SOLAR-DRIVEN WATER TREATMENT

SOLAR-DRIVEN WATER TREATMENT

Re-engineering and Accelerating Nature's Water Cycle

Edited by

Omid Mahian

School of Chemical Engineering and Technology,
Xi'an Jiaotong University, Xi'an, China

Jinjia Wei

School of Chemical Engineering and Technology,
Xi'an Jiaotong University, Xi'an, China

Robert A. Taylor

School of Mechanical and Manufacturing Engineering,
School of Photovoltaic and Renewable Energy Engineering,
University of New South Wales, Sydney, NSW, Australia

Somchai Wongwises

Department of Mechanical Engineering, Faculty of Engineering,
King Mongkut's University of Technology Thonburi, Bangkok, Thailand

ELSEVIER

ACADEMIC PRESS

An imprint of Elsevier

Academic Press is an imprint of Elsevier
125 London Wall, London EC2Y 5AS, United Kingdom
525 B Street, Suite 1650, San Diego, CA 92101, United States
50 Hampshire Street, 5th Floor, Cambridge, MA 02139, United States
The Boulevard, Langford Lane, Kidlington, Oxford OX5 1GB, United Kingdom

British Library Cataloguing-in-Publication Data
A catalogue record for this book is available from the British Library

Library of Congress Cataloging-in-Publication Data
A catalog record for this book is available from the Library of Congress

ISBN: 978-0-323-90991-4

For Information on all Academic Press publications
visit our website at https://www.elsevier.com/books-and-journals

Publisher: Joseph Hayton
Acquisitions Editor: Lisa Reading
Editorial Project Manager: Andrae Akeh
Production Project Manager: Prem Kumar Kaliamoorthi
Cover Designer: Christian Bilbow

Typeset by MPS Limited, Chennai, India

Contents

List of contributors

Avinash Alagumalai
Department of Mechanical Engineering, GMR Institute of Technology, Rajam, India

Simin Anvari
School of Chemical Engineering and Technology, Xi'an Jiaotong University, Xi'an, China

Mohamed M. Awad
Mechanical Power Engineering Department, Faculty of Engineering, Mansoura University, Mansoura, Egypt

Bac Van Dang
School of Mechanical and Manufacturing Engineering, University of New South Wales, Sydney, NSW, Australia

Maziar Dehghan
Department of Energy, Materials and Energy Research Center (MERC), Tehran, Iran

Balaji Devarajan
Department of Mechanical Engineering, KPR Institute of Engineering and Technology, Coimbatore, India

Mahsa Ghasemizadeh
Department of Energy, Materials and Energy Research Center (MERC), Tehran, Iran

Nipun Goel
Mechanical and Biomedical Engineering, Boise State University, Boise, ID, United States

Raja Murugadoss Jeyaraju
Department of Civil Engineering, Nadimpalli Satyanarayana Raju Institute of Technology, Visakhapatnam, India

Jegan Josephraj
Department of Civil Engineering, University College of Engineering Ramanathapuram, Anna University, Ramanathapuram, India

Rajeshkumar Lakshminarasimhan
Department of Mechanical Engineering, KPR Institute of Engineering and Technology, Coimbatore, India

Qiyuan Li
School of Chemical Engineering, University of New South Wales, Sydney, NSW, Australia

Amir Nashed
KarmSolar Group, Cairo, Egypt; School of Mechanical and Manufacturing Engineering, University of New South Wales, Sydney, NSW, Australia

Amr Omar
School of Mechanical and Manufacturing Engineering, University of New South Wales, Sydney, NSW, Australia

Todd P. Otanicar
Mechanical and Biomedical Engineering, Boise State University, Boise, ID, United States

Saman Rashidi
Department of Energy, Faculty of New Science and Technologies, Semnan University, Semnan, Iran

Gokulan Ravindiran
Department of Civil Engineering, GMR Institute of Technology, Rajam, India

David Saldivia
School of Photovoltaic and Renewable Energy, University of New South Wales, Sydney, NSW, Australia

Praveen Saravanan
Department of Civil Engineering, CMR Institute of Technology, Bengaluru, India

CHAPTER 1

Water: a global grand challenge and a path forward

Avinash Alagumalai[1], Simin Anvari[2] and Mohamed M. Awad[3]
[1]Department of Mechanical Engineering, GMR Institute of Technology, Rajam, India
[2]School of Chemical Engineering and Technology, Xi'an Jiaotong University, Xi'an, China
[3]Mechanical Power Engineering Department, Faculty of Engineering, Mansoura University, Mansoura, Egypt

1.1 Introduction

All living beings need water to live. However, humans take it for granted, and in recent years, everyone has been concerned about water shortages and contamination of drinking water. The two main concerns about water availability and quality are the decrease in groundwater levels and the rise in salinity of water. Freshwater supply is a major problem in both developing and developed countries. This is due to a lack of wastewater and industrial effluent management, which pollute natural water sources directly. Western countries are now focusing on the ecocide that has emerged from industrialization. The contamination of natural resources by industrial effluents is a major concern today, and it requires urgent action because it appears to pose a significant threat to human life. Industrial effluents are increasingly being mixed with natural resources in developing countries. Algeria, for example, has begun experiencing water contamination as a result of increasing population and poor waste management. Despite its reputation for industrialization, South Africa is currently facing a water crisis, with the issues being more about water quality than quantity [1]. All of the abovementioned concerns arise from the fact that many companies use obsolete waste management technology and a few industries do not adhere to regulations [2].

1.1.1 Water resources

Water is used for a number of purposes, including drinking, irrigation, and marine life. Water, for example, is used as a solvent for heating spaces and in batteries to conduct electricity through the dissociation effect. The real value of water should be recognized in this regard. It is important to note that groundwater availability is dictated not only by the storage

Solar-Driven Water Treatment
DOI: https://doi.org/10.1016/B978-0-323-90991-4.00005-0

spaces between soil, rocks, and clay but also by the nature's ability to supply it whenever needed. It is a well-known fact that water supply varies from one location to another. It is crucial to understand factors like an uneven distribution of precipitation on the earth's surface and water movement to fully comprehend why water is distributed unevenly from one place to another. Indeed, the amount of water available on this planet is finite and constant. However, there is an irregular distribution of water due to hydrological cycles and climatic changes. One of the main factors that cause irregular water distribution is the sun's radiation. Water serves as a natural thermal energy storage medium, affecting factors such as precipitation, air pressure, and temperature in the atmosphere. When the sun's radiation strikes the earth's surface, it heats the water up. It should be remembered, however, that the equator experiences net heat gain while the poles experience net heat loss. The surplus heat enters the poles due to movement of ocean and atmospheric air circulation. Cold air rises to the atmosphere from the earth's surface, while warm air containing evaporated water particles returns to the earth's surface in the form of rain. If we consider seawater cycle, water evaporates and form clouds. On contrary, water returns to the sea through rivers. Because of this cycle the climatic changes are envisaged and the uneven distribution of water is evident. It should also be noted that, in addition to the water cycle and climatic conditions the formation of aquifers is a major cause of unequal distribution of water supplies [3].

Surface water bodies, such as rivers, ponds, and lakes, are often seasonal and, in some cases, year-round. Seasonal water supplies necessitate the building of man-made dams or reservoirs to store and distribute water for irrigation and other uses. In the case of year-round water supplies, however, no such man-made storage is needed. Considering underground water reserves, they are formed because of oversaturation and the presence of underground water filters. The depth of groundwater level will vary from place to place. The water is stored in soil by molecular or capillary action.

1.1.2 Water quality

The source from which water is extracted determines the quality of the water. Human and industrial activities have the greatest effect on the quality of surface water compared to groundwater. Pathogens and suspended contaminants also pollute the surface water. The groundwater is often clear and odorless. Bacteria, viruses, and suspended solid particles are seldom found in groundwater. Groundwater, on the other hand, includes dissolved minerals known as total dissolved solids. In reality the concentration of minerals in groundwater determines its quality. If the concentration of dissolved minerals is higher in certain areas, pretreatment procedures must be used to clean the water before it is used [3].

Even though groundwater does not contain pathogens or suspended solids, it has the risk of contamination. Contaminants can seep into ground from the surface. Therefore the quality of water used for particular applications may vary. For example, the low-quality water with less salt content can be used for irrigation as it does not affect the crops much.

1.1.3 Water crisis

The Global Water Institute reports that 97.5% of the world's water is salt water found in oceans and seas. Land water, polar ice, and groundwater make up the remaining 2.5%. Some countries with sufficient water supplies can have an uneven distribution of these resources, resulting in widespread water scarcity [4]. The problem of water scarcity can be divided into three categories: water stress (water supply $<1700\ m^3$/person), water scarcity (water supply $<1000\ m^3$/person), and water total scarcity (water supply $<500\ m^3$/person) [5]. According to the World Health Organization, nearly 20% of the world's population lives in areas where there is insufficient water to meet their needs [6]. It is also projected that by 2025, more than 2.8 billion people in 48 countries will be affected by water stress or scarcity [4].

1.1.3.1 Potable water crisis

Potable water is important for all living things on the earth, and it is primarily used for drinking, domestic, agricultural, industrial, and other purposes. Potable water shortage is a severe disaster that poses a significant threat to human life. Potable water shortage occurs when the annual potable water supply per capita falls below $1000\ m^3$, while potable water stress occurs when the annual potable water supply per capita falls between $1000\ m^3$ and $1700\ m^3$ [7]. At the least, annual human need is around $1000\ m^3$ of potable water [8]. According to the Population Action International Institute, relative sufficiency, potable water tension, and scarcity will be 58%, 24%, and 18% in 2050, respectively [9]. By 2000, global demand for potable water had doubled from nearly 4.2 million cubic meters over the previous half-century, accounting for 30% of the available supply, with 70% projected by 2025 [8]. According to the United Nations Children's Fund more than 850 million people consume polluted water.

Water has been regarded as the source for establishing agricultural life throughout history, and it also plays a significant role in the shaping of nations. Drinking water supply has become a global issue and a threat. According to human development statistics, almost two million children die each year due to a lack of clean water and sanitation. Gathering or moving water over long distances takes a long time for women and young girls [10].

Water scarcity increases day by day due to population growth, which requires an increase in the rate of energy and water consumption.

Potable water is delivered to arid regions in one of the following ways: water distribution through a piping system and lift stations; and desalination of saline water. Air can be conceived as a limitless, pure, and sustainable supply of water. The amount of water vapor in the earth's atmosphere is estimated to be about 12,900 km^3. Every year the humidification and dehumidification processes cycle $4.95 \times 104\ km^3$ of water through the atmosphere, according to NASA [11].

1.2 History of water treatment

In the years between 4000 and 3000 BCE, human life style was affected by the *Bronze Age* and the *script* invention. The historic documents indicate that the first attempts for water treatment were also initiated in this period. Definitely, the aim of water treatment in ancient times was just the improvement of smell and taste in a way it can be drinkable even though with some impurities. One may ask what methods were used to improve the taste and smell of water 5000–6000 years ago. It was not a complicated approach: water heating. The humans in that time believed that heating the water was a way for water purification. Three basic schemes for water heating were boiling of water in a vessel using fire, putting a hot iron (or other metals) inside the water vessel, or putting the water vessel under sun radiation for heating.

Egyptians were probably the first in the world who used chemical alum to settle the impurities (relatively small but visible suspended particles) of water 3500 years ago. In 500 BCE, cloth bags were used to strain rainwater, an approach introduced by Hippocrates.

By beginning of the first industrial revolution in Europe (started about 1760) the human society experienced dramatic progress in all technological aspects, including water treatment. With the improvement of public health and consequently superior population growth, the development of industrialized methods for water purification in large scales got priority.

In 1804 the first plant for water purification designed based on slow sand filtration was built in Scotland. Slow sand filter can effectively strain the water by removing the turbidity and unhealthy impurities through physical, chemical, and biological approaches. At that time, there were not yet any advanced supply networks, so people had to use horses and carts for water transportation. Three years later, water supply networks were developed in Scotland to deliver the filtered water to consumers.

Some infectious diseases spread due to water pollution at the end of the 19th and in the early 20th century. Therefore scientists endeavored for finding effective approaches for water disinfection to remove bacteria and viruses; one of these methods was chlorination, which is still a common approach.

Although during the last 50 years, industries and technologies have been developed rapidly however, at the same time the number of sources of water pollution has increased. As a result, classic treatment approaches for water purification were not effective enough.

One of the well-developed technologies for water treatment during recent decades is membrane filtration. Compared to classic approaches for water treatment such as coagulation and sand filtration, membrane filtration technology provides higher separation capacity and generally lower chemical usage. It should be noted that membrane filtration technology is used not only for water treatment but also in food and beverage industries, chemical processing, gas separation, and medical fields. However, more than 50% of the market belongs to membrane filtrations used for water treatment.

1.3 Conventional water purification methods

Many hydrologists consider water that passes through a ground aquifer below 10 ft. to be fully filtered in the natural setting [12]. The majority of artificial filters are exact replicas of nature's filtration system. Some of these systems perform better than others depending on the raw water content. To choose the best treatment system for the job, one must first learn about the benefits and disadvantages of each choice.

When it comes to water purification, there are many factors to consider. Source of water supply, population density, water pollution, and topographic location are few of the factors. Depending on the application the cost of water purification can differ. Slow sand filters, for example, may be adequate in areas where water is in short supply.

1.3.1 Conventional methods for groundwater purification

When compared to surface water, groundwater is the more potable alternative. It should be remembered, however, that groundwater in some areas contains various minerals and other compounds, rendering it unfit for drinking. Surface water is unavoidable in this case. Groundwater is usually safe to drink. However, in some places the groundwater includes minerals and other substances that give it an unpleasant taste and odor. Surface water becomes inevitable in that situation.

Minerals and salts, on the other hand, can be removed using a number of basic techniques. Aeration and sand techniques, for example, can be used to reduce dissolved iron and manganese concentrations. The iron and manganese become insoluble as a result of the aeration, forming fine dark sediment that is easier to extract. It can be made without the use of a motorized pump in a simple manner. The mineral water can be pumped into the aeration device directly using a hand pump [3].

1.3.1.1 Slow sand filtration process

The best natural filters are clay and hardpan filters. Although with superior filter media the movement of water in this process is very slow unless external pressure is applied to this process as with reverse osmosis (RO). If the water is being forced through the filter media under pressure, then the purity may be compromised.

The slow sand filtration method of water treatment is the most ancient and is still widely used in many countries [13]. This method of treatment does not require pumps, depending on the topographic position. The treated water can be delivered to the delivery system by gravity. However, an external pump may be needed in some cases to transport the treated water. The representation of slow sand filtration systems is depicted in Fig. 1.1.

While the slow sand filtration method purifies water, it is difficult to know how many microorganisms still pass through the slow sand filter, necessitating disinfection. The use of chlorine gas for disinfection is advised. Electricity can be needed at the treatment plant to operate the external pump. The disinfection agent can be delivered by gravity. To purify water using a slow sand filtration process, two separate tanks are needed. The first tank is a sedimentation tank, which is used to remove dissolved and suspended substances, the second one is the slow sand

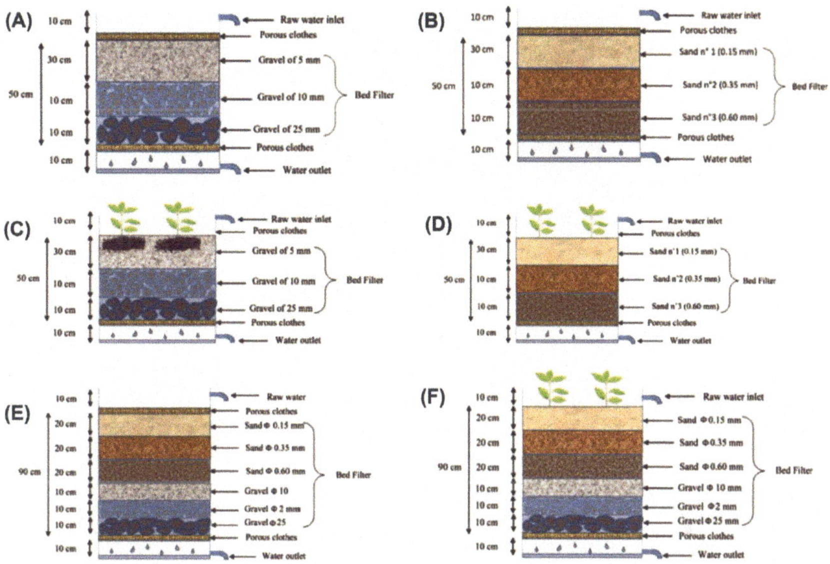

Figure 1.1 Slow sand filter: (A) gravel layers, (B) sand layers, (C) wetland plants and gravel layers, (D) wetland plants and sand layers, (E) mix aggregates, and (F) wetland plants and mix aggregates [14]. *Adapted from K.M. Sosthene, N. Gahi, Low cost filtration of domestic wastewater for irrigation purpose, World Journal of Engineering and Technology 6 (2018) 585–602.*

filtration tank where the pathogens and other substances are treated. The most important considerations in the design of water treatment plants are the water quality and demand. While treating highly turbid water, this system performance may not be good because dirt will obstruct the purification process. As a consequence, this treatment approach is suggested for low-demand situations. The filter tank is the most critical component of a slow sand filtration system's maintenance. For medium-sized community water sources, this method is one of the most cost-effective solutions. The device does not need a pump. In certain cases, however, an external pump is needed [3].

Slow sand filtration is the simplest process. Except for disinfection the device does not need chemicals. Since the filtration mechanism occurs naturally by gravity, this system cannot be used for communities with high water demands. For large water sources a more conservative water treatment system is recommended.

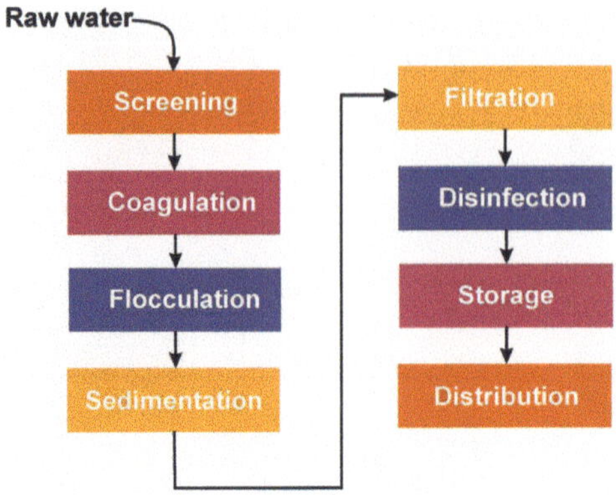

Figure 1.2 Various steps in traditional water treatment.

1.3.2 Conventional surface water treatment

Conventional methods to treat water are widely used for removing minerals and suspended solids from water in developing and developed nations. Many potentially harmful water pollutants can be removed using this process. Coagulants, coagulant aids, and filtration aids are the chemicals used in these systems. The toxicity and concentration of these chemicals are extremely low. After either of these procedures, chemical disinfection is needed for municipal water and wastewater treatment. Fig. 1.2 shows the various steps in traditional water treatment.

1.4 Water–energy nexus

There is a nexus (two-way relationship) between water and energy [15,16]. We need energy to transport water from specific sources to consumers or for straining nonpotable water in wastewater treatment plants. On the other hand, we can produce electrical energy using water, the same way that current is generated by dams (hydropower). Estimates show that approximately between 1% and 3% of the total electricity generated by countries is consumed by water treatment plants. In the United States, for example, 30 TW (trillion watts) hours per year is used by water treatment plants, which is equivalent to about $2 billion per year as reported in 2017. At present, a large part of the energy consumed by water treatment plants is supplied by fossil fuels. However, the use of fossil fuels is not a perfect choice due to some disadvantages such as the limitation of fossil fuels resources or their harmful effects on the environment and consequently human health.

So what is the solution? The best solution to combat pollution resulting from the use of fossil fuels is relying on renewable energy resources for cleaner production of power. The next section will provide a brief overview of the current status and potential of renewable energy.

1.5 An overview of renewable energy resources

Energy resources can be divided into renewable and nonrenewable energy resources. Renewable energy is simply a kind of energy that does not run out by consumption. Renewable energy resources can be divided into six groups: wind energy, geothermal energy, hydroenergy, marine energy, bioenergy, and solar energy. In some references, hydroenergy and marine energy are considered together as water energy. Note that nuclear energy is a nonrenewable energy resource since the amount of uranium sources as the main fuel of nuclear reactors is limited in the world. The various forms of renewable energy sources are depicted in Fig. 1.3.

In the following, each of the renewable energy resources will be described shortly.

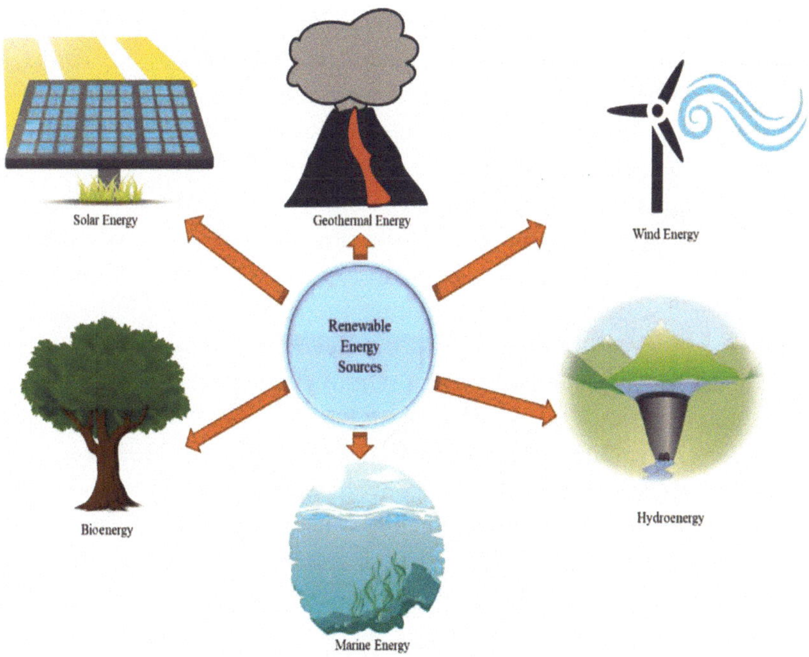

Figure 1.3 Different forms of renewable energy sources.

1.5.1 Wind energy

1.5.1.1 History and revolution

Wind energy originated from solar energy, how? The layers of air in the earth's atmosphere are heated up by solar radiation; however, there is a temperature difference between the layers. The temperature gradients across the air layers create pressure gradients that result in movement of air that is known as wind. A higher pressure difference means a higher driving force and, consequently, a stronger wind [17].

Wind energy applications have a long history. In ancient times, wind energy has been employed for driving sailing boats and milling wheat by windmills. Wind pumps and windmills were common facilities for food production in the Middle East by the 11th century [18].

A review of the history of wind turbines shows that there has always been controversy over who invented the first wind turbine to generate electricity. Although most available sources attribute the invention of the first electricity-generating wind turbine to Charles Brush, an American, in winter 1887, however, some other references give this credit to James Blyth, a Scottish Professor who performed similar experiments in July 1887, and the results were documented as a UK patent in 1891. Nevertheless, in the last decade of the 19th century, Poul la Cour, a Danish scientist, designed and constructed relatively advanced wind turbines for electricity generation, which were used later for hydrogen production. The range of electricity generation by wind turbines in the 19th century was between 1 and 12 kW [19].

The 20th century can be divided into two completely different periods in terms of development and attention to wind energy utilization for electricity generation in developed countries like the United States. In the early decades of the 20th century the attention to wind energy as a power source was limited due to the development of electricity networks in cities. The majority of the required electricity was supplied by power plants fed by coal and the use of wind energy for electricity generation was more highlighted in rural and remote areas. However, starting in the 1970s with increasing concerns about the shortage of fossil fuels resources and, on the other hand, environmental pollution considerations, the governments decided to invest much more on renewable energy resources including wind energy for power generation. Therefore it can be said that the 1970s was the beginning of modern history of wind turbines.

In 1990 wind turbines with electricity generation capacity of 500 kW (0.5 MW) were developed. It means that compared to 100 years ago the capacity increased by more than 40 times. In 2015 the wind turbine power generation capacity reached up to 9 MW, and it is expected to reach between 13 and 15 MW in 2025.

In 2020 despite the COVID-19 crisis, it is expected that the total global wind power capacity reaches about 700 GW, which reveals 8% growth compared to 2019. China and the United States have the most contribution to the global wind power capacity.

1.5.1.2 Wind turbines: types and specifications

Basically, wind turbines can be classified into two groups: horizontal-axis and vertical-axis wind turbines.

Nowadays, horizontal-axis wind turbines are used more often. Some advantages of horizontal-axis wind turbines are as follows [20]:

- *High power output*: The power output can be approximately between 2 and 8 MW depending on the application.
- *High efficiency*: The maximum efficiency of horizontal-axis wind turbines lies between 40% and 50%.
- *High reliability*: Enormous studies have been conducted on this type of wind turbine, so the challenges on design, construction, and exploitation can be handled easier.
- *High operational wind speed*: Since the height of this type of wind turbines is high, they can capture the wind at higher speeds, which means more efficient performance.

Despite the abovementioned advantages, horizontal-axis wind turbines have some disadvantages as follows:

- *Difficulties in setup and maintenance*: Transportation and setup of wind turbines with more than 100 m height and with a weight of several tons is not an easy task. Moreover, the maintenance engineers need to check and probably repair the main components of the wind turbine (such as gearbox and generator) at high altitudes that is difficult to do especially when wind speed is high.
- *Negative impact on ecosystem and environment*: Annually thousands birds are killed by blades of wind turbines. Another negative impact on the environment is the relatively noisy performance.
- *Severe rules for installation*: There are strict rules for installation of horizontal-axis wind turbines; so permissions must be obtained from

local government after checking the noise level and confirming the acceptable distance to neighbors.

1.5.2 Hydroenergy

1.5.2.1 What is Hydroenergy?

Hydroenergy is a renewable energy that is created from movements of running water or waterfalls. Water is a renewable energy source; it has a continuous cycle in nature. In this cycle the water evaporates due to solar energy from the lakes and oceans, and then due to distillation, it turns back to rivers, seas, ocean, dams, and other water bodies in the form of precipitation [21]. Therefore hydroenergy originates from solar energy. It is mostly known as hydropower because a large part of it is utilized for electricity generation globally.

By constructing the dam in rivers the possibility of converting the kinetic energy of flowing water as potential energy is achieved by the hydropower plants use for electricity generation. Therefore the amount of rainfall that flows into rivers and running water in a geographical area determines the amount of water available for hydropower generation. Seasonal changes in rainfall and long-term changes in rainfall patterns such as drought can significantly impact hydropower generation [22].

For thousands of years, humans have been using hydroenergy. More than 2000 years ago, Greeks utilized flowing water for moving mill wheels to turn wheat into flour. The first industrial utilization of hydropower to electricity generation was performed in the United States in 1880 [23].

1.5.2.2 Hydropower plants

Hydropower plants can be classified into three groups based on technology and application. They are usually categorized by size, head, and source of water. These classifications are described as follows [24–26]:

- *Classification based on size (capacity of power generation)*: Depending on the level of electricity generation capacity, hydropower plants can be divided into micro-, small, and large hydropower plants. Based on size difference, this division is different in various countries. According to a study [26], large hydropower has a capacity of more than 15–20 MW. Small hydropower has a capacity at the range of 0.1 to 15–20 MW, and microhydropower has a capacity of less than 100 kW.
- *Classification based on head*: This is the difference between the inlet and outlet levels of a hydropower installation. Since the head reveals the

pressure of water acting on turbines and the output power, it is an important parameter in the design. There is no standard for the classification of hydropower plants for the head, and it varies in different countries.

- *Classification based on source of water:* This includes the following: (1) Run of river, where a river's natural flow or drop in water height is utilized for electricity generation. (2) Storage hydropower is obtained from the water that is stored as a resource behind the dam to use its potential energy to generate electricity. One of the advantages of this power plant is the production of electricity during peak consumption times. (3) Pumped storage—storage pump plants are not an energy source, but they are a hydraulic energy storage system. In these plants, water is pumped from a lower source to a higher source using additional electricity generated from the power plant in off-peak or low consumption times.

A well-designed and adequately operated hydropower plant can be one of the cheapest power plants to generate electricity because its fuel (water) is naturally available and hence no need for fuel purchase. In addition, hydropower technology by electricity generation from direct mechanical work is an efficient energy conversion process [24]. Compared with other power generations, these power plants can behave much faster in response to the power demand fluctuations.

1.5.2.3 Advantages and disadvantages of hydropower

Today, hydropower is the largest source of renewables in electricity generation. According to the International Renewable Energy Agency report at the end of 2019, hydropower has the most considerable contribution of the capacity energy generation among renewables, and this energy provides about 47% of the global total [27]. It is expected, according to the Energy Information Administration (EIA) report, the hydropower growth will be relatively little in the world to electricity production by 2050. It is also expected that the most significant growth in hydropower generation in 2050 will be in countries such as China, Brazil, and OECD Europe, where there is a large access to hydropower resources [28].

Some advantages of hydropower plants are as follows [26]:

- It does not have CO_2 emission, radioactive waste problems, or air pollution.
- They are the best solution for supplying the peak demand, and their useful life is over 50 years.

- The efficiency of a hydropower plant is higher than thermal and gas plants.
- The average total cost of hydropower over the lifetime is lower than other energy sources.

Despite abovementioned advantages, there are some disadvantages for hydropower plants such as [26]:

- Access to hydropower plants is limited to the hilly or foothill regions because water and head are available there. In addition, seismic effects occur in large dams due to the heavy concentration of load behind the dam.
- The construction of dams causes a part of the dam's environment to go underwater, which causes the ecology of that area to be affected, destroyed, and changed. The lives of animals, plants, and even humans are endangered.
- By constructing a dam the downstream areas of the river no longer have access and control over the water flow.

1.5.3 Geothermal energy

1.5.3.1 What is geothermal energy?

The geothermal energy is coming from the Earth's solid crust. This thermal energy is obtained from hot molten rocks and natural liquids including the Earth's crust, which originates from stored solar energy and the decay of radioactive uranium, thorium, and potassium isotopes over many years in the depths of the Earth. These natural liquids are chiefly water that has been dissolved with varying amounts of salts. Most of the time, they are in the liquid phase but at times may include the saturated, superheated steam vapor or liquid—vapor mixture phase [29]. Hot springs, one of the geothermal sources, are used for cooking, washing, and bathing by humans for thousands of years. However, using geothermal energy for industrial application begun near Pisa, Italy, in the late 18th century [29].

Geothermal energy is used in many countries of the world to supply heating and electricity generation. Countries with volcanoes or prone to earthquakes have the most significant potential to use this energy [30]. The high temperature resource with a temperature above 200°C can be found in these countries. The moderate-low temperature source with a temperature between 50°C and 200°C is found in most locations of the world [29].

Geothermal energy is carbon-neutral, renewable, and sustainable energy that provides continuous heat because its use does not depend on weather conditions.

1.5.3.2 Utilization of geothermal energy resources

Geothermal sources are utilized in two ways, direct way to thermal energy generation and indirect way to electricity generation. However, recently, systems driven by geothermal energy are grouped into three categories [31,32]:

- *Region heating systems (direct use)*: The oldest way of geothermal energy utilization is the direct use of it. The moderate-low temperature reservoirs are suitable for the direct utilization of geothermal energy. The hot water of these reservoirs can be extracted by drilling rigs of water wells for a heating application. According to the table listed by Lindal (Iceland researcher), direct uses of geothermal energy can include heating of buildings, treatment of diseases, animal husbandry, agriculture, industrial applications, and others.
- *Geothermal power plants for electricity generation (indirect use)*: The high temperature geothermal reservoirs are mostly used for electrical generation by power plants. In other words, the hot water or steam extracted through drilled wells enters into the power plant's turbine for electrical generation. These power plants use three technologies: dry steam, flash steam, and binary for electricity generation.
- *Heat pumps based on geothermal energy*: Geothermal sources with a constant temperature near the earth's surface are usually used for heat pumps. Heat pumps use geothermal energy to heating and cooling the buildings/structures during winter and summer.

Generally, among renewable energies the contribution of geothermal energy production capacity is small; at the end of 2019, of the 2537-GW energy production capacity by renewables, about 14 GW was geothermal [27]. Among the countries of the world the United States, Indonesia, and Kenya, with 18,366, 15,315, and 9930 GWh energy generation, had the largest share of energy production from the geothermal sources in 2020, respectively [33].

In the world, geothermal energy has enormous potential for improvement and utilization. It is estimated that by 2050 the contribution of geothermal energy to the supply of world energy will be about 3.9% for heating and about 3.5% of annual electricity required. In this way the contribution of geothermal energy to thermal energy production and reduction of CO_2 emission will be about 1600 TWh and about 800 Mt per annum [31].

1.5.3.3 Advantages and disadvantages of geothermal energy

With a comparison of solar, wind, and geothermal energies, it is indicated that using geothermal energy in power plants is interesting because it is stable, independent of weather, and has higher thermal efficiency. The advantages of geothermal energy are as follows [34,35]:

- At a stable price, steady electricity is provided.
- The electricity is produced with lower environmental emissions.
- The geothermal energy systems by medium-high efficiency are obtained.
- It has a high capacity factor.
- It is weatherproof.
- It is used for heat production in industries, space heating, and agriculture.

However, this energy has some disadvantages: high initial investment, long construction time, long payback time, tough to assess resource, and tough to modularize [35].

As a result, there are both advantages and disadvantages to developing geothermal energy around the world. By reducing or eliminating disadvantages, it can help us provide sustainable energy in the future.

1.5.4 Marine energy

1.5.4.1 What is marine energy?

Based on the United States National Oceanic and Atmospheric Association, oceans cover about 71% of the Earth. The oceans have a vast energy source and are available in most coastal countries. The energy extracted from oceans and seas is called marine energy. The origin of marine energy, like all renewable energies, is the sun. It is renewable, carbon-free, and one of the large high-energy sources. It is expected that marine energy will significantly contribute to the world's energy sources in the next decade. Extractable energies from the oceans include wave, tidal, ocean thermal energy, marine current energy, and salinity gradient (osmotic energy) resources, and their development in today's world is briefly described in the following sections.

1.5.4.2 Different types of marine energy

1.5.4.2.1 Wave energy

More than 80% of marine energy contains wave energy. It is one of the most promising and widespread fuels in the world [36]. Waves are created by the wind moving on the oceans' surface, so they can be considered an indirect solar source. For hundreds of years, humans have used wave energy. The first patented device for wave energy conversion was recorded in England in 1855 [37]. Wave energy has the most significant potential in coastal countries such as the west coasts of North and South Americas, the west coast of Europe, and the south coast of Australia. One of the advantages of wave energy is high power density because it can be considered a combination of wind and solar energies. Usually, a wave's power density is 30 kW/m, which is 10 times larger than solar energy and five times larger than wind energy [38].

One of the most critical challenges in using wave energy is the high cost of converting this energy into electricity. Therefore it is predicted that by 2050 about 10% of the global electricity demand will be generated by wave energy [39].

1.5.4.2.2 Tidal energy

Tidal energy can be defined as the energy caused by the gravitational energy of the moon and sun on the seawater. Because the behavior of this energy is predictable, it has an advantage over wind and solar energy. It is predicted that the power of tidal energy is about 100 GW in the world's waters, but only part of it is exploitable, which happens in estuaries and gulfs [40]. The first tidal power plant was built in 1966 with a production capacity of 240 MW on the Rance River in northwestern France [41]. Today, tidal energy is not yet widespread due to its high cost and limitations, such as it is mostly available in places with extreme lethality or high water flow velocities. However, it could be one of the promising sources in future electricity generation. Leading countries in converting this energy to electricity are Canada, Britain, France, and the United States.

1.5.4.2.3 Ocean thermal energy

Ocean thermal energy conversion (OTEC) technology is a renewable energy production technology that converts sunlight into power. The temperature difference between hot and cold water levels in the oceans is about 20°C–25°C. Using the OTEC method, this thermal gradient of different water layers is used to drive a power cycle [36]. As regards, the ocean water temperature changes slightly during the night and day and in various seasons, so OTEC energy has a small termination behavior; therefore the use of inverter and additional storage systems is eliminated. The first OTEC power plant was built in 1881. Afterward, the use of this energy as renewable energy expanded. It is expected that OTEC systems can generate about 5×10^6 MW power in the world [42]. One of the advantages of OTEC power plants is that in addition to generating electricity, they also can produce products such as freshwater and microalgae streams and are also usable for air conditioning applications.

1.5.4.2.4 Marine current energy

Ocean current energy, including Gulf Stream, is obtained from the kinetic energy of the ocean current. It is used to drive turbines that are typically installed in the flow path. This energy has two constant and directional currents, which is its advantage over tidal currents. It is expected that the global ocean current power is 5000 GW, with power densities of up to 15 kW/m^2 [43].

1.5.4.2.5 Osmotic energy

Osmotic energy is a source of clean energy produced by a phenomenon in nature called transmutation (the passage of water through a semipermeable membrane). When fresh water is placed in contact with salt water, as when a river reaches the sea, a vast amount of energy is released, and this energy can be applied to move the turbine to electricity generation. The global potential of electricity generated by these power plants is expected to be around 2.6×106 MW [42].

1.5.5 Biomass energy

1.5.5.1 What is biomass?

Biomass energy, one of the sustainable renewable energy sources at the time of the industrial revolution, became important after exploring coal [44]. Biomass is made up of living species such as plants, animals, or anything currently live or have been living for a little time. In the world the largest source of biomass is the forestry section. Charcoal, pellets, fuelwood, and wood chips are obtained from it. They are composed of about 85% of all biomass sources consumed for energy objectives. Historically, humans used fuelwood, the oldest form of biomass [45]. For traditional cooking and heating the fuelwood is still used in Africa and Asia [46].

Every year a part of the sun's radiation that arrives in Earth's atmosphere is absorbed and stored by plants through the photosynthesis process. Plants absorb CO_2 from the atmosphere and keep it as carbon in themselves; they are a source of CO_2 saving. Animals also grow by receiving food from the biomass. In this way, a large amount of biomass grows annually. Therefore for its development, it does not need millions of years, same as fossil fuels.

By burning the biomass the additional CO_2 does not add to the atmosphere because the stored CO_2 in plants is released. The main attraction of biomass is regeneration; for this, it is considered renewable [45].

1.5.5.2 Applications of biomass energy

Direct and indirect applications can use biomass for energy. Direct combustion or cofiring with fossil fuels is placed in direct biomass applications. For indirect biomass applications processes the raw biomass converts into various gaseous fuels [syngas (CO and H_2), biogas (CH_4 and CO_2), producer gas (H_2, CO, CO_2, CH_4, and H_2), and alternative natural gas (CH_4)], liquid fuels (ethanol, methanol, biodiesel, pyrolysis oil, and vegetable oil), or solid fuels (biochar, biocoke disturbed biomass, and charcoal), which can be utilized directly for energy production [44]. This conversion can be achieved through three methods, including biochemical, chemical, and thermochemical, which are briefly described further [47]:

- *Biochemical*: Probably, it is the oldest means of gasification of biomass. In this method, bacteria or enzymes decompose the biomass molecules for production of methane and CO_2.
- *Chemical*: In chemical conversion, biomass oils are converted into liquid fuel and gasoline additives. For example, biodiesel is constructed by mixing alcohol (typically methanol) with animal fat, vegetable oil, or recycled cooking oil.
- *Thermochemical*: In this method, gases are obtained by converting all biomass that can be used directly or synthesized into the wanted chemicals. Three technologies are used for thermochemical conversion: direct liquefaction, gasification, and pyrolysis.

These biomass products are applied to meet some of the energy needs, including electricity generation, heating buildings, and fueling vehicles. In the following sections the advantages and share of biomass energy in the world are expressed.

1.5.5.3 Extracting energy from biomass

When comparing biomass energy with other energies, it is more economical and can produce more energy. Both rich and developing countries have an enormous potential from biomass sources or can create it. However, the contribution of developing countries is higher in supplying energy from biomass. Mostly, they use it in the domestic section [48].

Biomass is a flexible energy source for producing different energies such as power, heat, gaseous fuels, and liquid. It is often used to supply heating energy for buildings and industries. For example, by burning wood, that is, the largest biomass, the required energy for household cooking, lighting, and space heating is supplied in developing countries. In addition, for the production of power and heat by biomass fuel on a large scale, the combined heat and power plants are used. Still, it is expected that about 3% of the electricity generation and about 5% of the energy needed for transportation will be supplied from bioenergy in 2023 [49].

The International Energy Agency in 2019 reported that, among renewable energies, bioenergy has about 70% of use and about 9.5% of total primary energy is supplied by it [7]. Moreover, based on the special report on global warming in 2018, it is found that bioenergy will make up about 26% of primary energy in 2050, up from 10% in 2020 [50].

Bioenergy has extensive potential to supply most energy required in the world because of the following reasons [51,52]:

- In the developing countries, by supplying about 35% of energy demand, biomass helps reduce poverty there.
- It can supply energy at any time without the need for using costly devices to convert.
- It can carry out the consumer's energy in all required forms such as gaseous or liquid and heat or electricity.
- Carbon dioxide emitted from the burning of biomass is neutral.

In summary, it is notable that the traditional biomass production can be improved to the modern biomass production by using efficient, economic, and clean technologies. In other words, modern biomass production is generated in a sustainable way for heat and power production that is mostly used in industrial countries [53].

1.5.6 Solar energy

1.5.6.1 What is solar energy?

The primary energy source on the earth's surface is sunlight, which can be harnessed in several ways [54]. Fossil fuels containing coal, oil, and natural gas have been produced through photosynthesis and chemical processes over the years.

The significant advantage of solar energy compared to fossil fuels is that solar energy is a clean energy and causes no pollution to the environment. This energy is one of the most promising and sustainable choices for electricity generation [55]. It can also replace conventional fuels for both today's and the future world.

The oldest energy source that humans use is solar energy [56]. The burning of the Roman fleet in the Gulf of Syracuse by the famous mathematician and philosopher Archimedes is one of the oldest known solar energy uses. Between 1864 and 1878, August Monchot was pioneered in the construction and operation of several solar steam engines to convert solar energy into other forms of energy in Europe and North Africa [57]. Solar energy can be directly or indirectly harnessed and converted into thermal and electrical energies. Collectors are the chief technology used to harness solar energy, which is explained in the following sections, along with their types and applications.

1.5.6.2 Solar energy collectors

The collector is the main component in solar systems. A collector is an instrument for absorbing solar rays and transforming its heat into the fluid (typically air, oil, or water) circulating in the collector. In this way, solar thermal energy is transferred directly to intended equipment or the energy storage tank (for cloudy days or night time use). Solar collectors can be categorized into two types: the nonconcentrating or the stationary and the concentrating [58]:

- *Nonconcentrating collectors*: In these collectors the surface for absorbing and repulsing solar radiation is the same. They cannot follow the sunshine because they are fixed in position. The three major types of these collectors are flat plate collectors [including a dark plate type to absorbing solar energy, a transparent cover to pass solar energy and reduce heat loss, a fluid (usually air, water, or antifreeze) to transfer heat from the absorber, and a thermal insulator in the back], fixed parabolic collectors (which can reflect all the received beams in a wide range and are of two main types, namely, symmetric and asymmetric), and evacuated tube collectors (which are typical solar flat panel collectors expanded for winter and summer use).
- *Concentrating collectors*: In these collectors the daily motion of the sun is followed by collectors. They usually have concave surfaces. They receive the sun's beams that are focused in a small area. Collectors that are categorized into this classification are as follows: (1) parabolic trough collector (devices for heating applications in temperatures between 500°C and 4000°C that have light construction and low-cost technology); (2) linear Fresnel reflectors that are of two types—Fresnel lens collectors, made of plastic and is designed to focus the sun's rays at the receiving point, and Fresnel linear collectors, which focus the sun's rays on the

receiver in a row of linear mirror strips; (3) parabolic dish, which follows the sunshine in two axes and can reach temperatures above 15,000°C, and the receiver at the focal point of dish can receive the concentrated solar energy; and (4) central receiver, which is a series of flat mirrors or heliostats that are used to receive very high solar radiation. This heliostat field is called the central receiver collector. Heliostats can be used to collect solar energy by directly reflecting solar radiation to a joint target. Absorbed thermal energy by the central receiver is transferred to the flowing fluid to be stored for later use for electricity generation.

1.5.6.3 Solar energy applications

Since solar energy in the form of radiation is plentiful and clean, our society has many options for producing energy for their everyday needs from this renewable energy source. The solar energy that falls on the earth can be collected in two ways: (1) electrical energy (using solar panels) and (2) thermal energy (using concentrated solar power systems). Solar panels transform sunlight directly into electricity [59,60]. Light is concentrated from a wide area to a small area in solar thermal energy plants using reflectors such as mirrors or lenses, and then this concentrated light is transformed into heat, which is used to power turbines or energizes a thermochemical reaction, which produces electricity. Solar thermal plants have a higher power conversion efficiency than solar panels, but they have the disadvantage of high maintenance costs. Furthermore, solar thermal plants need a large amount of direct radiation to operate properly, and their efficiency plummets when the sun is obscured by clouds. Solar panels, on the other hand, can produce electricity from diffused sun radiation. Solar panels have also been used on the roofs of buildings, homes, and now roads to improve energy efficiency and ensure an adequate supply of electricity to buildings, houses, and street lights, among other things. More importantly, there are no CO_2 emissions when electricity is produced this way. As a result, solar-generated electricity is promising, dependable, and environmentally sustainable. The photovoltaic market is rapidly expanding, with solar panel installations producing renewable, environmentally friendly, and low cost electricity. While thin film solar cells have reduced manufacturing costs in this emerging technology, we must compromise on radiation absorption, which lowers solar cell performance. Various light trapping technologies have been used to increase absorption, but absorption enhancement with plasmonic metal nanoparticles has emerged as a strong candidate for cost-effective and efficient thin film solar cells.

- *Nonpower plant applications*: These systems operate based on low temperature thermal concentrators and use the solar source for final consumption. Nonpower plant applications include many types. Some of them include the following: (1) Solar water heating systems, which are used for water heating. Solar water heaters consist of three main parts: the collector, the piping circuit, and the heat storage tank. The collector absorbs the sun's radiant energy and transmits it to working fluid. Fluid, by receiving thermal heating, heats up and moves to the storage tank. After passing through the heat exchanger, it transfers its heat to the water inside the tank and then returns to the collector. In the same way, it continues to move in this cycle permanently [61]. (2) Solar space heating and cooling—heating and cooling buildings by solar energy was a new idea introduced in the 1930s and achieved significant progress in less than a decade. Supplying the thermal energy to buildings using the sun energy is done in two ways, passive and active. In the passive state the building's quality and architecture affect the receipt and storage of solar energy. In active state the use of solar collectors and other energy source is necessary to transfer the heated fluid into the building [62]. (3) Solar desalination, in which when the heat received from the sun acts on the salt water at a low temperature, only the water evaporates, and the solutes remain. Then the evaporated water can be adjusted by different methods, and thus fresh water can be produced. Solar water desalination can be constructed for both domestic and industrial sizes [63]. (4) Solar dryer—drying materials use the sun's energy directly or indirectly, and the air flows naturally or forcibly, then both of them causing to dry out the product. (5) Solar cooker are of three common types: parabolic, heat pipes, and boxes. The working principle of solar stoves is to collect direct solar rays at a focal point and increase the temperature. (6) Solar furnace usually uses a large number of mirrors to focus the rays of sunlight, and their collection is reflected on the furnace to increase its temperature and be heated.
- *Power plant applications*: In power plant applications the solar radiant energy absorbed by solar collectors is used to drive power generation plants [64]. Some of these plants also have a thermal energy storage system to store excess solar energy during peak solar radiation hours. Energy stored can run the power plant on cloudy days or at nighttime when solar energy is not available. The leading technologies used in

centralized solar power systems are as follows [64]: (1) Parabolic trough collectors focus the solar beams over the entire receiving tube in which synthetic oil is circulating. It enters the heat exchanger through piping and generates the energy needed to drive the power plant's turbine. These types of solar technology systems are mostly used to generate heat at temperatures above 4000°C for electricity generation or thermal applications. (2) Power towers use thousands of mirrors that follow the solar radiation. Power tower plants are generally large (usually 10 MWe or more). (3) Dish/stirling engine systems, in which the sun's rays radiate on the parabolic concentrating surface and are concentrated in its center. By a thermal energy stirling engine the concentrated radiant energy is converted into mechanical energy. Then, this mechanical energy, with the help of an alternator, is converted to electricity.

- *Photovoltaic systems*: Photovoltaic is a system that can convert solar energy directly into electricity. The main element in solar cells' structure is semiconductors such as silicon and gallium arsenide [65]. The foundation work of solar cells can be justified based on electrons theory in the atomic circuits. Photovoltaic systems are used as power plants independent of the grid or systems connected to the grid with fixed or mobile installation structures in small units with low power to supply the electricity required by small calculators to large power plant systems. Among the most critical applications of these systems are the following: Off-grid power supply systems (to supply electricity for residential houses and rural areas and, generally, to meet the electricity needs of areas without grid electricity), solar pumping (to supply drinking water to houses, irrigation of lands, animal husbandry, fish farming, drinking water for animals, etc.), solar lighting (to provide lighting for residential houses and schools, waypoints, tunnels, lighthouses, park lights, etc.), and portable power supply system (are used in solar vehicles; they have nonindustrial electrical applications and are used in tools such as toys, solar calculators, etc.).

1.5.6.4 Solar energy status

One of the renewable energy types used in the energy—water nexus is solar energy; for example, toward a sustainable energy transition in the Mediterranean where this region will be facing 62% increase in energy demand by 2040. Solar energy and wind energy will be the cheapest way to generate power by 2040. Based on the data of solar atlas, it is recommended to use solar energy in energy–water nexus in the sun-belt countries where these countries are endowed with high intensity of direct solar radiation. For example, Egypt has 2900–3200 hours of sunshine per year, with direct normal intensity of 1970–3200 kWh/m^2 per year and a total radiation intensity of 2000 and 3200 kWh/m^2 per year [66].

After the 1973 energy crisis the international society's attention increased to solar energy use. In most countries of the world, especially industrialized countries, much capital was assigned to researching and achieving optimal solar energy application designs. In industrialized countries, considerable attention is paid to concentrated power plants and photovoltaic systems for power generation. In addition, nonpower plant systems are of interest in many countries worldwide, especially the United States.

According to a study [67], about 629 GW of electricity were generated by converting solar energy into electricity worldwide by the end of 2019. China has generated one-third of the world's solar electricity by generating about 208 GW of electricity from solar energy.

From 2016 to 2019, China, the United States, and India were pioneered in electricity supply by photovoltaic systems. Moreover, the world's biggest photovoltaic farm with a capacity of 850 MW of electricity generation is located in China.

On the other hand, since 2007, implementing and using concentrated power systems to provide power has been improved after a two-decade delay. However, because photovoltaic systems were becoming cheaper, there was no speedy development in their implementation. Hence, photovoltaic farms are widespread in most countries. The largest concentrated power plants are located in Spain and the United States. The largest concentrated power plant with a capacity of 377 MW is located in the United States.

It is expected, utilization of solar energy for electricity production will be significantly grown by 2030 globally. China will have a significant share in electricity generation by solar energy and will produce about 1200 GW of electricity by combination of solar and wind capacity by 2030 [67].

1.6 An overview of solar energy in water treatment

It is apparent that readily available fossil reserves are depleting at an unprecedented pace, necessitating an urgent shift to alternative and renewable energy sources. Photovoltaic systems may be combined with water/wastewater treatment systems in this scenario. There are different water treatment options using solar energy.

Considering water extraction from humid air using desiccants, we depend on atmospheric air as a renewable and huge water reservoir. This endless water source is available everywhere on the surface of earth. The water amount in atmospheric air is estimated as 14000 km^3, while the fresh water amount in lakes and rivers on the surface of earth is only about 1200 km^3 [68]. Water extraction from humid air has many advantages in comparison with the other ways. Water extraction from humid air can be done in various ways. The most common ways are cooling humid air to a temperature below than the dew point of air, fog collection, and water vapor absorption from humid air utilizing a liquid or a solid desiccant. One of the most commonly used desiccant materials is solution of calcium chloride ($CaCl_2$) due to its cheap price and being nontoxic. It is recommended to use $CaCl_2$ solution at initial concentration of 30% [68–70].

Fathy et al. [70] compared the water productivity in seven studies using $CaCl_2$ solution at different cities in Egypt, India, and China. In literature, there are many different desiccants for water extraction from humid air [16]. Salehi et al. [16] showed that hydrogels and metal organic framework as the adsorbents that are appropriate for water harvest.

There are two processes in water extraction from air using desiccant materials. They are the absorption process and the regeneration process. The absorption process takes place during the night while the regeneration process takes place during the day. In the absorption process, mass transfer of water vapor between desiccant surface and atmospheric air. The partial pressure difference of water is the driving force, which mainly has an effect on the absorbed water quantity from the air during the absorption period. When the partial pressure difference of water is equal to zero, there is no mass transfer of water vapor occurring between desiccant surface and atmospheric air. In the regeneration process the solar energy is used as the heating source to heat and dry the wet desiccant materials.

Solar desalination is the process of making fresh water/potable water from saline or brackish water. Solar desalination is particularly beneficial in areas where there is a high demand for drinking water. Because of the high demand for fresh water, industrial desalination has become a viable alternative. Desalination's biggest benefit is that it is a drought-resistant technology [71]. The simplest desalination system can be built using solar

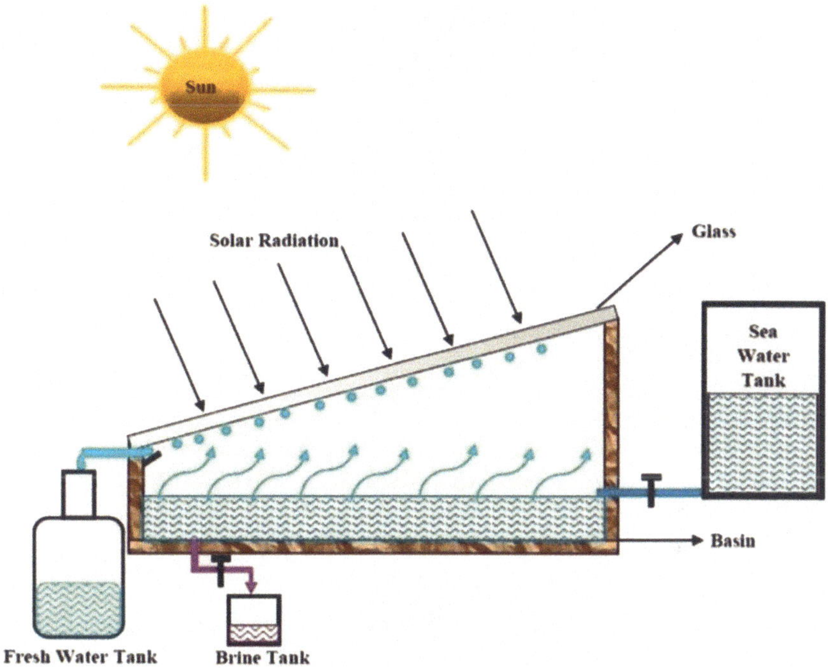

Figure 1.4 Representation of simple solar still [72]. *With permission from Elsevier, LN: 5039170877523.*

still. There are different arrangements of solar stills, and the simplest arrangement is depicted in Fig. 1.4.

As mentioned earlier, in the desalination method, we depend on the existence of salty water resources such as seas and oceans. The energy–water nexus in desalination systems are dependent on applying renewable energy to produce water. Desalination can be classified into two general kinds (the existence and the absence of membrane technology). In their review on solar-powered desalination, Ahmed et al. [73] reported that the half of the cost is related to consuming energy. One of the most used equipment kinds in water desalination is solar still. In this approach the water evaporates with solar energy while the steam condensates to produce water. Salehi et al. [16] reviewed different system forms of solar still with various conditions.

The humidification–dehumidification (HDH) system is considered as one of the most used equipment kinds in solar desalination. Both solar still and HDH can be classified as direct solar desalination. On the other hand, indirect solar desalination can be classified into two general kinds (solar

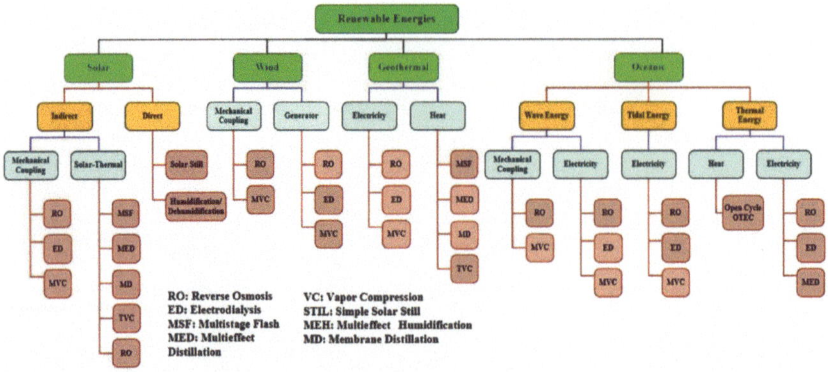

Figure 1.5 Renewable sources for different desalination techniques. *From A.A. Salehi, M. Ghannadi-Maragheh, M. Torab-Mostaedi, R. Torkaman, M. Asadollahzadeh, A review on the water-energy nexus for drinking water production from humid air, Renewable and Sustainable Energy Reviews 120 (2020), with permission from Elsevier, LN: 5039170708102.*

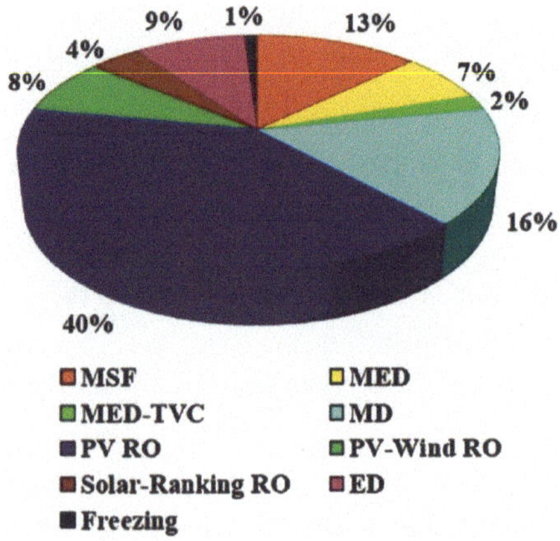

Figure 1.6 The different desalination process contribution by using renewable energy. *MSF*, Multistage flash; *MED*, multi effect distillation; *MED-TVC*, multi effect desalination with thermal vapor compression; *MD*, membrane distillation; *PV RO*, photovoltaic reverse osmosis; *ED*, electrodialysis. *From A.A. Salehi, M. Ghannadi-Maragheh, M. Torab-Mostaedi, R. Torkaman, M. Asadollahzadeh, A review on the water-energy nexus for drinking water production from humid air, Renewable and Sustainable Energy Reviews 120 (2020), with Permission from Elsevier, LN: 5039170708102.*

thermal and mechanical coupling). Salehi et al. [16] showed the different processes shared and the desalination plants distribution with the use of renewable energy. As shown by Salehi et al. [16] in Fig. 1.5, there are five solar thermal types.

As shown by Salehi et al. [16] in Fig. 1.6, photovoltaic RO represents the largest contribution by using renewable energy with 40%.

Recently, new solar desalination systems have been introduced. For example, Abu El-Maaty et al. [74,75] introduced a fog desalination system integrated with evacuated tube solar collectors. They proposed this new solar desalination system as the HDH system development. In this system the fogged droplets passed through a solar powered heater to help fogged droplets to evaporate, and the air was heated and humidified. Then, this air rose and passed through a condenser. Finally, the condensate was collected. The maximum productivity reached 5.83 kg/m^2 per day. The cost was 0.035 $/L.

Solar energy, in the form of photovoltaic modules, is the most suitable power source for RO systems. Solar-powered RO systems are ideal for remote areas where pure water is in high demand [76]. To make the system economically feasible for large-scale water supply, the RO system should be combined with energy recovery turbines.

Similar to desalination technology, electrodialysis is a type of desalination process which separates salts by using membranes [77]. Membranes are used to draw these salts by making a potential difference and transferring ions. After electrodialysis the salt-free water is left behind. It should be mentioned that electrodialysis is better suited to water that is less saline or brackish (less than 6000 ppm). Furthermore, the electrodialysis process is ineffective for water with dissolved solids below 400 ppm. Solar energy using photovoltaic modules may be the best source of power to make electrodialysis a cost-effective choice in remote locations.

In addition to water treatment systems, solar energy can be integrated with water disinfection options. It is well known that chlorine, chloramines, and ozone are widely used disinfectants. In addition, ultraviolet (UV) light, mixed-oxidant gases, and photocatalysis are other disinfection options. If we consider UV light, it is a widely used disinfection option in rural areas and health care centers. This method uses UV lamp to disinfect water. In this method a thin layer of water is allowed to pass through a tube exposed to UV light. As a result the bacteria are destroyed. However, when there are dust and mud particles in water, they can shield the bacteria, and bacteria can escape before being disinfected. Furthermore, certain microorganisms that have hard cysts cannot be destroyed by UV technology. In remote locations a parabolic trough solar concentrator is a viable system to treat water. By integrating a solar pump with this system, disinfected water can be circulated in the tank. In the Florida solar energy

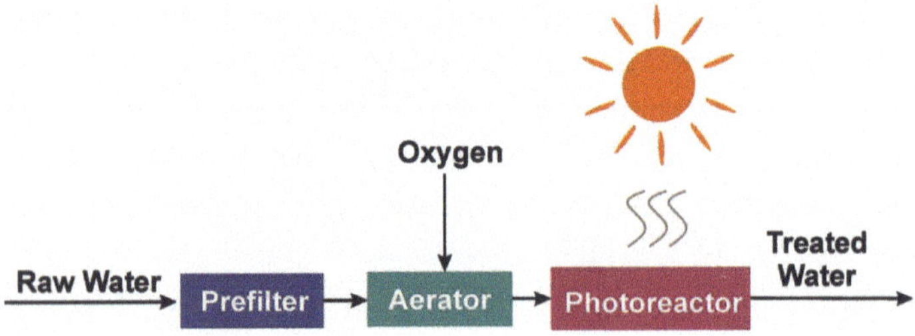

Figure 1.7 Solar water detoxification system.

center, this disinfection method was used to disinfect about 2500 m^3/day of water using a 28 m^2 solar concentrator [78].

Mixed-oxidant gases technology is a most effective technology to destroy bacteria and viruses [79]. There are numerous designs available for the oxidant gas process. The input energy requirements to this system are dependent on the oxidant solution quantity. When the higher oxidant solution dosage is used, the flow rate of the treated water is less. The photovoltaic arrays can be used to power this disinfection system, and this system is much suitable for large rural areas.

A photocatalytic system with a specific series reaction is used in a solar water detoxification system [80]. A parabolic trough concentrator is the most common design for this device. The TiO_2 catalyst in this photoreactor is activated by the near-UV portion of the solar spectrum, which creates hydroxyl radicals [81]. As a result organic compounds are attacked and broken down into nontoxic materials including carbon dioxide and water. The representation of the solar water detoxification device is shown in Fig. 1.7.

1.7 Conclusion

Wastewater needs treatment before it comes into contact with surface or groundwater. Similarly, domestic water should be free of contaminants and safe to drink. Water treatment uses and will use renewable energy sources, either directly or indirectly. Since the power needed to treat water in rural areas is relatively low, renewable energy systems are now used in rural areas to treat water. However, because of the high capital cost, there are many problems with using renewable energy to treat water on a large scale.

Among different renewable energy resources, solar energy is the simplest method to treat saline water and disinfect drinking water. Solar energy in the form of photovoltaic modules/arrays can be used to power the water treatment and disinfection systems including, RO, UV technology, and photocatalysis.

However, the issues and risks associated with solar-driven water purification technologies such as economic viability of solar-driven systems, access to finance, system installation, operation and maintenance, standardization and quality control need to be assessed. Some other issues that need to be discussed related to water purification are the effect of different parameters that affect water treatment and disinfection. In addition, the assessment of solar-driven water treatment technologies based on several factors like resources and labor availability, suitability to different geographical locations, favorability and suitability to conditions and needs, technology infrastructure, performance, and suitability of the technology should be addressed.

References

[1] S.O. Fakayode, Impact assessment of industrial effluent on water quality of the receiving Alaro River in Ibadan, Nigeria, African Journal of Environment Assessment and Management (2005).

[2] T. Debièche, J. Mania, J. Mudry, Pollution d'une nappe alluviale par le chrome et l'étain à partir d'un stockage de résidus métallurgiques. Application à la basse plaine de la Seybouse, Nord-Est algérien, Africa Geoscience Review 8 (2003) 425–435.

[3] N. Argaw, Renewable energy in water and wastewater treatment applications, 2003.

[4] Future global water security in a changing environment human and environmental security issues, 2013.

[5] UN Water, Water stress versus water scarcity, 2017 (n.d.). Available from: <http://www.unwater.org/water-facts/scarcity>.

[6] UN Water, Managing water under uncertainty and risk, The United Nations world water development report 4, UN Water Reports, World Water Assessment Programme, UNESCO, Paris, France, 2012.

[7] R. Engelman, R.P. Cincotta, B. Dye, T. Gardner-Outlaw, J. Wisnewski, People in the balance: Population and Natural Resources at the Turn of the Millennium (2000).

[8] K. Komoto, K. Kurokawa, M. Ito, P. Van Der Vleuten, Energy from the desert: very large scale photovoltaic systems: socio-economic, financial, technical and environmental aspects, 2009 (n.d.).

[9] S. Liburd, Solar-driven humidification dehumidification desalination for potable use in Haiti, 2010 (n.d.).

[10] WHO [online] (n.d.). Available from: <https://wholives.org/our-mission/mission/#>.

[11] NASA, 1 October 2010 (n.d.). Available from: <https://earthobservatory.nasa.gov/features/Water>.

[12] S. Campbell, The Home Water Supply: How to Find, Filter, Store, and Conserve It, Storey Publishing, LLC; Later Printing edition, 1983.

[13] C. Joel, L.A. Mwamburi, E.K. Kiprop, Use of slow sand filtration technique to improve wastewater effluent for crop irrigation, Microbiology Research (Pavia) (2018). Available from: https://doi.org/10.4081/mr.2018.7269.

[14] K.M. Sosthene, N. Gahi, Low cost filtration of domestic wastewater for irrigation purpose, World Journal of Engineering and Technology (2018). Available from: https://doi.org/10.4236/wjet.2018.63036.

[15] J. Dai, S. Wu, G. Han, J. Weinberg, X. Xie, X. Wu, et al., Water-energy nexus: a review of methods and tools for macro-assessment, Applied Energy (2018). Available from: https://doi.org/10.1016/j.apenergy.2017.08.243.

[16] A.A. Salehi, M. Ghannadi-Maragheh, M. Torab-Mostaedi, R. Torkaman, M. Asadollahzadeh, A review on the water-energy nexus for drinking water production from humid air, Renewable and Sustainable Energy Reviews 120 (2020). Available from: https://doi.org/10.1016/j.rser.2019.109627.

[17] P. Breeze, Chapter 11—Wind power, in: P. Breeze (Ed.), Power Generation Technologies, third ed., Newnes, 2019, pp. 251–273. <https://doi.org/10.1016/B978-0-08-102631-1.00011-0>.

[18] EIA (n.d.). Available from: <https://www.eia.gov/>.

[19] T.J. Price, James Blyth—Britain's first modern wind power pioneer, Wind Engineering 29 (2005) 191–200. Available from: https://doi.org/10.1260/030952405774354921.

[20] Horizontal Axis Wind Turbines (HAWT): Advantages and Disadvantages (2020). Available from: <https://www.luvside.de/en/hawt-advantage-disadvantage/>.

[21] IPCC, Special report on renewable energy sources and climate change mitigation, Chapter 5: Hydropower, 2012.

[22] J.S. Hecht, G. Lacombe, M.E. Arias, T.D. Dang, T. Piman, Hydropower dams of the Mekong River basin: a review of their hydrological impacts, Journal of Hydrology (2019). Available from: https://doi.org/10.1016/j.jhydrol.2018.10.045.

[23] P. Breeze, Chapter 1—an introduction to hydropower, in: P. Breeze (Ed.), Hydropower, Academic Press, 2018, pp. 1–12. Available from: https://doi.org/10.1016/B978-0-12-812906-7.00001-6.

[24] C.S. Kaunda, C.Z. Kimambo, T.K. Nielsen, Hydropower in the context of sustainable energy supply: a review of technologies and challenges, ISRN Renewable Energy (2012). Available from: https://doi.org/10.5402/2012/730631.

[25] H.J. Wagner, J. Mathur, Introduction to hydro energy systems: basics, technology and operation, Green Energy and Technology (2011). Available from: https://doi.org/10.1007/978-3-642-20709-9.

[26] M. Kötter, J. Mathur, Introduction to Hydro Energy Systems: Basics, Technology and Operation, 2011.

[27] IRENA, Renewable capacity highlights (n.d.). Available from: <http://www.IRENA.org>.

[28] EIA (n.d.). Available from: <https://www.eia.gov/todayinenergy/detail.php?id = 41533>.

[29] H. Gupta, S. Roy, Geothermal Energy: An Alternative Resource for the 21st Century, 2007. <https://doi.org/10.1016/B978-0-444-52875-9.X5000-X>.

[30] P. Stelling, L. Shevenell, N. Hinz, M. Coolbaugh, G. Melosh, W. Cumming, Geothermal systems in volcanic arcs: volcanic characteristics and surface manifestations as indicators of geothermal potential and favorability worldwide, Journal of Volcanology and Geothermal Research. (2016). Available from: https://doi.org/10.1016/j.jvolgeores.2016.05.018.

[31] M. Melikoglu, Geothermal energy in Turkey and around the World: a review of the literature and an analysis based on Turkey's Vision 2023 energy targets, Renewable and Sustainable Energy Reviews (2017). Available from: https://doi.org/10.1016/j.rser.2017.03.082.

[32] O.D. Elvan, Y.O. Turker, Geothermal energy capacity and legislation in Turkey, The Journal of World Energy Law and Business (2013). Available from: https://doi.org/10.1093/jwelb/jwt015.

[33] Geothermal energy generation worldwide in 2020, by country (n.d.). Available from: <https://www.statista.com/statistics/514488/geothermal-generation-worldwide-by-key-country/#statisticContainer>.
[34] Y. Noorollahi, M.S. Shabbir, A.F. Siddiqi, L.K. Ilyashenko, E. Ahmadi, Review of two decade geothermal energy development in Iran, benefits, challenges, and future policy, Geothermics (2019). Available from: https://doi.org/10.1016/j.geothermics.2018.10.004.
[35] K. Li, H. Bian, C. Liu, D. Zhang, Y. Yang, Comparison of geothermal with solar and wind power generation systems, Renewable and Sustainable Energy Reviews (2015). Available from: https://doi.org/10.1016/j.rser.2014.10.049.
[36] Y. Sang, H.B. Karayaka, Y. Yan, N. Yilmaz, D. Souders, 1.18 Ocean (marine) energy, in: I. Dincer (Ed.), Comprehensive Energy Systems, Elsevier, Oxford, 2018, pp. 733–769. Available from: https://doi.org/10.1016/B978-0-12-809597-3.00120-6.
[37] A. Clément, P. McCullen, A. Falcão, A. Fiorentino, F. Gardner, K. Hammarlund, et al., Wave energy in Europe: current status and perspectives, Renewable and Sustainable Energy Reviews 6 (2002) 405–431. Available from: https://doi.org/10.1016/S1364-0321(02)00009-6.
[38] T. Brekken, Fundamentals of ocean wave energy conversion, modeling, and control, in: IEEE International Symposium on Industrial Electronics, 2010. Available from: <https://doi.org/10.1109/ISIE.2010.5637668>.
[39] Wave power's share of global electricity demand to reach 10% by 2050 (2017). Available from: <https://www.climateaction.org/news/wave-powers-share-of-global-electricity-demand-to-reach-10-by-2050>.
[40] R. Pelc, R.M. Fujita, Renewable energy from the ocean, Marine Policy (2002). Available from: https://doi.org/10.1016/S0308-597X(02)00045-3.
[41] Tidal Energy (n.d.). Available from: <https://ei.lehigh.edu/learners/energy/tidal/tidal4.html>.
[42] F. Zabihian, A.S. Fung, Review of marine renewable energies: case study of Iran, Renewable and Sustainable Energy Reviews (2011). Available from: https://doi.org/10.1016/j.rser.2011.02.006.
[43] United States Department of the Interior, Technology white paper on ocean current energy potential on the United States outer continental shelf, 2006.
[44] P. Basu, Chapter 1—introduction, in: P. Basu (Ed.), Biomass Gasification, Pyrolysis Torrefaction, third ed., Academic Press, 2018, pp. 1–27. Available from: https://doi.org/10.1016/B978-0-12-812992-0.00001-7.
[45] M. Balat, G. Ayar, Biomass energy in the world, use of biomass and potential trends, Energy Sources (2005). Available from: https://doi.org/10.1080/00908310490449045.
[46] Global Bioenergy Statistics (2019). Available from: <https://worldbioenergy.org/191129WBAGBS2019_HQ>.
[47] Y. Zhang, G. Riskowski, T. Funk, Thermochemical conversion of swine manure to produce fuel and reduce waste (1999). Available from: <https://citeseerx.ist.psu.edu/viewdoc/download?doi = 10.1.1.505.715&rep = rep1&type = pdf>.
[48] S. Bhattacharya, Biomass energy in developing countries, 2006.
[49] IEA (2018). Available from: <https://www.iea.org/fuels-and-technologies/renewables>.
[50] W.V. Reid, M.K. Ali, C.B. Field, The future of bioenergy, Global Change Biology (2020). Available from: https://doi.org/10.1111/gcb.14883.
[51] M.F. Demirbas, M. Balat, H. Balat, Potential contribution of biomass to the sustainable energy development, Energy Conversion and Management (2009). Available from: https://doi.org/10.1016/j.enconman.2009.03.013.
[52] A. Demirbaş, Global renewable energy resources, Energy Sources, Part A: Recovery, Utilization and Environmental Effects (2006). Available from: https://doi.org/10.1080/00908310600718742.

[53] A. Demirbas, Modernization of biomass energy conversion facilities, Energy Sources, Part B: Economics, Planning, and Policy (2007). Available from: https://doi.org/10.1080/15567240500402784.
[54] I. Vinoth Kanna, D. Pinky, Solar research—a review and recommendations for the most important supplier of energy for the earth with solar systems, International Journal of Ambient Energy. (2020). Available from: https://doi.org/10.1080/01430750.2018.1472658.
[55] J. Gong, C. Li, M.R. Wasielewski, Advances in solar energy conversion, Chemical Society Reviews (2019). Available from: https://doi.org/10.1039/c9cs90020a.
[56] S.A. Kalogirou, Solar thermal collectors and applications, Progress in Energy and Combustion Science (2004). Available from: https://doi.org/10.1016/j.pecs.2004.02.001.
[57] J.Y. Yoo, H.J. Kim, E.J. Woo, C.J. Park, On solar energy utilization for drying technology, International Journal of Environmental Science and Development (2017). Available from: https://doi.org/10.18178/ijesd.2017.8.4.968.
[58] S.A. Kalogirou, Chapter one—introduction, in: S.A. Kalogirou (Ed.), Solar Energy Engineering, Academic Press, Boston, 2009, pp. 1−48. Available from: https://doi.org/10.1016/B978-0-12-374501-9.00001-7.
[59] S.K. Sansaniwal, V. Sharma, J. Mathur, Energy and exergy analyses of various typical solar energy applications: a comprehensive review, Renewable and Sustainable Energy Reviews (2018). Available from: https://doi.org/10.1016/j.rser.2017.07.003.
[60] M.R.S. Shaikh, A review paper on electricity generation from solar energy, International Journal for Research In Applied Science and Engineering Technology (2017). Available from: https://doi.org/10.22214/ijraset.2017.9272.
[61] A. Jamar, Z.A.A. Majid, W.H. Azmi, M. Norhafana, A.A. Razak, A review of water heating system for solar energy applications, International Communications in Heat and Mass Transfer (2016). Available from: https://doi.org/10.1016/j.icheatmasstransfer.2016.05.028.
[62] T. Pauschinger, 5—Solar thermal energy for district heating, in: R. Wiltshire (Ed.), Advanced District Heating and Cooling Systems, Woodhead Publishing, Oxford, 2016, pp. 99−120. Available from: https://doi.org/10.1016/B978-1-78242-374-4.00005-7.
[63] I. Ullah, M.G. Rasul, Recent developments in solar thermal desalination technologies: a review, Energies (2019). Available from: https://doi.org/10.3390/en12010119.
[64] S.A. Kalogirou, Solar Energy Engineering: Processes and Systems, second ed., 2014. <https://doi.org/10.1016/C2011-0-07038-2>.
[65] L. Hernández-Callejo, S. Gallardo-Saavedra, V. Alonso-Gómez, A review of photovoltaic systems: design, operation and maintenance, Solar Energy (2019). Available from: https://doi.org/10.1016/j.solener.2019.06.017.
[66] A. Lashin, Review of the geothermal resources of Egypt: 2015−2020, in: Proceedings of the World Geothermal Congress 2020, Reykjavik, Iceland, n.d.
[67] China's new 2030 climate commitments: beyond peak emissions, n.d.
[68] K. Awad, M. Awad, A. Kandel, Extraction of water from atmospheric air using double slope condensation surface, Bulletin of the Faculty of Engineering Mansoura University (2020). Available from: https://doi.org/10.21608/bfemu.2020.97651.
[69] M.A. Talaat, M.M. Awad, E.B. Zeidan, A.M. Hamed, Solar-powered portable apparatus for extracting water from air using desiccant solution, Renewable Energy (2018). Available from: https://doi.org/10.1016/j.renene.2017.12.050.
[70] M.H. Fathy, M.M. Awad, E.S.B. Zeidan, A.M. Hamed, Solar powered foldable apparatus for extracting water from atmospheric air, Renewable Energy. (2020). Available from: https://doi.org/10.1016/j.renene.2020.07.020.
[71] V. Martínez-Alvarez, M.J. González-Ortega, B. Martin-Gorriz, M. Soto-García, J.F. Maestre-Valero, 14: Seawater desalination for crop irrigation—Current status and

perspectives, in: V.G. Gude (Ed.), Emerging Technologies for Sustainable Desalination Handbook., Butterworth-Heinemann, 2018, pp. 461–492. <https://doi.org/10.1016/B978-0-12-815818-0.00014-X>.

[72] S.W. Sharshir, N. Yang, G. Peng, A.E. Kabeel, Factors affecting solar stills productivity and improvement techniques: a detailed review, Applied Thermal Engineering. (2016). Available from: https://doi.org/10.1016/j.applthermaleng.2015.11.041.

[73] F.E. Ahmed, R. Hashaikeh, N. Hilal, Solar powered desalination—Technology, energy and future outlook, Desalination 453 (2019) 54–76. Available from: https://doi.org/10.1016/j.desal.2018.12.002.

[74] A.E. Abu El-Maaty, M.M. Awad, G.I. Sultan, A.M. Hamed, Solar powered fog desalination system, Desalination (2019). Available from: https://doi.org/10.1016/j.desal.2019.114130.

[75] A.E. Abu El-Maaty, M.M. Awad, G.I. Sultan, A.M. Hamed, Performance study of fog desalination system coupled with evacuated tube solar collector, Desalination (2021). Available from: https://doi.org/10.1016/j.desal.2021.114960.

[76] A.M. Delgado-Torres, L. García-Rodríguez, B. Peñate, J.A. de la Fuente, G. Melián, Chapter 3—water desalination by solar-powered RO systems, in: A. Basile, A. Cassano, A. Figoli (Eds.), Current Trends and Future Developments on Membranes, Elsevier, 2019, pp. 45–84. Available from: https://doi.org/10.1016/B978-0-12-813545-7.00003-9.

[77] F. Macedonio, E. Drioli, 4.09: Membrane systems for seawater and brackish water desalination, in: D. Enrico, G. Lidietta (Eds.), Comprehensive Membrane Science and Engineering, Elsevier, 2010, pp. 241–257. ISBN 9780080932507. <https://doi.org/10.1016/B978-0-08-093250-7.00056-6>.

[78] R. Anderson, K. Collier, Solar water disinfection, in: American Solar Energy Society 1996 Annual Conference Proceedings, 13–18 April 1996, Asheville, North Carolina, 1996.

[79] H.B. Carmona, A. Benitez, J. Zarco, E. Sánchez, I. Mascher, Efficiency of oxidant gas generator cells powered by electricity or solar energy, Pan American Journal of Public Health (1998). Available from: https://doi.org/10.1590/s1020-49891998000200007.

[80] A.G. Gutierrez-Mata, S. Velazquez-Martínez, A. Álvarez-Gallegos, M. Ahmadi, J.A. Hernández-Pérez, F. Ghanbari, et al., Recent overview of solar photocatalysis and solar photo-fenton processes for wastewater treatment, International Journal of Photoenergy. (2017). Available from: https://doi.org/10.1155/2017/8528063.

[81] O. Ola, M.M. Maroto-Valer, Review of material design and reactor engineering on TiO_2 photocatalysis for CO_2 reduction, Journal of Photochemistry and Photobiology C: Photochemistry Reviews (2015). Available from: https://doi.org/10.1016/j.jphotochemrev.2015.06.001.

CHAPTER 2

Water—Conventional and novel treatment methods

Gokulan Ravindiran[1], Praveen Saravanan[2], Raja Murugadoss Jeyaraju[3] and Jegan Josephraj[4]
[1]Department of Civil Engineering, GMR Institute of Technology, Rajam, India
[2]Department of Civil Engineering, CMR Institute of Technology, Bengaluru, India
[3]Department of Civil Engineering, Nadimpalli Satyanarayana Raju Institute of Technology, Visakhapatnam, India
[4]Department of Civil Engineering, University College of Engineering Ramanathapuram, Anna University, Ramanathapuram, India

2.1 Introduction

In general, human civilization has placed itself in areas with locally sustainable water sources, in the form of runoff, and/or rivers and streams [1]. Water is an essential life-sustaining element; human life on the earth has become sustainable due to the presence of water, and agricultural productivity on the earth is also possible only in the presence of water. Water has been discussed in terms of quantity and quality, wherein the quantity measures the total available form of water that can be used for human and agricultural practice and the quality indicates the suitability of the water for drinking, domestic, and irrigation purposes [2].

Since the 1980s, water demand has increased by about 1% annually worldwide because of the cumulative population increase, socio-economic progress, and changing lifestyle patterns. Global water demand is projected to rise at the same rate until 2050, reflecting 20%–30% higher levels of water usage than at present, largely because the demand in the manufacturing and domestic sectors is increasing. More than two billion people around the world have high water scarcity and over four billion have a serious water deficit for at least one month a year. Stress levels will continue to rise as water demand increases and climate change impacts increase. Fig. 2.1 summarizes the role of the ecosystem in the water cycle.

Water and electricity are intertwined intricately. Although there are various electricity sources, water requirement has become important since its wide range of utilization, such as raw material production, thermal process cooling, cleaning materials, the cultivation of biofuels crops, as well as

Solar-Driven Water Treatment
DOI: https://doi.org/10.1016/B978-0-323-90991-4.00003-7

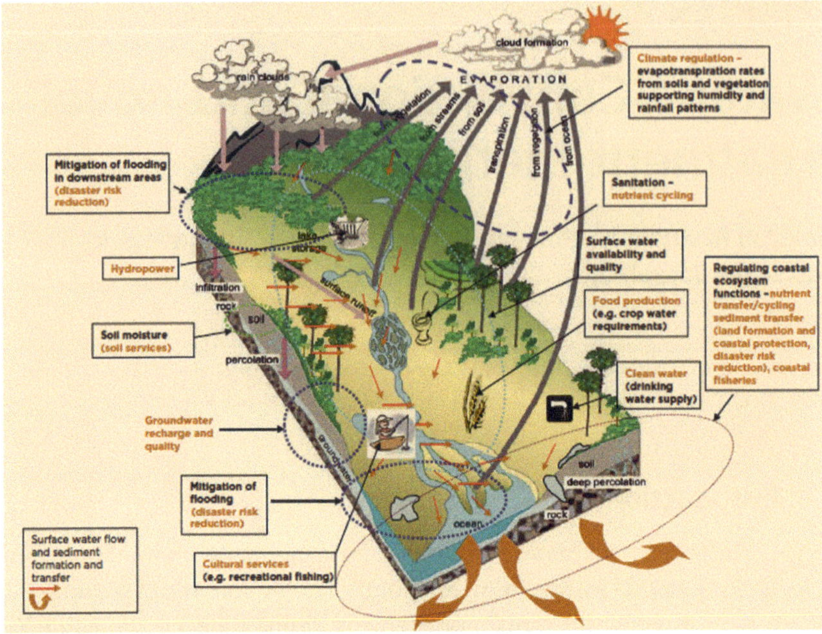

Figure 2.1 Role of the ecosystem in the water cycle [3].

turbines for power supply. In comparison, energy is required for pumping, transport, treatment, desalination, and irrigation of water supplies and human consumption. This two-sided interdependence of the two services is a coined relation between water power and the major cross-sector vulnerabilities.

Water is not spread uniformly across the world, and certain areas would be exposed to higher water-for-energy strains. The WEC (2010) [4] predicts that it will become increasingly important to explore new technology to process primary energy and produce electricity in Asia and the Middle East, where water scarcity and demand for water increased due to expectations to raise their electrical output by five times. External drivers are the struggle of managing this connection, whose effect can only be measured but never totally expected. Climate change is a key external driver, impacting specifically water and electricity; steps for reducing energy use and carbon emissions are centralized, while preparation includes planning to improve hydrological diversity and severe weather conditions, including flooding, droughts, and storms.

Further external pressure is generated by population growth and migration, and increased economic activity and living conditions which would produce increased energy demand, in particular in non-OECD nations. Further external pressures emerge from demographic change. Finally, government policy decisions also aggravate the burden by seeking higher water and higher cost energy. Energy needs to be drained (surface water, groundwater), treated (drinking water treatment standard, desalination), supplied with water services (municipal, commercial, and agricultural supplies), reconditioned (wastewater treatment), and released. Yet few countries are currently researching and commenting on water energy needs.

In terms of procurement of water and wastewater system, EPRI (2002) [5] reports 2%–4% of overall US energy usage. Including end-user consumption, US national water energy use is around 10% [6]. Surface water injection energy needs are normally 30% less than in groundwater pumping (EPRI, 2002) [7]. Groundwater will be predicted to rise as the energy use rates decrease in many areas. Moreover, where surface water is not plentiful, it could be more energy-intensive to bring water into the area than to drain available groundwater supplies.

Water is commonly treated using physical, chemical, and biological methods to remove the pollutants and to meet the drinking water standards and these standards are varying based on the requirements. But in many countries, fresh water is not available and they depend on the seawater for drinking purpose and they are using membrane technologies for the removal of dissolved salts present in the water. So, desalination plants with high capacity are installed and these resulted in the consumption of more energy for the operation and treatment of the saline water [7,8]. Many types of research have been carried out in the prediction of the total energy consumption for the treatment of saltwater in a desalination plant. Strokes and Horvath (2009) [9], from their research, concluded that the average requirement of the electricity for the desalination plant was found to be 0.38 kWh per year per m^3 for the conventional treatment plant, and for the brackish desalination plant, it requires 0.26 kWh per year per m^3 for the treatment. These increased the cost of the treated water and this was gradually rising with an increase in the energy demand and water demand in the past decade. Desalination also creates heavily concentrated waste sources of salt which have to be disposed of. Coastal waste plants dump solely into the surrounding waters and have detrimental effects on marine coastal ecology.

As seen in Table 2.1, wastes often require significant quantities of energy for treatment [10]. Stricter discharge laws in high-income countries install more energy-intensive technology of treatment. Filter treatment for aerobic

Table 2.1 Energy requirement for the treatment of water for different source [10].

Treatment type		Energy use (kWh/million L)
Water	Surface water	60
	Groundwater	160
	Brackish groundwater	1000–2600
	Seawater	2600–4400
Wastewater	Trickling filter	250
	Activated sludge	340
	Advanced treatment without nitrification	400
	Advanced treatment with nitrification	500

treatment absorbs an average of 250 kWh/mL [5,11]. Diffused aeration is a more energetic wastewater system as part of the activated slot method, and requires 340 kWh/mL [5,11]. Further processing of wastewater with filtration and nitrification option includes 400–500 kWh per mL [5,11]. The recent methods for sludge treatment will potentially absorb 30%–80% of the overall energy consumption of wastewater plants [12]. A green fuel, namely, biogas or biomethane can be produced from the sludge that was generated from wastewater treatment plants using an anaerobic digestion tank and it can give back 50% of the total consumed energy as a by-product [13,14].

As wastewater treatment plants require more treatment facilities while comparing with the water treatment plants, it is likely to raise wastewater treatment units' energy requirements in the future for income-enhancing countries [15]. It is therefore possible to mitigate the expected energy consumption changes for tighter treatment conditions by implementing better energy conservation and reducing the predicted rise in the usage of energy in treatment plants.

Energy is now being used for field irrigation. A limited proportion of the overall resources embedded in water in OECD countries is irrigation. However, irrigation absorbs a comparatively greater share of water resources in non-OECD countries where treatment and heating are less common. Power demands for the promotion of rising populations are growing and water shortages would push countries to increasingly pursue unconventional water supplies with increased energy demands. Thereby, innovations are increasingly more effective in terms of electricity [8], the increased energy use of water supply from sources and sources increasingly disadvantageous because of the location or the handling of water of less quality is offset by this benefit inefficiency risk.

2.2 Water: uncertainty

Water quality becomes an increasingly significant global issue, as the possibility of decline directly translates into socio-economic implications. The consistency of the water is inextricably related to the volume of water as these are the main factors in the supply. In recent decades, water quantity, research, and public attention have attracted significantly more funding compared with water quality. However, the quality of water is as critical for fulfilling basic human and environmental standards as the volume of water. Besides, the Deux are intertwined inextricably, and water quality in different ways influences water quantities.

For instance, the amount of water available for use in a given environment, which is polluted and not used to drink, bathe, or farm, can effectively be decreased [16]. Further contamination of the water will increase the treatment costs [16]. According to Stellar (2010) [17], in the cases of groundwater and surface water, the relationship between quality and quantity will take different forms. There is an apparent connection between overuse and depletion of the content of underground aquifers. Excessive groundwater pumping over time will reduce water quality in two ways. First, groundwater pumping impacted with a decrease in volume of water, as in India with fluorosis that may theoretically endanger or impact millions of persons, raises naturally occurring compounds. Second, as in the case of Cyprus and the Gaza Strip, consistency may be influenced by increasing salinity by saltwater injection into coastal aquifers [17]. The common problem created by saltwater pollution is a decrease in water available for human use, but it also specifically affects other applications such as agriculture and industry. Surface water can contribute to increased water contamination or mineral leaching concentrations of hazardous material. There is a good example of this with regard to the Rio Grande River, where decreasing flows correlate with major water quality reductions in the summer months.

The concentration of disease is about 100 times higher during the dry season [17]. Politicians must work together to further incorporate water quantity and water quality concerns into their answers. On the other hand, they need the assistance of the scientific community, which will increase the quantification of the challenges and the creation of solutions. The quality of water depends on socio-economic growth. Human and ecosystem health threats have been correlated with low quality of water, which in turn threatens socio-economic growth. The wealthier, more developed countries and their conservation campaign have traditionally had responsibility for ecosystem health. The work of marine environments degrades through the removal of varied products and services that are accessible through water contaminated with harmful pollutants, such as inorganic compounds and untreated sewage. Because people below poverty rely on these products and services, this attempts to alleviate poverty further [18].

As regards the reactions, cost-effective processing, storage, and disposal of human waste are important. More than 80% of the water used globally is projected not to be obtained or processed [19]. Solutions in the treatment of wastewater can also be paired with efforts in public safety such as personal hygiene and protection in the environment. Studies have shown that proper hygiene and healthy drinking water will minimize diarrheal diseases by almost 90% [20]. Efforts must therefore be rendered to the companies that consume or manufacture hazardous chemicals.

The priority factor is the production of renewable technologies and replacement technologies combined with economical treatment solutions. The management of nonpoint pollution sources, particularly eutrophication nutrients, poses an ever-growing global challenge. Regulations and effective regulatory compliance are important—in addition to operational attempts to improve emergency responses, particularly if the protection of potable water sources in natural disasters is compromised. Toxic water is less prevalent, but in many regions, especially the developing economies, it remains an important problem. The leakage of hazardous waste from waste disposal and commercial undertakings also poses a danger and cost to clean water supplies in the developing world.

2.3 The water–energy nexus

Water and electricity are interconnected tightly and interdependent. Choices and decisions taken in one area can have an important, positive or negative effect on the other. Trade-offs must be controlled to limit harmful repercussions and to promote synergies. Water and energy have key consequences for alleviating poverty directly. As many of the Millennium Development Goals, depend on substantial increases in water, sanitation, and energy infrastructure, - which is essentially the ultimate prospect of a widespread poverty reduction.

Over the next few decades, global demand for fresh water and electricity will continue to rise dramatically to satisfy the demands of rising demographics, markets, shifting habits, and emerging patterns of use. The stresses on scarce natural resources and habitats are significantly intensified. The problem would be undesirable in countries that experience rapid economic transition and expansion, especially in those countries that have insufficient water supplies or insufficient water-related infrastructure and services and modern energy systems continue to be significantly underdeveloped.

Water and energy are intertwined strongly: to a certain degree, the water needs to be generated, transported, and utilized in all categories of energy; and energy is needed to capture, process, and deliver water and its collection and treatment after use. Water and energy are also strongly interdependent, with direct and indirect implications for one domain. Forms of energy generation can be used to assess the amounts of water used to create energy. Freshwater supplies will be distributed, consumed, and controlled to determine how much (or little) water for energy production is usable.

Many businesses can also be influenced by water and energy decisions, and vice versa. These interconnections and interdependencies, along with the negative and positive externalities, are fundamental to the so-called "water–energy nexus". Water threats contribute to energy risks. The demand for water-intensive energy producers raises the pressure to follow alternative approaches, particularly in areas where energy competes with other big water users (e.g., irrigation, construction, drinking water, and urban sanitation) and where the use of water can be minimized to preserve healthy habitats. With the demand for water leading to increased shortages, trade-offs are being handled more urgently and cobenefits across several industries like electricity being maximized.

Dependence on the water in the energy field produces vulnerabilities. Water shortages and high temperatures cycles will lead to power plants shutting off or decreasing their output. With more severe weather conditions due to climate change, the power industry will become vulnerable to higher risk. In the electricity market worldwide, such effects have existed, with various cases on all continents. Water limits are one of the limiting factors for the design of power stations and the option of cooling systems. These options will become constrained and vital as the water-related risks increase. In comparison, climate change expected will also reduce some threats to hydroelectricity production in certain regions [21].

Likewise, uncertainty related to global energy demand growth and development could pose new and substantial threats to water supplies and other users. Unconventional gas and oil supplies (e.g., shale and bitumen) are emerging and there will be substantial consequences for the state, demand for, use of and management of fresh water, as well as recurring concern about the generation, and policy changes towards renewables. For certain energy managers, the efficient usage of water supplies is becoming a business risk. The World Economic Forum ranks in its 2013 Global Risks Report the "water supply crisis" as the fourth and second most likely crisis, a significant increase from its position in previous reports. Multinationals and other large companies are more interested in their water footprint and how they are minimized [22].

2.4 Challenges in water–energy nexus

As previously mentioned, over the next two decades global energy demand is anticipated to rise significantly. In developed countries, this pattern depends mainly on demographic and economic development. Water and electricity will be challenged mainly by supplying water services to ensure an improvement in energy requirements. Policymakers have to encourage more efficient and optimized oil–water usage and vice versa. Comprehensive evaluations of water supply at the national level would be the first step in these policies. Second, water and energy policy that is frequently followed in multiple government agencies or ministries must be more tightly coordinated and coordinated with policymakers.

All the above developments indicate a likely turn towards increasingly energy-intensive water-processing methods. Many countries are in the need of more energy for water treatment since the quality of the water was very low and this has become a challenge in the treatment cost of the water. In addition to handling water to better cleanliness levels, communities will now spend more time moving fresh water from their sources to dense urban areas. These efforts involve drilling into ever deeper lakes or transporting water across broad long-distance ventures [14].

The available water and its extremes of drought and flood and quality are supported by habitats. Water management often includes trade-offs between environmental resources—and also risk transfers. The need for water in ecosystems depends on the need for water to maintain or restore the advantages to citizens (services) which we wish to provide to ecosystems. Improved water conservation and sustainable growth need a stronger emphasis on how to overcome rising competition between habitats and the socio-economic industries for water.

For decades, requests for water from humans and the "environment" or "ecosystem" have been debated. The idea that these issues are distinct in certain respects, which encourages tensions between growth and the environment or environmental interests, was a major source of early dispute. In recent years, greater alignment of interest has been reached by better understanding that the water that is used to protect the atmosphere or the dignity of habitats is actually also a way of serving human interests by maintaining the benefits to humans that a balanced ecosystem provides. Ecosystems have been traditionally considered by others as an unproductive water user. This is not valid, since water is not used by ecosystems—they recover it. But expectations change to the control of human relations with environments to facilitate growth objectives relevant to water. Water and efficiency resources are also ecosystem services in terms of human direct application, as are the advantages of wetlands for the prevention of severe drought and water events. Most of the ecosystem resources, particularly through water, are interrelated. Decisions that facilitate or at the cost of one service invariably require compromise.

Importantly, environment-based trading can also contribute to the transition of risks through associated ecology-related changes, and biodiversity is often considered to be an ecosystem asset because it is directly valued (i.e., cultural/esthetic/recreating benefits; importance for the life of the species). Increasingly, wetlands are viewed as water cure rather than simply as a survivor. The shift in ecosystem perception as yet another field of "demand" stems from an improvement in awareness, appreciation and, if not required, the ability to support services rendered by ecosystems. Practically, the "competition" between and discussion of sectors and ecosystem needs has inevitably intensified more and more. However, this is a welcome and optimistic trend, as it represents a stronger dialog and a drive towards a better balanced use of water supplies and thus more sustainable growth.

2.5 Differences and divergences

Water is, in the simplest words, rare, irreplaceable, and impossible to transfer beyond the pull of gravity, a sustainable natural resource. Power, on the other hand, comes in various forms from a range of sources. It is normally powered by the demand and can be spread over large distances (e.g., via transmission lines for electricity or pipelines for fuels). When considering the role of water in the nexus, and how they are handled, it is important to differentiate between water supplies and water services. The conservation of water supply is a management of the water cycle in which water flows in both quality and quantity through the ecosystem, as a natural resource (i.e., rivers, streams, estuaries, and others, soils and aquifers). The maintenance of water supplies includes the construction and management of infrastructures for storage, transport, and delivery of water to end-users, and the processing of the waste streams by reticulation for treatment and secure reuse. Although energy is primarily needed for water services, the development of energy involves water resources.

2.6 Water treatment technologies

The water treatment technologies are broadly categorized into physical, chemical, and energy-intensive technologies. Each treatment method is having an advantage and a disadvantage. The physical method is one in which the physical forces are used for the removal of pollutants from water and it is based on solid-in-liquid interaction. One of the commonly used physical methods for the removal of pollutants from water is the filtration process and these processes are based on the utilization of filter media for the removal of suspended solids. The filtration process is classified into conventional and advanced filtration. The conventional treatment units utilize sand as the filter media and the removal efficiency of upto 90%–99% can be achieved. The advanced filtration units refer to a pressure filter where the media are placed in a closed pressure vessel and the inlet water is given with pressure and the filtration rate is very high when compared with the conventional filtration units. The chemical methods refer to the treatment units where the suspended solids are removed based on the addition of a chemical to increase the interaction of the pollutants in the water, thereby increasing the floc size that will result in settling of the pollutants under gravity. The commonly used methods are coagulation and sedimentation aided with coagulation. The energy-intensive methods are more efficient and it was used to kill the pathogens present in the water and some of the commonly used methods are sterilization techniques. These methods are very costly and will increase the cost of the treatment rather than its high removal efficiency. Fig. 2.2 summarizes the most commonly used methods for the treatment of water.

The water treatment plants will have all these three treatment technologies in the combined form or, in some cases, it may be standalone treatment

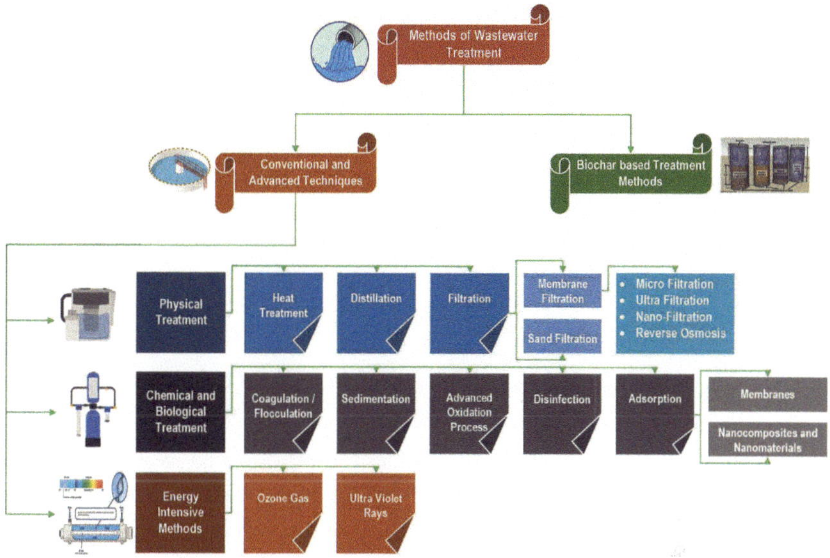

Figure 2.2 Commonly used methods for the treatment of water.

units based on the type of the pollutants. The selection of the treatment methods is based on the following consideration:

1. characteristics of the water,
2. quality of the water,
3. treatment efficiency required to meet the drinking water standard,
4. type of pollutants present in the water, and
5. treatment cost for the removal of the pollutants.

2.6.1 Water purification

Water purification is a general term that refers to the removal of pollutants from the water. So, it is important to know about the characteristics of the water and the major pollutants that pollute the water. The water that is used for drinking water or domestics purposes is taken from groundwater or surface water. Surface water is known to be highly polluted due to the discharge of industrial wastewater and due to leaching of soil during stormwater runoff. The pesticide that is used for the agricultural practices will be washed away with the rainwater and reaches the surface water. These will increase the pollutant level that again increases the treatment cost. Groundwater is known to be highly affected by heavy metals and it is based on the locality of the places. In the places where the industries are more, there will be the possibility of groundwater pollution. Fig. 2.3 summarizes the sources of water pollution.

2.6.2 Physical treatment methods

The most commonly used physical methods for the treatment of the water are listed below:

1. Heat treatment
2. Reverse osmosis
3. Distillation
4. Microfilter
5. Slow sand filter
6. Activated charcoal filter

Heat treatment: It is one of the ancient methods that was commonly used for removing pathogens. Besides killing the pathogens present in the water, it is also capable of removing temporary hardness. If the water is rich in temporary hardness, during boiling, the carbonate hardness will be converted into insoluble salt and it will settle down as residue. Heating the water can be done by the solar heater or electric driven heaters. On average, it costs $1 for treating 1000 gallons of water. Commonly, the

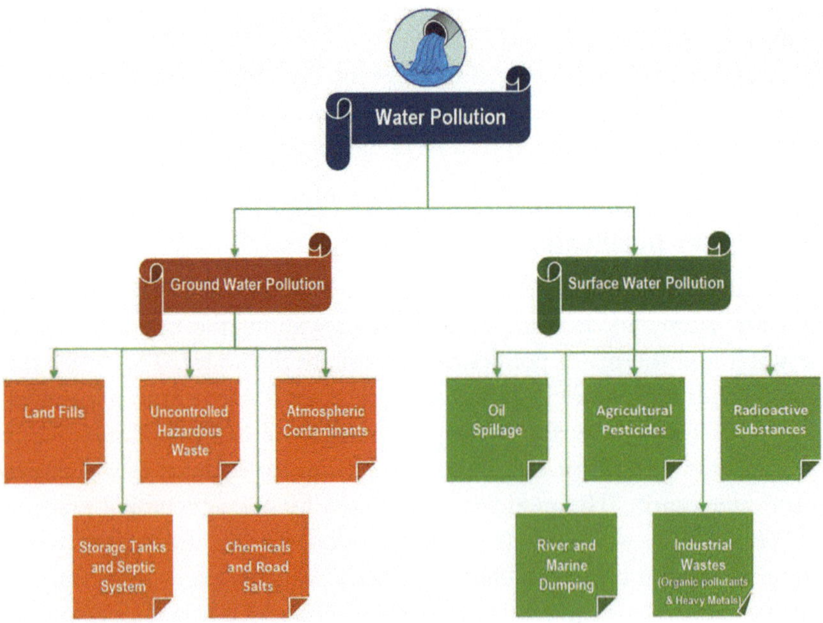

Figure 2.3 Sources of water pollution.

water is heated to 161°F for 15 seconds to kill the pathogens present in the water.

Reverse osmosis: It is one of the new and advanced technologies that was based on the principle of filtration, where a filter media is used to remove the pollutants. Reverse osmosis utilizes a membrane for filtration. The membranes are rolled and fixed inside the pressure vessel around the inlet pipe and the water to be treated will be given a feed with high pressure and this water molecule will move from successive layers of the membrane. The impurities/pollutants present in the water molecules will get trapped in the membranes and the filtered water is 99% free from the pollutants. The pollutants that are trapped in the membrane will be backwashed with the desired solvents and the membranes were used for the next treatment cycles. The filtered water will be entirely free from the dissolved solids and this is one of the emerging problems. Recent research concluded that a trace amount of minerals is required in the water for the regular metabolism in humans. So, a reverse osmosis plant is equipped with the additional unit for the artificial feeding of the required amount of the minerals as per the drinking water standards.

Distillation: It is also one of the ancient methods; due to its disadvantages, its application in water treatment is not common. The main disadvantage of the distillation method is the requirement of the energy is very high and the chemical that is having boiling point less than the water will get escaped with the water and it is again tough to remove the pollutants. Distillation involves two steps: First, evaporation of the water and then condensation of the vapors. Based on the quantity of the water to be treated, there are single, multiple, and flash evaporators used for the water treatment.

Microfilters: These filters are designed at a small-scale level to remove the bacteria and other suspended solids present in the water. These filters are used with pressure, gravity, or by siphon action. The filters are placed in the tank in such a way that water can enter via gravity forces and hence can be filtered; the only disadvantage of the filter is, over the filter media, biomass growth will be increased, which will reduce the efficiency of the treatment. To overcome the biomass growth, the filters are pretreated with silver and other disinfecting agents to avoid the growth of the microbes over the microfilters. One such commonly used microfilter was activated carbon.

Sand filtration: It is the most commonly used treatment unit for the removal of pollutants. Due to its ease in construction, operation, and maintenance, its application becomes wide. There are different types of sand filtration, namely, slow sand filter, rapid sand filter, and pressure filter. The classification is also done based on the type of flow, namely, gravity and pressure filters. The difference among the filters was based on the efficiency of the filter, rate of filtration, and quantity of the water treated. The efficiency was very high for the slow rate filter with a removal efficiency of 95%–99%, whereas the rapid sand filter will have a removal efficiency of 90%–95%. But the pressure filter will have a removal efficiency of 90% and below. The characteristics of the water decide the type of filter needed for the treatment. The maintenance of the filter plays a very important role in the slow sand filter; since the void ratio is very low, it is most commonly got trapped by pollutants, and in regular practice, the pollutants over the sand filter are to be removed, and the formation of mud balls within the filer media is another main problem.

Activated charcoal filter: The activated charcoal filter works based on the principle of the adsorption process. Adsorption is a surface phenomenon where the pollutants to be removed will come in contact with the adsorbent and it gets bonded over the surface of the adsorbents. It is very easy in working, construction, and operation of the plant. It can be designed as a prototype or for small-scale treatment or even for large-scale treatment. Activated carbon is used to remove the chemical that is used in the treatment process and it is capable of removing bacteria from the water. The activated carbon is to be fixed in such a way that all the water molecules are in contact with the filter. These filters will give additional taste to the water and the activated carbon generally has a pore size of 0.6–0.9 mm and these pores will be reduced due to the growth of the microbial biomass. Since activated carbon favors the growth of bacteria and even though these bacteria may not be harmful to humans, it can be removed by using silica pretreated with the activated carbon.

2.6.3 Chemical treatment methods

Some of the commonly used chemical methods in wastewater treatment are

1. Coagulation/flocculation
2. Sedimentation
3. Disinfection

2.6.4 Coagulation/flocculation

The impurities/pollutants are classified into settleable solids, suspended solids, dissolved solids, and pathogens. Easily settleable solids will be removed in the collection well or in a primary settling tank. Initially, the suspended solids present in water are very small and will not settle easily. To increase the floc size, chemicals are added to the water and this will result in the formation of big floc and it will settle down under gravity and it is easy to remove from the water in a sedimentation tank. Coagulants are the chemical used to increase the floc size. Commonly used coagulants are alum and lime, and many chemicals are recently used based on the characteristics of the water. The coagulants are added and it is stirred for 1–2 minutes and the water is allowed to coagulation tank for 30 minutes to increase the floc size.

Sedimentation tank: The sedimentation tank is based on the principle of settling. The sedimentation tanks are generally classified into rectangular tank and circular tank. Nowadays sedimentation tanks are named clarifiers, where the flocculation, coagulation, and sedimentation are carried out in the clarifier. The water that is coming from coagulation will be feed into the sedimentation tank and the detention time is maintained as 4 hours. The flocs will settle down and it will be collected from the bottom of the sedimentation tank as sludge. This sludge needs to be disposed of carefully; since all the pollutants are in this sludge, if not disposed of properly, it will affect the environment.

Disinfection: After the treatment of water in physical and chemical methods, all the impurities present in the water will be removed to a maximum extent of 99%. But the presence of bacteria and other microbes will not be completely removed. These microbes need to be removed before it is supplied to the distribution system since these microbes are the carrier of waterborne disease. To kill these bacteria, a commonly used method is called the disinfection process. Many chemicals namely chlorine in gas and powder form, iodine, bromide, silver, and potassium permanganate are used for pathogen removal. The quantity of the chemical used depends on the quantity of the water that needs to be treated and the time taken for the water to reach the public. The residual chlorine of 0.2 mg/L was maintained as per the drinking water standards.

2.6.5 Energy-intensive treatment methods

The commonly used energy-intensive treatment methods are

1. Ozone
2. Ultraviolet rays

Ozone: Ozone is considered as one of the recent and advanced treatment methods for the removal of microorganisms from the water. Ozone gas is formed when air is exposed to a high electric voltage. It consists of three atoms of oxygen and it has a half-life period of 120 minutes in water at a temperature of 20°C. These ozone gasses are unstable and after 120 minutes, it will not be in the water and there is no residual ozone so there won't be any additional taste created by ozone gas treatment. Since there is no residual ozone present in the water, there may be a chance of microbial growth if the water is stored for a long time. During treatment, some of the molecules will also be interchanged with the ions and this will affect the flocculation and coagulation process if this unit is present before the primary treatment of the water.

Ultraviolet rays: Ultraviolet lights are used for killing the pathogens present in the water. It is produced at a wavelength of 253.7 nm, and like ozone, there is no residual left in the ultraviolet treatment. These will result in the growth of microbes if the water is stored for a long time. The main disadvantage of this method is after treatment, if the water is exposed to visible range, then there is a chance of growth of the microbes.

2.7 Novel treatment methods—biochar and its applications

2.7.1 Introduction

Adsorption is the most commonly used method in the recent decade due to its wide application in water treatment. Any type of pollutants can be removed using the adsorption process. Adsorption is a surface phenomenon in which a solid–liquid interface will occur and results in the removal of ions from the water. The adsorbent is the general term that refers to the solid material that was used for the adsorption process and the substance which gets attached to the adsorbent is called adsorbate. The adsorption of the pollutants will occur based on the unbalanced force occurring on the surface of the adsorbate and this may be due to the physical force like weak van der Waals force, so-called physisorption, or it may occur due to the chemical bonding called chemisorption. If an adsorbent is to become successful in industrial treatment, it should have a more specific surface area, pore volume, and pore size. The adsorbent should also have good mechanical properties, showing its resistance during the treatment process [23]. Fig. 2.4 summarizes the various adsorption mechanism involved in the removal of pollutants.

Any adsorption process will have some adsorption mechanism. As discussed earlier, other than physisorption or chemisorption, other factors governing the adsorption mechanism are ion exchange, precipitation, and

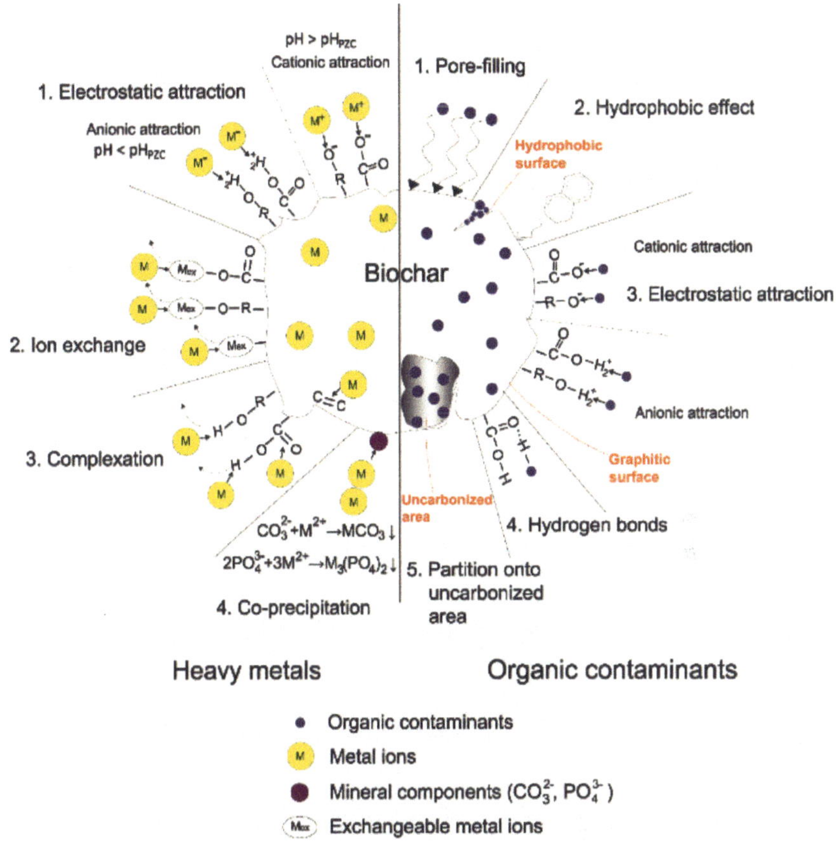

Figure 2.4 Adsorption mechanism for the removal of pollutants [24].

complexation. Ion exchange mechanisms are favored by the types of ions present on the adsorbate and the targeted pollutants. During the ion exchange process, the surface chemistry of the biochar and the pollutants plays a very important role. The positively charged adsorbent will have more adsorption of negatively charged reactive dyes or heavy metals. So, it depends on the size of the molecules that are present. The positive and negative charges of the adsorbent are influenced by the solution pH and point of zero charges [25,26]. For instance, at a pH of 2, the solution is acidic and it is protonated with excess hydrogen ions that are positively charged ions. So, these will enhance the removal of the pollutants that are negatively charged and vice versa. Complexation is based on the formation of multiatom structures or complexes involving the formation of metal–ligand interaction. Precipitation favors the adsorption by forming a solid substance in the liquid, and thereby increasing the attraction, and increases the removal efficiency of the adsorbents [27,28]. There are different types of adsorbents used in water and wastewater treatment. Based on the pollutants, most of the adsorbents are specifically produced for specific treatment, namely, activated carbon and some adsorbents are naturally available, namely, clay, silicates, zeolites, activated alumina, and polymeric adsorbents [28–30]. But these adsorbents are available in small quantity or the production cost is more, which will increase the treatment cost. So, alternate adsorbents are produced in the last decade and many modifications are carried out for the specific types of pollutants.

Biochar is a carbon-rich material produced in the absence of oxygen using thermochemical decomposition of the waste biomass. The agricultural and other waste biomass is used as a feedstock for the production of biochar using slow pyrolysis, thermal decomposition, and gasification. Carbon-rich biochar has a wide application in both organic and inorganic pollution removal and its characteristics are similar to that of activated carbon. This new adsorbent will lower the treatment cost and also acts as the best tool for environmental tool in solid waste management [31,32]. The production cost of biochar will be varying between $0.2 and $0.6 per kg. The characteristics of the biochar that favor the adsorption of pollutants include surface area and porosity, pH, surface charges, functional groups, and chemical composition.

2.7.2 Feedstocks for biochar

The selection of feedstock for the production of biochar plays a very crucial role in the economy of the treatment cost. Generally, the feedstock is classified as dry and wet feedstocks. The biomass with a moisture content less than 30% is called a dry feedstock and the biomass with a moisture content greater than 30% is called wet biomass [33–35]. For the effective production of biochar, it was recommended to select the feedstock that is grouped under dry feedstocks. Since the moisture content of the biomass plays a very important role in the production of biochar, the higher the moisture content, higher the production cost. The feedstocks, namely, sludge and algae come under wet feedstock, and wood chips and agricultural residues come under dry feedstocks. Biomass is further classified as energy-intensive crops and waste biomass. Waste biomass is generally overgrown without any application and utilization of this waste biomass will give an alternate solution for the climate change mitigation measure

[36,37]. For example, seaweeds grown naturally and the application of the seaweed is very much limited. The biochar from this waste seaweed biomass will have a great impact on the environment. On the other hand, energy-intensive crops have a wide application in fuel industries and the residues of these industries will be used for the production of biomass.

2.7.3 Synthesis of biochar

Production of biochar via pyrolysis is a crucial factor that decides the overall properties and characteristics of the biochar. The chemical reaction involving during the thermal decomposition is still the main challenge for the researchers to understand the mechanism involved in the conversion process. The known chemical reactions that are occurring during decomposition are aromatization, condensation, decarboxylation, decarbonylation, dehydration, demethoxylation, and intermolecular rearrangement. The temperature for the thermal decomposition plays a very important role and it is difficult to identify the type of reaction involved at desired temperatures. The mechanism of biochar production from biomass under pyrolysis is based on cellulose, hemicellulose, and lignin because the thermal stability of each component is varying. The thermal decomposition of cellulose will happen at a temperature of 240°C–350°C, decomposition of hemicellulose will happen between 200°C and 260°C, and the decomposition of lignin will happen between 280°C and 500°C. Table 2.2 summarizes the different operating conditions for the synthesis of the biochar and Fig. 2.5 summarizes the overall process involved in the conversion of feedstock to the biochar and its application in the environment [40].

2.7.4 Characteristics of biochar

The properties of the biochar that influences the adsorption mechanism of the pollutants are depended on surface area and porosity, functional

Table 2.2 Classification of pyrolysis in terms of operating conditions and biochar yield [38,39].

Process		Operation temp. (°C)	Heating rate (°C/min)	Residence time	Biochar yield (wt.%)
Pyrolysis	Slow	300–800	5–7	>1 h	35–50
	Fast	400–600	300–800	0.5–10 s	15–35
	Flash	400–1000	~1000	<2 s	10–20

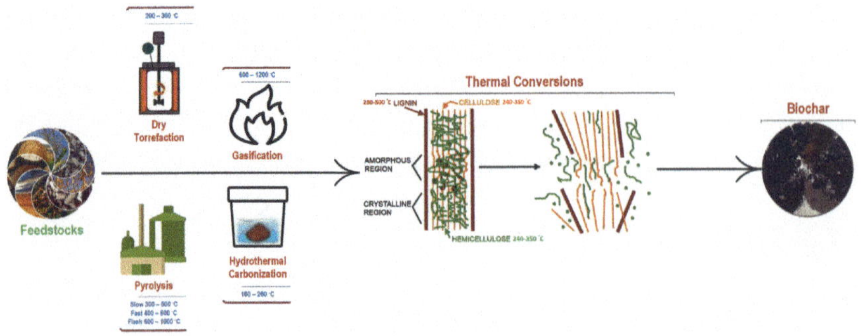

Figure 2.5 Overall process involved in the conversion of feedstock to the biochar.

groups present in the biochar, pH and point of zero charges, and mineral content. The surface area and porosity of the material are considered as one of the important parameters that decide the sorption capacity of the biochar [41].

Surface area and porosity: During thermal decomposition, the cellulose, hemicellulose, and lignin will break down and the water molecules present in the biomass will be lost and this will create pores in the residues and these pores are micropore and the size of the pore is very less. The biochar will have a greater potential only when the pore size is very high and these will be achieved at an increasing temperature of 500°C–900°C and the micropore will be converted to macropore. For instance, the porosity of the biochar is depended on the type of biomass used. Biochar produced from coconut shell and bamboo will have macropores and it is constitutes of more cellulose, that will result in the formation of micropore [42,43].

pH and surface charge: Pyrolysis temperature of the feedstock decides the charges on the biochar. An increase in temperature will increase the pH since the acidic groups present in the biomass will be converted into ash content with an increase in temperature. So elevated temperature will favor the biochar with more negative ions present on the biochar. The pH of the solution and point of zero charge plays an important role. If the ions present in a solution become zero, it is called a point of zero charges. If the pH of the solution is greater than point of zero charges, the biochar will have more negative charges and favors the pollutants that have positive ions and vice versa [44,45].

Functional groups: The pyrolysis temperature of the biomass will result in the variation of carbon, hydrogen, and oxygen content present in the biochar. This variation will result in variation of the different ratios, namely, H/C, O/C, and N/C into the biochar. These ratios have a major role in the polarity and aromaticity of the biochar. For instance, at increased temperature, these ratios will decrease, and at decreased pyrolysis temperature, these ratios will increase and show more polarity and aromaticity.

Mineral content: The mineral content, namely, calcium, magnesium, potassium, and phosphate, present in the biochar will favor the adsorption of the metal ion from water. Like other parameters, mineral content is also affected by the pyrolysis temperature. With an increase in temperature, the mineral content present in the biomass will be converted into ash or some other chemical compound. On the contrary, increase in temperature favors the adsorption by increasing the surface area and the porosity of the biochar [46,47].

2.7.5 Activation of biochar

The raw biochar produced through pyrolysis will have the several disadvantages. Since the biochar size is very small its potential in the adsorption of pollutants will be challenging. The pyrolysis process may result in the degradation of the functional groups that favor the adsorption process. But in some situation the produced biochar will have very less surface area and pore volume. These may result in poor adsorption capacity when the pollutant's concentration is very high. So, much research has been carried out to increase the properties of the biochar and to enhance the adsorption capacity of the biochar. Such modifications involve physical activation and chemical reactions [48]. Some of the commonly used methods for the modification of the biochar are steam-enhanced biochar, heat-treated biochar, acid-treated biochar, alkaline-treated biochar, nitrogen-based modified biochar, and biochar-based composites. These methods will increase the surface characteristic and physical properties of the biochar. Fig. 2.6 summarizes the different methods for the activation of biochar for pollutant removal.

Steam-enhanced biochar: The biochar produced from the pyrolysis process is again modified in the gasification process at a temperature of 800°C–900°C. These will result in partial devolatilization, and the formation of carbon over the surface of the biochar will appear. During this process, the pore size will also increase and it will increase the porosity of the biochar and the surface of the biochar will also increase. The additional

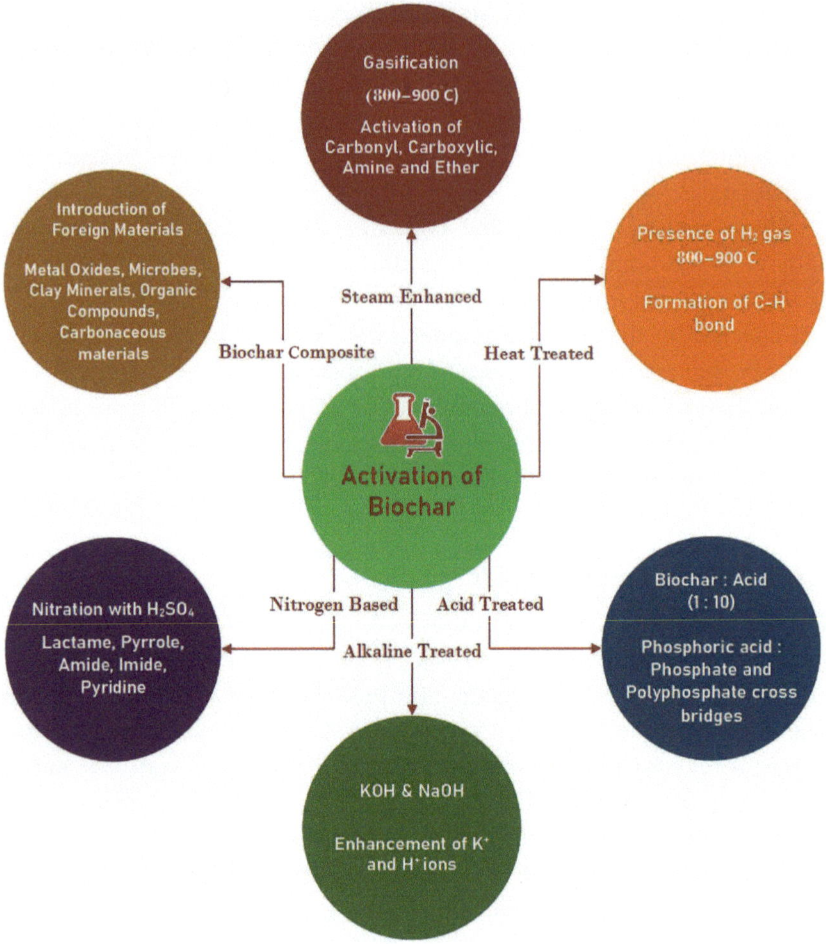

Figure 2.6 Different methods for the activation of biochar for pollutant removal.

functional groups namely carbonyl, carboxylic, amine, and ether will be formed on the surface of the biochar. These modifications will enhance the adsorption capacity of the biochar [49,50].

Heat-treated biochar: In the presence of hydrogen gas when the biochar is heated at a temperature of 800°C–900°C, the porous carbon will be formed on the biochar surface and the heteroatoms of the carbon content will disappear from the surface and result in the formation of C–H bonds at the edges of the crystalline structure of the biochar [51,52].

Acid-treated biochar: Biochar when treated with strong acids, it will increase the functional groups of the biochar and this will enhance the metal sorption capacity or pollutant removal efficiency. Acids, namely, sulfuric acid, hydrochloric acid, nitric acid, and phosphoric acid are used to increase the surface characteristics of the biochar and it is added in the ratio of 1:10. Phosphoric acid is the most commonly used acid over its advantage. Phosphoric acid will give phosphate and polyphosphate cross-bridges that will increase the sorption of the pollutants [53,54].

Alkaline-treated biochar: Alkaline treatment of the biochar will increase the positive ions present on the surface of the biochar and thereby increasing the sorption capacity of the biochar. The commonly used chemicals are sodium hydroxide and potassium hydroxide. But when compared with potassium hydroxide, sodium hydroxide is less expensive and less corrosive. But when the biochar is treated with potassium hydroxide, it will enhance the biochar surface with K^+ ions and this increases the surface properties of the biochar. It will also increase some of the potassium species like K_2O and K_2CO_3 and this will result in the modification of the internal structure of the biochar and the porosity of the biochar will be increased [55,56].

Nitrogen-based modified biochar: Functional groups, namely, lactame, pyrrole, amide, imide, and pyridine are rich in *N*-containing functionalities and this compound will show greater affinity towards metals copper, zinc, and cadmium that are considered as base. These groups are introduced to the surface of the biochar through nitration, simultaneously adding sulfuric acid and followed by reduction [57,58].

Biochar-based composites: Biochar-based composite is a new emerging technology that increases the physical and chemical properties of the biochar. It combines all the advantages of the modification techniques and this method has the advantage of all other methods. The incorporation of foreign materials over the biochar will increase the biochar properties such as surface area, new functional groups, crystalline nature, porosity, and internal structural modifications. This modification enhances the sorption capacity of the pollutants. These foreign materials will be introduced to the biochar either by metal oxides, microbes, clay minerals, organic compounds, or carbonaceous material. There are different types of composites, namely, metal oxide-based biochar, magnetic biochar-based nanocomposites, and functional nanoparticle-based biochar [59].

Biochar for pollutant removal: Water pollution is increasing day by day and it is estimated that 10 million tons of chemicals were discharged annually to the environment and over 700 contaminants are contributing more to the water pollution. These pollutants are mostly anionic and heavy metals. The anionic contaminants will easily mix with the water and create a harmful effect to the human and the environment. For instance, domestic sewage and industrial effluent are very rich in nitrogen and phosphorus. These chemicals when mixed with the river, lake, or the sea will result in the eutrophication process. This process will also affect the ecosystem and the environment. There are physical, chemical, or biological methods, namely, ion exchange, membrane technology, and microbial treatment. But the treatment cost associated with conventional and advanced technology is very high, and in the biological method, the cost of the treatment is very low but the time for the treatment is very high. So, to overcome these disadvantages, adsorption was considered as one of the cost-effective methods. Activated carbons are long being used but the production cost is very high. So, recently, biochar is playing a very important role in pollutants removal. Several types of research have been conducted in the past and it is concluded that biochar has the potential of sorption of all types of contaminants from water. The adsorption mechanism will be based on pore filling, hydrogen bonding, surface complexation/precipitation, electrostatic attraction, and π–π interaction. The influencing parameters, namely, solution pH, biochar dose, initial pollutant concentration, contact time, and temperature play a very important role in the removal of pollutants using biochar. Fig. 2.7 summarizes the

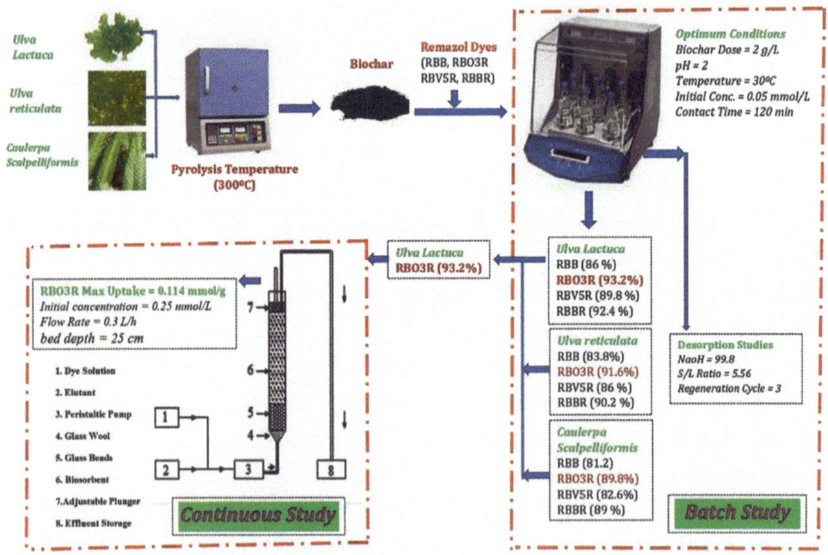

Figure 2.7 Synthesis of biochar from seaweeds for the sorption of the reactive dyes.

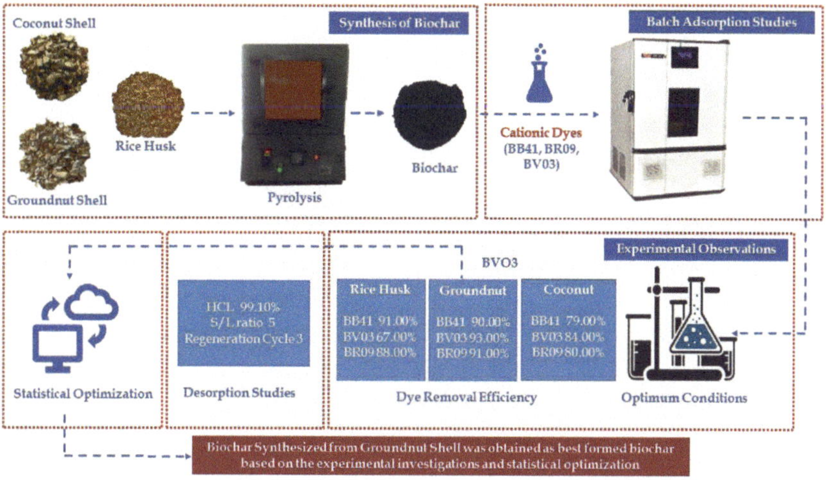

Figure 2.8 Application of biochar produced from agricultural waste for cationic dye removal.

biochar synthesis from different feedstock and its potential in the sorption of reactive dyes in batch and column studies. The results concluded that the *Ulva lactuca* was having high potential in the sorption of reactive dyes [32]. Fig. 2.8 summarizes the synthesis of biochar from agricultural waste, namely, coconut shell, groundnut shell, and rice husk for the sorption of dyes in batch-mode operation. The results concluded that the biochar derived from groundnut shell has high potential in the removal of cationic dyes [60,61].

2.8 Need for solar-driven water treatment techniques

Compared with water, energy can be produced in many ways, with individual water source specifications and results. Water treatment systems powered by solar energy represent sustainable solutions to solve the world water crisis and the adverse consequences of fossil fuel burning in the sense of a global water shortfall and potential power crisis. Solar water technology can be picked rather locally. There are easy, low-tech, low-investment technologies like a solar still and SODIS, among other numerous technologies, which are suitable especially for remote areas in the developing countries which lack comprehensive financial support, high-tech, and skilled worker's access. While some are not commercially viable, the reliability and the technological sophistication of indirect desalination technologies and solar photocatalysis technologies are growing as a result of both solar and water treatment technologies and progress in technology.

The cost of generating water for solar-based desalination is also reasonably high compared with the traditional fossil-based desalination plants because of the inefficient solar collectors, which is a significant factor for restricting the pace of commercialization. In most cases, however, nonrenewable electricity environment prices have been overlooked poorly. In preparation for a prosperous world, the loss of fossil fuels, greenhouse gas emissions, and air pollution should all be taken into account in the energy market. In comparison, the projected cost of large-scale solar desalination plants indicated economic comparability with traditional plants, although there was still no actual large-scale project. Besides, even more research is needed to investigate the effect of solar and solar-based desalination technology on the sustainable future.

2.9 Conclusion

The quality and quantity of the water are decreasing with an increase in industrial development and the population. The present technologies and the treatment cost associated with the removal of pollutants are very high. So, it has become an emerging problem, and the treatment of this pollution more economically needs to be focused. The solar-driven water treatment technologies and some new novel low-cost treatment methods will contribute to a new revolution in conventional waste treatment technologies. Biochar and biochar-based nanocomposites-based adsorbents are the recent focus and these adsorbents can be applied to all types of contaminants as the sorption capacity of the adsorbents is high. But the characteristics of the biochar decides the efficiency of the treatment. Similarly, selection of feedstock for the synthesis of the biochar is a major concern since a unique biochar cannot be applied to several types of contaminant. So, investigations have to be carried out to modify the biochar for the treatment of all type of contaminants. Techno-economic and life cycle assessment of the biochar is also an important factor to be considered in future research.

References

[1] C.J. Vorosmarty, E.M. Douglas, P.A. Green, C. Revenga, Geospatial indicators of emerging water stress: an application to Africa, AMBIO: A Journal of the Human Environment 34 (3) (2005) 230–236.

[2] D.G. Neary, Hydrologic values, in: J. Richardson, R. Bjorheden, P. Hakkila, A.T. Lowe, C.T. Smith (Eds.), Bioenergy from Sustainable Forestry: Guiding Principles and Practice, Kluwer Academic Publishers, Dordrecht, 2002, pp. 190–215.
[3] WWAP (World Water Assessment Programme), The United Nations World Water Development Report 4: Managing Water Under Uncertainty and Risk, UNESCO, Paris, 2012.
[4] WEC (World Energy Council), Water for Energy, WEC, London, 2010. Available from:. Available from: http://www.worldenergy.org/documents/water_energy_1.pdf.
[5] EPRI (Electric Power Research Institute), Water and Sustainability (Volume 4): US Electricity Consumption for Water Supply and Treatment – The Next Half Century, EPRI, Palo Alto, CA, 2002. Available from:. Available from: http://www.circleofblue.org/waternews/wp-content/uploads/2010/08/EPRI-Volume-4.pdf.
[6] K.M. Twomey, M.E. Webber, Evaluating the energy intensity of the United States public water system, in: ASME 2011 5th International Conference on Energy Sustainability, ES 2011, 2011.
[7] WEF (World Economic Forum), Water Security: The Water–Food–Energy–Climate Nexus: The World Economic Forum initiative, Island Press, Washington, DC, 2011.
[8] J.R. Strokes, A. Horvath, Energy and air emission effects of water supply, Environmental Science and Technology 43 (8) (2009) 2680–2687.
[9] EIA (US Energy Information Administration), International Energy Outlook 2010: Highlights, Office of Integrated Analysis and Forecasting, EIA, US Department of Energy, Washington, DC, 2010. Available from:. Available from: http://www.eia.gov/oiaf/archive/ieo10/highlights.html.
[10] WEF (Water Environment Federation), Energy Conservation in Wastewater Treatment Facilities Manual of Practice, WEF, Alexandria, VA, 1997.
[11] A.S. Stillwell, C.W. King, M.E. Webber, I.J. Duncan, A. Hardberger, The energy–-water nexus in Texas, Ecology and Society (Special Feature: The Energy-Water Nexus: Managing the Links between Energy and Water for a Sustainable Future) 16 (1) (2011) 2.
[12] Center for Sustainable Systems. United States Wastewater Treatment Factsheet. University of Michigan. Available from: <http://css.snre.umich.edu/facts/factsheets.html>, 2008.
[13] R.B. Sieger, D. Whitlock, Session for the CHP and Bioenergy for Landfills and Wastewater Treatment Plants Workshop, Salt Lake City, UT, 11 August 2005, 2005. Available from: <http://www.docstoc.com/docs/22892252/CHP-and-Bioenergy-for-Landfills-and-Wastewater-Treatment-Plants>.
[14] A.S. Stillwell, C.W. King, M.E. Webber, Desalination and long-haul water transfer as a water supply for Dallas, Texas: a case study of the energy-water nexus in Texas, Texas Water Journal 1 (1) (2010) 33–41.
[15] B. Applebaum, Water and Sustainability, Vol. 4: US Electricity Consumption for Water Supply and Treatment – The Next Half Century. Report 1006787, Electric Power Research Institute (EPRI), Palo Alto, CA, 2000. Available from:. Available from: http://dc213.4shared.com/doc/BJ17AQaE/preview.html.
[16] UNEP (United Nations Environment Programme), UNEP Yearbook: New Science and Developments in Our Changing Environment, UNEP, Nairobi, 2010.
[17] D. Stellar, Can We Have Our Water And Drink It, Too? Exploring the Water Quality–Quantity Nexus. State of the Planet Blog, Earth Institute, Columbia University, New York, 2010. Available from:. Available from: http://blogs.ei.columbia.edu/2010/10/28/can-we-haveour-water-and-drink-it-too-exploring-the-water-qualityquantity-nexus/.
[18] MA (UN Millennium Ecosystem Assessment), Living Beyond our Means. Natural Assets and Human Well-being. Statement from the Board, March 2005.

[19] E. Corcoran, C. Nellemann, E. Baker, R. Bos, D. Osborn, H. Savelli (Eds.), Sick Water? The Central Role of Wastewater Management in Sustainable Development. A Rapid Response Assessment, United Nations Environment Programme (UNEP), UN-HABITAT, GRID-Arendal, 2010. Available from:. Available from: http://www.grida.no/files/publications/sickwater/poster1_SickWater.pdf.
[20] WHO (World Health Organization) and UN-Water, GLAAS 2010: UN-Water Global Annual Assessment of Sanitation and Drinking Water: Targeting Resources for Better Results, WHO, Geneva, 2010. Available from:. Available from: http://www.unwater.org/activities_GLAAS2010.html.
[21] Hydro-Québec, Plan stratégique 2006−2010, Montreal, Hydro-Québec, 2006.
[22] WEF (World Economic Forum), Global Risks 2013, eighth ed., WE, Geneva, 2013.
[23] W.J. Thomas, B. Crittenden, Adsorbents. Adsorption Technology & Design, Butterworth-Heinemann, Oxford, 1998.
[24] X. Wang, Z. Guo, Z. Hu, J. Zhang, Recent advances in biochar application for water and wastewater treatment: A review, Peer Journal 8 (2020) e9164. Available from: https://doi.org/10.7717/peerj.9164/fig-1.
[25] X. Dong, L.Q. Ma, Y. Li, Characteristics and mechanisms of hexavalent chromium removal by biocharfrom sugar beet tailing, Jounal of Hazardous Materials. 190 (2011) 909915.
[26] A. Mukherjee, A.R. Zimmerman, W. Harris, Surface chemistry variations among a series of laboratoryproducedbiochars, Geoderma 163 (2011) 247−255.
[27] R.H. Crabtree, Metalligand multiple bonds, The Organometallic Chemistry of the Transition Metals, John Wiley & Sons, Inc, 2005.
[28] R.T. Yang, Silica gel, MCM, and activated alumina, Adsorbents: Fundamentals and Applications, John Wiley & Sons, Inc, 2003.
[29] R.T. Yang, Zeolites and molecular sieves, Adsorbents: Fundamentals and Applications, John Wiley & Sons, Inc, 2003.
[30] Jaroniec, M. Adsorbents: Fundamentals and Applications By Ralph T. Yang (University of Michigan). Wiley-Interscience, Hoboken, NJ, xii 1410, pp. $89.95. ISBN 0-471-29741-0. Journal of American Chemical Society. 125 (2003) 1205912059.
[31] A.K. Priya, R. Gokulan, A. Vijay Kumar, S. Praveen, Biodecolorization of remazol dyes using biochar derived from *Ulva reticulata*: isotherm, kinetics, desorption and thermodynamic studies, Desalination and Water Treatment 200 (2020) 286−295. Available from: https://doi.org/10.5004/dwt.2020.26098.
[32] R. Gokulan, G. Ganesh Prabhu, A. Avinash, J. Jegan, Experimental and chemometric analysis of bioremediation of remazol dyes using biochar derived from green seaweeds, Desalination and Water Treatment 184 (2020) 340−353.
[33] S. Mani, S. Sokhansanj, X. Bi, A. Turhollow, Economics of producing fuel pellets from biomass, Applied Engineering in Agriculture. 22 (2006) 421-−426.
[34] S. Sokhansanj, J. Fenton, Cost Benefit of Biomass Supply and Pre-Processing, BIOCAP Canada Foundation, 2006.
[35] J. Lehmann, J. Gaunt, M. Rondon, Bio-char sequestration in terrestrial ecosystems—a review, Mitigatigation and Adaption Strategies for Global Change 11 (2006) 395−419.
[36] S. Brick, S. Lyutse, Biochar: Assessing the Promise and Risks to Guide US Policy, Natural Resource Defence Council, USA, 2010.
[37] D. Mohan, C.U. Pittman, P.H. Steele, Pyrolysis of wood/biomass for bio-oil: A critical review, Energy Fuels 20 (2006) 848−889.
[38] W.-J. Liu, H. Jiang, H.-Q. Yu, Development of biochar-based functional materials: Toward a sustainableplatform carbon material, Chemical Reviews. 115 (2015) 12251−12285.
[39] X. Cao, W. Harris, Properties of dairy-manure-derived biochar pertinent to its potential use in remediation, Bioresource Technology 101 (2010) 5222−5228.

[40] H.S. Kambo, A. Dutta, A comparative review of biochar and hydrochar in terms of production, physicochemical properties and applications, Renewable and Sustainable Energy Reviews. 45 (2015) 359–378.
[41] T. Chen, Y. Zhang, H. Wang, W. Lu, Z. Zhou, Y. Zhang, et al., Influence of pyrolysis temperature oncharacteristics and heavy metal adsorptive performance of biochar derived from municipal sewage sludge, Bioresource Technology 164 (2014) 4754.
[42] S.D. Joseph, A. Downie, A. Crosky, J. Lehmann, P. Munroe, Biochar for carbon sequestration, reductionof greenhouse gas emissions and enhancement of soil fertility; a review of the materials science, Rendiconti del Circolo Matematico Palermo Supplement. 48 (2007) 101–106.
[43] J. Jin, Y. Li, J. Zhang, S. Wu, Y. Cao, P. Liang, et al., Influence of pyrolysis temperature on properties andenvironmental safety of heavy metals in biochars derived from municipal sewage sludge, Jounal of Hazardous Materials 320 (2016) 417–426.
[44] B.T. Nguyen, J. Lehmann, W.C. Hockaday, S. Joseph, C.A. Masiello, Temperature sensitivity of black carbon decomposition and oxidation, Environmental Science and Technology. 44 (2010) 3324–3331.
[45] M. Uchimiya, I.M. Lima, K. Thomas Klasson, S. Chang, L.H. Wartelle, J.E. Rodgers, Immobilization ofheavy metal ions (CuII, CdII, NiII, and PbII) by broiler litter-derived biochars in water and soil, Journal of Agricultural and Food Chemistry. 58 (2010) 5538–5544.
[46] X. Cao, L. Ma, B. Gao, W. Harris, Dairy-manure derived biochar effectively sorbs lead and atrazine, Environmental Science and Technology 43 (2009) 3285–3291.
[47] Y. Yao, B. Gao, J. Chen, L. Yang, Engineered biochar reclaiming phosphate from aqueous solutions:mechanisms and potential application as a slow-release fertilizer, Environmental Science and Technology 47 (2013) 8700–8708.
[48] A.U. Rajapaksha, M. Vithanage, M. Ahmad, D.-C. Seo, J.-S. Cho, S.-E. Lee, et al., Enhanced sulfamethazine removal by steam-activated invasive plant-derived biochar, Journal of Hazardous Materials. 290 (2015) 4350.
[49] J.A. Ippolito, D.G. Strawn, K.G. Scheckel, J.M. Novak, M. Ahmedna, M.A. Niandou, Macroscopic andmolecular investigations of copper sorption by a steam-activated biochar, Journal of Environmental Quality. 41 (2012) 1150–1156.
[50] W. Shen, Z. Li, Y. Liu, Surface chemical functional groups modification of porous carbon, Recent Patents on Chemical Engineering (Discontinued) 1 (2010) 27. Available from:. Available from: https://doi.org/10.2174/2211334710801010027.
[51] L.R. Radovic, I.F. Silva, J.I. Ume, J.A. Menéndez, C.A.L.Y. Leon, A.W. Scaroni, An experimental and theoretical study of the adsorption of aromatics possessing electron-withdrawing and electron-donating functional groups by chemically modified activated carbons, Carbon (New York) 35 (1997) 1339–1348.
[52] M.-M. Zhang, Y.-G. Liu, T.-T. Li, W.-H. Xu, B.-H. Zheng, X.-F. Tan, et al., Chitosan modification of magnetic biochar produced from *Eichhornia crassipes* for enhanced sorption of Cr(vi) from aqueous solution, RSC Advances 5 (2015) 46955–46964.
[53] Y. Zhou, B. Gao, A.R. Zimmerman, J. Fang, Y. Sun, X. Cao, Sorption of heavy metals on chitosanmodifiedbiochars and its biological effects, Chemical Engineering Journal. 231 (2013) 512–518.
[54] Y. Li, J. Shao, X. Wang, Y. Deng, H. Yang, H. Chen, Characterization of modified biochars derived frombamboo pyrolysis and their utilization for target component (furfural) adsorption, Energy & Fuels 28 (2014) 5119–5127.
[55] Y. Ma, W.J. Liu, N. Zhang, Y.S. Li, H. Jiang, G.P. Sheng, Polyethylenimine modified biochar adsorbentfor hexavalent chromium removal from the aqueous solution, Bioresource Technology. 169 (2014) 403–408.

[56] A.U. Rajapaksha, S.S. Chen, D.C. Tsang, M. Zhang, M. Vithanage, S. Mandal, et al., Engineered/designer biochar for contaminant removal/immobilization from soil and water: potential and implication ofbiochar modification, Chemosphere 148 (2016) 276–291.

[57] G.X. Yang, H. Jiang, Amino modification of biochar for enhanced adsorption of copper ions from syntheticwastewater, Water Research 48 (2014) 396–405.

[58] X.F. Tan, Y.G. Liu, Y.L. Gu, Y. Xu, G.M. Zeng, X.J. Hu, et al., Biochar-based nano-composites for thedecontamination of wastewater: a review, Bioresource Technology. 212 (2016) 318–333.

[59] P.S. Jegan Josephraj, G.R. Bhagavathi Pushpa Thillainayagam, Evaluation of the adsorption capacity of *Cocos nucifera* shell derived biochar for basic dyes sequestration from aqueous solution, Energy Sources, Part A: Recovery, Utilization, and Environmental Effects (2020). Available from: https://doi.org/10.1080/15567036.2020.1800142.

[60] J. Jegan, S. Praveen, T. Bhagavathi Pushpa, R. Gokulan,), Sorption kinetics and isotherm studies of cationic dyes by *Arachis hypogaea* shell derived biochar as low-cost adsorbent, Applied Ecology and Environmental Research 18 (1) (2020) 1925–1939.

[61] J. Jegan, S. Praveen, T. Bhagavathi Pushpa, R. Gokulan, Biodecolorization of Basic Violet 3 using biochar derived from agricultural waste: isotherm and kinetics, Journal of Bio Based Materials and Bioenergy 14 (3) (2020) 316–326.

CHAPTER 3

Solar and water: high-technology readiness technologies

Amir Nashed and Bac Van Dang
School of Mechanical and Manufacturing Engineering, University of New South Wales, Sydney, NSW, Australia

3.1 Introduction

This chapter focuses on matured renewable energy-driven water applications, which includes water desalination and water-pumping systems (WPS). Maturity here is defined by both market share and the reliability of the technology.

There are various well-established technologies in the market that can be classified by their consumed/generated energy type (Fig. 3.1). WPS typically use electricity to drive a motor-pump unit unless they are driven by muscle (hands or animals) or in some cases wind if windmills are used.

Conversely, water desalination technologies typically use thermal and/or electrical energy. The main thermal desalination technologies are multieffect distillation (MED) and multistage flash distillation (MSF); and the main electricity-powered desalination technologies are reverse osmosis (RO) and mechanical vapor compression[1] (MVC). Other well-established technologies with very low market shares such as electro-dialysis and capacitive de-ionization will be covered in this chapter. It should be also noted that all thermal-based desalination technologies require electricity to run auxiliary pumps and various instrumentations.

The main solar-based energy generation technologies are photovoltaics (PV), flat plate collectors (FPC), and evacuated tube collectors. There are other solar technologies that can be considered in a maturing stage such as concentrated solar power; yet their use in water applications is still very limited and mostly in early commercialization stage [1]. Wind energy is also another well-established and widely spread renewable energy technology. However, it has limited use in water application which is because

[1] There are also thermal vapor compression (TVC) systems, but they are typically integrated with MED systems to raise their gain output ratios.

Solar-Driven Water Treatment
DOI: https://doi.org/10.1016/B978-0-323-90991-4.00007-4

they are site-specific [2,3] and are typically used for large-scale energy generation (MW scale) because of their low cost of generation. It should be noted that this chapter focuses on direct coupling of renewable energy systems with water applications. In other words, indirect coupling, where large-scale renewable energy systems are connected to the grid and used to offset the carbon footprint of various energy applications, is not covered in this chapter.

Solar PV technology is increasingly being applied to gradually replace conventional polluting and hazardous power plants such as coal, natural gas, and nuclear energy in addition to diesel generators in remote/rural applications.

The levelized cost of electricity of solar energy together with wind energy is currently the lowest [4]. The once regarded niche and expensive technology experienced dramatic reduction in costs from 2009 because of overcapacity and market competition [5]. For example, the average price of PV modules in 2009 was $4.34 per Watt peak (Wp) [6], then it dropped to less than 0.4$/Wp in 2020, a staggering 11-fold reduction in a decade [7].

On top of that, as most areas with water scarcity are also blessed with high solar resources and are typically remote areas [8], PV has gained strong ground in solar WPSs and solar desalination, replacing either diesel generators in remote areas or the grid in areas with high electricity prices or with unreliable electricity network.

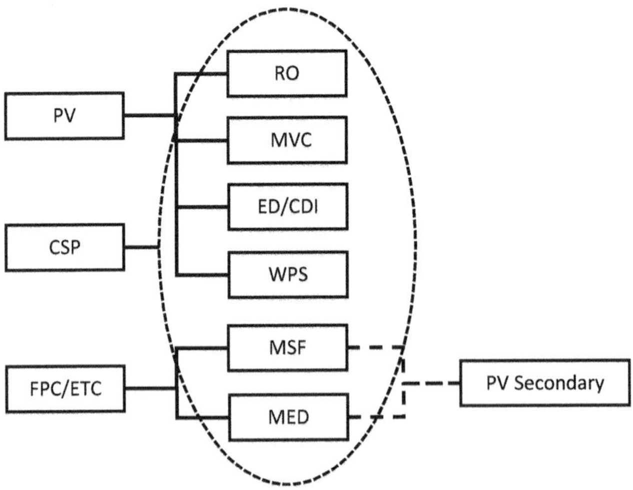

Figure 3.1 Classification of solar-driven water applications.

Solar PV energy has various other important advantages such as simple operation, low maintenance [9–11], and modularity. This makes their application ideal for decentralized applications in remote and rural areas where the plant can be installed and operated by locals with sufficient training [12]. Solar energy is also a local energy resource with high availability, making it ideal to create resilient and sustainable communities [13].

Conversely, while FPC and ETC technologies established itself themselves as the dominant technology for solar water heating, they have little application in water pumping and desalination application. This is attributed to two main reasons: first, all water applications require electricity that cannot be supplied by a low-grade thermal heating; and second, all thermal desalination technologies such as MSF and MED are energy intensive and found very little use in renewable application, which will be discussed in more details in this chapter.

For this reason, this chapter will focus on solar PV-driven RO and WPS as they are currently the most widely used renewable energy driven water applications.

3.2 Solar water pumping

3.2.1 Historical background

Perhaps the oldest device used in moving water from a low-level location to a higher level was the "Shaduf" invented by the ancient Egyptian civilization in 2500 BCE. Another breakthrough was the use of water wheels with attached water compartments that are typically moved by animals or humans (muscle force) in 200 BCE; followed by windmills invented by the Chinese in CE 1200. Since then, several inventions for different kinds of modern pump concepts came to light such as steam vacuum pumps (CE 1698), gear pumps (CE 1580), and piston vacuum pump (CE 1650). The first true centrifugal pump was developed by Denis Pain in CE 1687. Screw pumps were then developed in CE 1830 and vane pumps in CE 1874. The first deep-well vertical pump was developed by Byron Jackson in CE 1901; then the multistage centrifugal pump was developed in CE 1905. The first electrical motor for underwater use was invented in 1908 by Hayward Tyler [14].

To this day, windmills, water wheels, and other muscle-operated pumps such as treadmills [15] and hand pumps are still used in many rural areas in developing countries due to their low cost; even the "Shaduf" is still used in some locations in Egypt.

With generators being developed and the increasing need for higher water flow rates and heads especially for irrigation, diesel generators or grid-connected WPS were increasingly being used and became the most commercially WPSs used [16]. The next step in this evolution of WPS was the use of renewable energy, which is discussed in more detail in the following section.

3.2.2 Application and global status

Solar WPSs are mainly used for domestic, irrigation, livestock, and water distribution applications. They are particularly used in rural and/or remote areas to replace conventional diesel power or grid-connected WPS [17].

Grid availability in many remote/rural areas where WPS are typically used is either nonexistent or unreliable. The alternative solution is using diesel generators which require heavy maintenance and are costly to run mainly due to rising fuel prices. For example, in the last decade diesel fuel prices have risen by over 250% [18]. On top of that and against conventional knowledge, relative to the recent drop in PV prices, diesel generators indeed have moderate to high initial costs [19]. They are also very polluting not just to atmosphere, but also to the surrounding environment because of the fuel and lubricants used which pollute the wells and soil.

Solar-powered WPS, therefore, started to become more economical than those driven by diesel generators in all size ranges [20] and are now the most widely used systems for domestic and agricultural applications [21]. Moreover, as the energy source is locally available, solar WPSs are more reliable compared with transporting fuel to rural areas with weak or expensive supply network when diesel generator WPSs are used.

Another important advantage for solar WPSs is that they can be portable, which is appealing because an increasing number of agriculture users want systems that can move among various locations [19].

Some governments already have schemes to use solar pumping to replace both grid-connected and diesel-powered pumps used in agriculture with financial subsidies to encourage the farmers to adopt such systems [22]. The use of solar WPS in irrigation was also found to significantly improve household income and health through better nutrition [23], and, in general, resulted in economic development, especially in rural areas [24]. The solar WPS technology, particularly pumps and motors, has also progressed allowing their use in deep-well (500 m) applications and to pump larger volumes of water reaching 1500 m^3/d [20].

Other untraditional applications for solar WPS include mining industries [8] and the use of such systems to generate surplus electricity to meet electrical load demand i.e., cogeneration [25]. They are also suggested to be used for community scale applications, for example, large village and institutions [20]. Perhaps, another application that, in our opinion, is worth considering is the use of solar pumping systems in oil applications. In our experience, many oil wells rely on electrical submersible pumps, a type of artificial lifting devices, to extract oil. All these systems are typically powered by diesel generators as they are mostly located in remote area but also due to their large head requirements, flow rate, and power use which can reach 500 kW or more. The use of solar can significantly reduce the cost of diesel use and once battery costs are low enough, then solar energy can totally replace the use of diesel generators.

The main limitation of solar WPS, however, is its high initial cost, but this can be solved with innovative financing options, especially due to the short pay pack time of the system which will be discussed in more detail in the economics section. The lack of local expertise to maintain the system could be also a challenge which will require creating local training centers to provide skilled professionals [26]. Theft and vandalism is also another risk, but this could be tackled by raising awareness for the whole community served by the WPS and by simple antitheft measures [20].

3.2.3 Design and operation

3.2.3.1 System components

Fig. 3.2 shows the typical components in a solar pumping system which can be classified as follows:

1. Solar modules

   The solar modules convert solar energy into electrical energy, that is, photovoltaic effect. Modules are connected in series (to from strings) and in parallel to obtain the required voltages and currents, respectively. The entire set of series and parallel modules is referred to as the solar array. More information on the operation concept of the solar cells can be found in [27]. The solar system can also have single-axis trackers (vertical axis) to track the sun which can increase the water production by more than 40% [28]. Solar trackers are, however, expensive and require extensive maintenance and reduce the reliability of the system. Alternatively, periodic manual adjustment of the module orientation to take advantage of morning and afternoon sun can boost water production by 20%–30% compared with fixed tilt arrays [29,30].

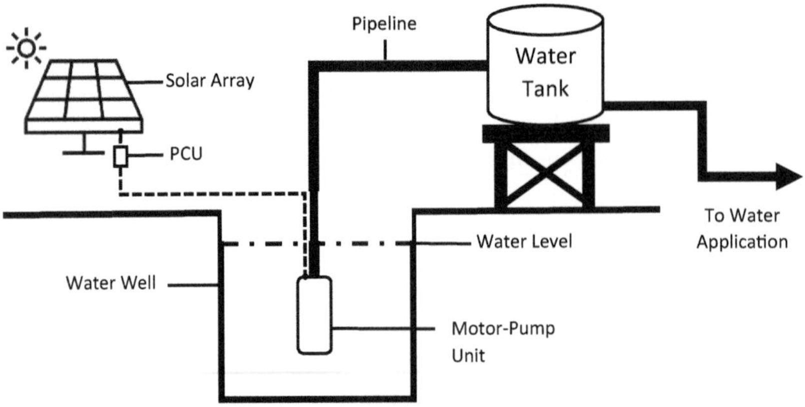

Figure 3.2 Typical solar water pumping system.

2. Power conditioning unit

   The power-conditioning units (PCUs) are required to modify the voltages to suit the motor requirements. They are usually DC–DC converters or DC–AC inverters. Maximum power point trackers (MPPTs) are the mostly used type of DC–DC converters as they are essential to ensure the solar array always operates at the maximum efficiency point. Variable speed drives are also another type of PCUs, which can control the speed of the pump, particularly with AC motors.
3. Water storage tank

   Water storage tanks are typically used in solar pumping systems. The tank increases the availability of water by supplying water in times when the solar pump is not operating or cannot fully meet the water demand. The tank is usually installed at a high level to use gravity when the water is discharged to the load [8,22].
4. Batteries

   Batteries, that is, chemical storage, are used to store the electrical energy to be used in periods of low solar irradiation such as cloudy days and nighttime. Typically, deep cycle lead acid batteries are used in standalone[2] solar applications due to their low cost and simple maintenance, making them ideal for remote and rural applications. However, they have short lifetimes, typically 5 years [9,31–33], though some manufacturers report 15 years for their batteries [34].

[2] Not connected to the power grid.

They also have low maximum depth of discharge,[3] which necessitates oversizing the battery bank. In addition, they require regular maintenance, and their capacity is highly impacted by their discharge rates. In this decade, however, lithium-ion batteries became widely available with lifetimes that can be double that of a lead acid battery [35]. Another main advantage is that their capacities are nearly independent of the discharge rate [36].

Now all types of batteries require a charge controller to avoid overcharging or overdischarging which can damage them or significantly reduce their lifetime and capacity.

5. Motors and pumps

   Pumps are the core of a WPS and are used to deliver water with the required flow rate and pressure/head. Electrical motors are used to drive the pumps. Both are discussed in more detail in Section 3.2.3.3.

6. Control system

   Typically, solar WPS's will have a control system to manage the start and stop of the pump depending on solar availability and water demand [26]. There are also control systems that take into consideration the water supply availability to avoid groundwater overextraction, for example [37]. More recent systems include soil moisture sensors to control the water supply [26].

7. Balance of system

This includes DC isolators, DC & AC cables, junction boxes, circuit breakers, valves, pipes, and instrumentation. The sizing of cables, DC isolators, and circuit breakers should follow the relevant codes and standards.

3.2.3.2 System configuration

There are various solar pumping system configurations as shown in Fig. 3.3.

3.2.3.2.1 Direct coupling

There are various configurations of solar pumping systems. Perhaps the most applied is direct systems where the PV system is directly connected to a DC motor-pump unit. Such system is the cheapest and the simplest in terms of operation and maintenance. For this reason, it is ideal for remote areas with limited access to skilled operators and spare parts.

However, it is the least efficient as it does not maximize the amount of power that could be obtained from the PV system which can only be

[3] Not all available battery capacity can be used. The amount that can be discharged is described by the depth of discharge, which is a percentage of the battery capacity.

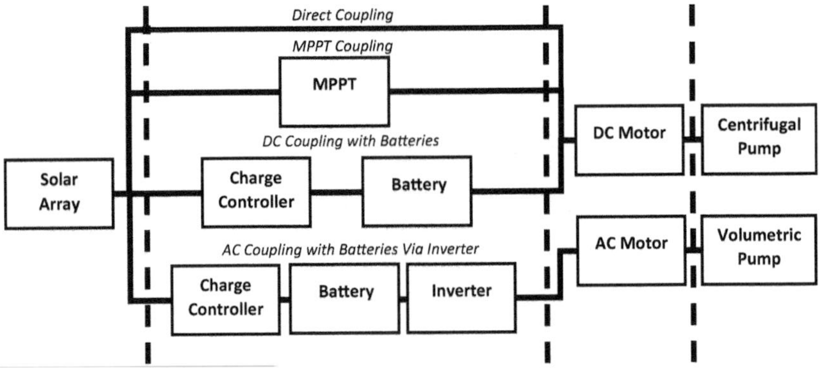

Figure 3.3 Typical solar water-pumping system configurations.

achieved through an MPPT. In addition, in such systems, it is nearly impossible to match the water demand with the supply as the latter depends on the available intermittent solar resources.

For this reason, direct solar pumping systems will typically require water storage to ensure the use of all available energy, especially in periods with high solar irradiation in addition to storing water to meet the demand when the solar pump is not working or have low production, for example, during nighttime, cloudy days, early mornings, and late afternoons.

3.2.3.2.2 Coupling through power-conditioning units

Adding an MPPT significantly boosts the energy generated from the solar system and increases the overall energy efficiency of the system, and hence its economics, that is, more water can be generated from the same capital. The main purpose of the MPPT is to keep the solar panel operating near the maximum efficiency point.

In a directly coupled system, the generated current of the solar modules falls proportionally with the incident solar irradiance. This results in a reduction in the DC motor voltage moving the operating point further from its maximum power point (MPP) as shown in Fig. 3.4. This is exacerbated by the fact that both the motor and the pump now operate away from their best efficiency points, which reduces the overall system efficiency. Conversely, the MPPT boosts the voltage on the module side to force it to operate near the MPP, and this increase in voltage is translated into more current on the load side. This in turn ensures the pump can operate once more near its maximum efficiency point [38].

A decade ago, MPPT's were considered unreliable and expensive but that is not the case anymore; and for this reason, it is highly advisable to

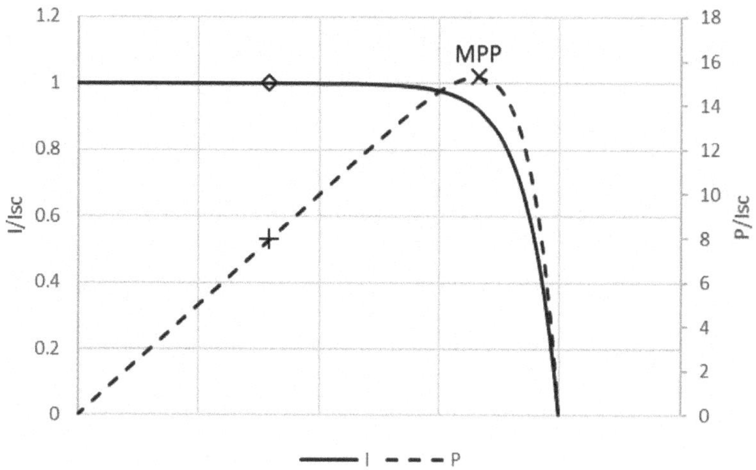

Figure 3.4 Typical solar module *I*–*V* curve.

use MPPTs in solar pumping units, as it makes them more economically feasible than direct coupled system [39].

3.2.3.2.3 Battery storage

Battery storage has the advantage of allowing the solar pumping system to produce water for longer periods, for example, nighttime, depending on the size of the battery used. It is, therefore, essential if high water availability is required.

It also allows constant operation of the motor-pump unit at their maximum efficiency point regardless of the amount of incident light, which is not achievable by direct or MPPT-coupled systems. Batteries are also necessary for portable systems where water storage is not applicable.

However, batteries reduce the reliability of the system, and increases maintenance requirements and cost of the solar system [40]; and it may eventually be less economical than a conventional diesel power pumping system. Nevertheless, as the cost of batteries keeps dropping, this may no longer be the case within this decade.

Water storage tanks can be used in conjunction with battery storage to meet the demand during cloudy days to avoid oversizing the capacity of the latter (less days of autonomy[4] can be used in the design). This

[4] For standalone solar systems, the battery bank is sized based on selected days of autonomy which accounts for potential periods of consecutive cloudy days.

significantly reduces the costs of the system. Water storage, in general, should be the first kind of storage to be considered due to its low cost. Table 3.1 shows a summary of main advantages and disadvantages of the different solar WPS configurations.

Table 3.1 Solar WPS configuration comparison.

Configuration	Advantages	Disadvantages
Direct coupling	• Lowest initial cost • Lowest maintenance • Simplest operation • Long life • Reliability	• Water storage typically required • Inefficient use of available solar energy • Difficulty to match the demand with water supply • Not suitable for large pumping head requirements • Care must be taken that the pumping rate during noon time (i.e., maximum solar irradiance) does not exceed the well replenishment rate • Not suitable for locations with cloudy weather (system will operate mostly at part-load, reducing its efficiency)
Coupling through MPPT	• Maximizes the energy output of the solar system • Longer operation of the solar pumping system and more pumped volume of water	• More expensive than direct coupled system • More components added to the system increasing complexity and reducing reliability
Systems with battery storage	• Constant operation • Maximum operating efficiency for motor and pump • Reliable water supply regardless of weather conditions • Provides high starting current which is essential for some pumps and AC motors	• Significant increase in capital and operating costs (e.g., battery replacement) • Increased maintenance

3.2.3.3 Design considerations

3.2.3.3.1 Water demand

Water demand depends on the type of application/end–user, for example, domestic, irrigation, or livestock water supply. Typical domestic water demand is 30 L/person [20]. For livestock, some typical values for various animal species can be found in the literature [19,41]. Water demand for crops can be obtained from various sources such as the food and agriculture organization [42]. It should be noted that the water demand for crops is also highly dependent on the type of irrigation used. Generally, drip irrigation is the most suitable for solar WPSs due to its minimal water use, followed by sprinkler irrigation which requires higher flow rates and pressures. Due to its high water requirements, flood irrigation is the least feasible to use with solar WPSs.

The daily water demand may vary with seasons or months; therefore, if necessary, daily water requirements for each month will need to be specified. A multiplier from 0% to 20% might be also added to account for various water losses, for example, evaporation, and to account for future growth.

3.2.3.3.2 Water source

Another important design aspect is the characteristics of the water source. The water supply can be surface or underground water. While underground water pumping is the most common use for solar pumping, surface water applications is finding more application for livestock use. For example, pumping water away from the water surface is useful to avoid polluting the latter with animals' urine and feces. Otherwise, such polluted water can reduce animal growth and health, and eventually, the water can become too polluted for the animals to drink [19].

The other important aspects to consider, especially for underground water applications, are the well depth, drawdown level, replenishment rate, and the impact of seasonal variations particularly on the water level. This is important to calculate the head requirements that must be achieved by the pump. On top that, such data is essential to estimate the flow rate limit of the well to avoid excessive pumping ensuring a sustainable use of the well without reducing the water table to uneconomical levels.

3.2.3.3.3 Solar resources

Solar irradiation is a crucial factor as it directly determines the amount of water that can be pumped as it is basically the energy source running the motor-pump unit. A solar pumping system is, therefore, more economically feasible in areas with high solar resources, which also coincide with areas with limited water availability that can be mainly accessed from underground aquifers such as the MENA region. As a rule of thumb, solar resources must be greater than 3 kWh/m^2 per day for a solar pumping system to be feasible [19].

For this reason, a solar resource assessment and accessibility need to be conducted. Accessibility assessment means analyzing the location where the solar system will be installed and ensuring it is free from any natural or man-made structures that can shade the solar modules. Shading has a significant impact on the amount of solar energy generated knowing that the current generated from the solar system is nearly directly proportionate to the amount of incident light.

The other important factor to consider is the orientation and tilt angle of the solar modules which impact the energy generated. Ideally, the solar system must face north and south in southern and northern hemispheres, respectively, to maximize the overall yearly yield. However, the optimal orientation will also depend on the time of water demand, especially if no kind of storage is used. For example, for the pump to operate during morning period, east-facing solar modules would maximize the production during this period.

The tilt angle mainly depends on the seasonal variation of the water demand. If high water demand is expected during summer, then the solar modules need to have low tilt angle as the sun will be most of the time at high altitudes. Conversely, a high water demand in winter requires a large tilt angle because the sun will be at low altitudes most of the time in winter. It should be noted that, ideally, the solar modules need to be perpendicular to the incident solar irradiance to maximize the yield.

3.2.3.3.4 Type of pumps

Pumps can be classified based on their placement, type, and the type of motor they are connected to.

In terms of placement, pumps in solar water-pumping applications can either be surface pumps or submersible, which will depend on the water source type and the well depth.

Regarding type, pumps are generally classified as centrifugal or positive displacement pumps. Centrifugal pumps have impellers that push the water in radial direction towards the casing, generating the required

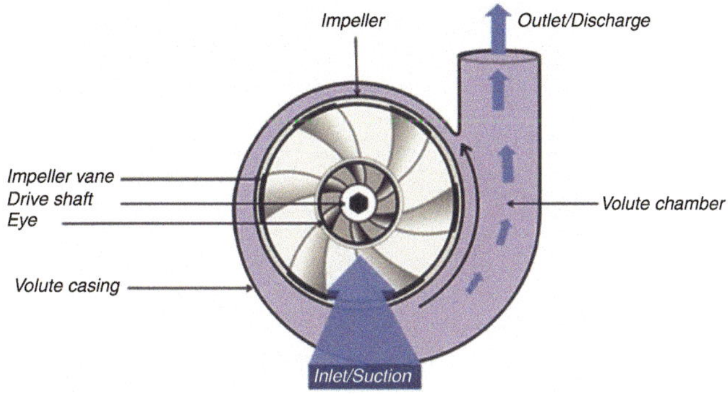

Figure 3.5 Centrifugal pump. *From Michael Smith Engineer Ltd. Useful information on centrifugal pumps. Michael Smith Engineer Ltd, 2021.*

pressure to raise the water from one level to the other (Fig. 3.5). They are typically used in direct coupled systems with low head (<25 m) requirements at which they have maximum efficiency. If larger heads are required, then multistage centrifugal pumps can be used. Another limitation for this type is that they are not self-priming; that is, the pump needs to be filled with water to start and cannot evacuate the trapped air by itself, and for this reason, centrifugal pumps are typically submerged in water (Fig. 3.5).

Positive displacement, or volumetric pumps, is also used in water pumping applications, particularly when low flow rates and large heads are required, that is, water from deep wells. This type of pumps includes[5] progressive cavity, helical rotor, piston, and rotary screw pumps. This type of pumps, however, is typically not suitable for direct coupling applications, as the pumping flow rate is mostly proportional to the rotation speed which in turn is proportional to the power (Fig. 3.6), hence the energy available from the sun. This contrasts with centrifugal pumps where the flow rate is less sensitive to changes in the input power. Volumetric pumps also require high starting torque, which is another main reason to be unsuitable for direct coupling. Their main advantage, however, is that the head is almost independent of the input power which means that speed control may not be required.

[5] Diaphragm pumps are also volumetric pumps, yet it is not recommended for water pumping applications as it is sensitive to particulates in the water such as sand which increases its maintenance requirements.

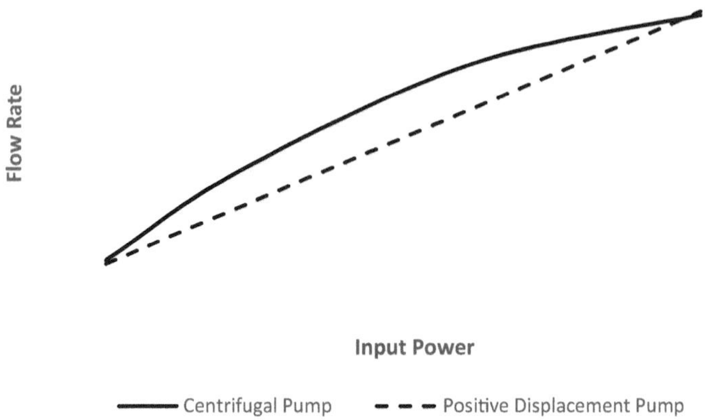

Figure 3.6 Effect of input power on flow rate.

Pumps can be driven by either DC or AC motors. DC motor-pump units are more widely used as they have the main advantage of the capability to be directly connected to the solar system without the need for any type of electrical conversion, that is DC motors are also more efficient using up to half the energy needed by AC motor-pumps units. They also can handle low voltages, which are expected to occur when the solar irradiance is low, without overheating or stalling.

For this reason, DC-powered motor-pumps are typically used in direct solar water-pumping applications, especially the submersible type. They, however, have low flow rates ranging from 0.2 to 1.1 m^3/h [19,22] as they are limited to up to 5 kW size motors [20]. This may change, however, with the progress in electric vehicles which require much larger DC motors.

AC motors are reliable and more available in the market making them ideal for remote and rural areas with lack of skilled personnel. Moreover, for all retrofitted systems, in most situation the client will be already using an AC pump as it is usually coupled with a diesel generator. However, they need solar pumping inverters which include a variable speed drive, to work, which adds to the cost. The variable speed drive controls the frequency of the voltage, hence the speed of the motor, to ensure the required head is satisfied regardless of the input power.

Finally, with submersible pumps, typically, the motor is integrated with the pump [22] and sold as one unit, which makes it easier to design the WPS. Table 3.2 shows a summary of main advantages and disadvantages of different motor/pump configurations.

Table 3.2 Motor and pump configuration comparison.

Configuration	Advantages	Disadvantages
Centrifugal	• Simplicity • Reliability • Low cost • Low starting torque. • Robust • Suitable for high flow rates	• Suitable only for low head applications unless multistage pumps are used • Not self-priming • Operating at high speeds for maximum efficiencies may shorten pump lifetime
Positive displacement	• High efficiency at large heads (>20 m) • Head is less sensitive to available power • Self-priming	• Not suitable for direct coupling • MPPT or battery storage required • High starting torque required, which could be an issue at low light conditions
DC motor-pump	• High efficiency • Suitable for direct coupling	• Low flow rates • Not all DC motors are suitable for direct coupling (only permanent magnet and potentially series DC motors though may not work at low light conditions) • Needs to be brushless especially if it will be submerged (requires less maintenance)
AC motor-pump	• More available in market • Reliable	• Requires inverters which adds to the costs • High starting current

3.2.3.4 Solar pumping system design

The design method discussed in this section is based on design methods discussed in Refs. [19] and [29]. They are for direct coupled systems but can also be used for MPPT-coupled systems. For battery-based systems, the same steps are used except for the part of the solar system design where in that case, typical standalone solar system design methods are to be followed such as those discussed in [36].

1. Determine the average daily water requirements for each month.
2. Determine the source of water, that is, surface or underground water. In both cases, determine the total vertical and horizontal distances the

water needs to be pumped along. This includes water level and drawdown for well applications. These distances are needed to estimate the piping length and the head requirements. Recharge rates must be also determined to make sure the flow rate never exceeds these values. Finally, the condition of the water needs to be examined, for example, too much sand, algae, and grit which can affect pump performance

3. Determine the optimal system configuration as discussed in Section 3.2.3.2.
4. Decide whether water storage is needed. Typical values of 3–10 days of storage are recommended where the maximum value is necessary for domestic use in cloudy climates.
5. Determine the daily solar irradiation in the location. Average daily global horizontal irradiation can be easily found from typical meteorological year data.
6. Select the optimal tilt angle for the solar modules. Table 3.3 shows the recommended tilt angles based on the water demand periods; otherwise, solar design simulation tools such as PVSYST can be used to optimize the tilt angle.
7. Calculate the required pump flow rate. This is obtained by dividing the monthly/seasonal water requirement by the expected operating hours of the solar pumping system. For directly or MPPT-coupled systems, the operating hours can be assumed to be equal to the peak sun hours (PSH) during the same period [19,20]. The PSH is the same as the daily global horizontal solar irradiation.[6]

$$Q^{\cdot} = \frac{V_{prd}}{Hr_{op} \cdot \mathrm{Days}_{prd} \cdot 3600} \tag{3.1}$$

Q is the water flow rate in m^3/s, V_{prd} is the period required water volume in m^3, and $Days_{prd}$ is the number of days in the selected period.

This method is typically used in practice due to its simplicity. However, it can underestimate the required flow rate, especially for direct coupled systems, as it fails to account for periods of time when the weather is cloudy. During cloudy days, a direct coupled system

[6] Peak suns hours are the equivalent hours of the day at which the solar irradiance is 1 kW/m^2 which is the standard test conditions. In other words, it is the number of hours that gives the same solar irradiation/energy if the irradiance is assumed to be constant at 1 kW/m^2. Therefore, dividing solar irradiation in kWh/m^2 by 1 kW/m^2 gives the number of hours.

Table 3.3 Recommended tilt angles based on water demand profile [29,44].

Tilt angle	Description	Water profile
Latitude +20°	Module tilt optimized for wintertime when the sun is at low altitude to ensure uniform energy generation all over the year (less generation in summer and more in winter)	Water demand is constant throughout the year, for example, domestic water supply
Latitude +10°	Module optimized for wintertime with slightly more generation in summer	Water demand is fairly constant throughout the year but with some increase during summer
Latitude −10°	Annual energy generation is maximized	Water supply is to be maximized. This is applicable when a large water storage is used so excess water in summer can be stored and then used in periods of low solar irradiation
Latitude −20°	Summer generation is maximized	Water demand in larger during summer, for example, summer crop irrigation

Source: From S.R. Wenham, M.A. Green, M.E. Watt, R. Corkish, Applied Photovoltaics, second ed., Centre for Photovoltaic Engineering, UNSW, 2006; G. Li, Y. Jin, M.W. Akram, X. Chen, Research and current status of the solar photovoltaic water pumping system – A review, Renewable & Sustainable Energy Reviews 79 (2017) 440–458.

will likely not work and therefore the actual total monthly operating hours will be less than the value calculated using Eq. (3.1). Nevertheless, this is compensated by the fact that global horizontal irradiation is used rather than the tilted plan of array value. A more accurate method for estimation of the operating hours is shown in the Appendix, which is obtained from [29].

8. Specify the pipe diameter and the number and type of valves and pipe bends used. The pipe diameter should be carefully considered. Very large pipes will help reducing friction losses; yet will increase the cost and, according to Jenkins [19], may have low water velocities that could result in the accumulation of suspended solids at the bottom the pipe. Typically, 40–50 mm diameter are sufficient for most direct coupled solar pumping applications due to their typically low flow rates. Otherwise, a design velocity of 1 m/s can be used to estimate the pipe diameter [45].

9. Calculate the friction losses in the pipes and valves. This can be assumed to be 2%–5% of total head required in a good design. Otherwise, it can be calculated as follows:

a. Determine the friction loss coefficient of the pipe bends and valves (K values). Table 3.4 shows the K values for some pipe bends and valves obtained from [47].

b. Calculate the pipe friction factor using the formulas below, obtained from [48], which are used for laminar and turbulent flows, respectively:

$$f = \frac{64}{Re} \tag{3.2}$$

$$f = \left[-1.8\log\left[\frac{6.9}{Re} + \left(\frac{\varepsilon/D}{3.7}\right)^{1.11}\right]\right]^{-0.5} \tag{3.3}$$

Re is the Reynolds number, and ε and D are the pipe surface roughness and diameter, respectively, both in meters. Some typical values of ε obtained from [49] are shown in Table 3.5.

c. Determine the total length of the pipe which includes any horizontal distance, for example, distance from well to storage tank or load; and the total vertical lift required. The total vertical lift includes both the static head, that is, the distance between the water level when the well is still to the level at which the water is discharged, and the drawdown level, that is, the drop distance in the water level once pumping has started. The latter depends on reservoir properties and the flow rate.

d. The total friction loss in the pipe can be then calculated as follows:

$$H_{\text{loss}} = \frac{v_{av}^2}{2g}\left[\sum K + \frac{fL}{D}\right] \tag{3.4}$$

where v_{av} is the average water velocity inside the pipe, L is the total length of the pipe in m, and g is the gravitational constant in m/s^2.

e. Calculate the total dynamic head (TDH).

$$TDH = H_s + H_{dd} + H_{\text{loss}} \tag{3.5}$$

where H_s and H_{dd} are the static head and drawdown level, respectively, both in meters.

Table 3.4 Typical fitting *K* values.

Description	L/D	Nominal pipe size (in.)											
		$\frac{1}{2}$	$\frac{3}{4}$	1	$1\frac{1}{4}$	$1\frac{1}{2}$	2	$2\frac{1}{2}-3$	4	6	8–10	12–16	18–24
Gate valve	8	0.22	0.20	0.18	0.18	0.15	0.15	0.14	0.14	0.12	0.11	0.10	0.10
Globe valve	340	9.20	8.50	7.80	7.50	7.10	6.50	6.10	5.80	5.10	4.80	4.40	4.10
Angle valve	55	1.48	1.38	1.27	1.21	1.16	1.05	0.99	0.94	0.83	0.77	0.72	0.66
Ball valve	3	0.08	0.08	0.07	0.07	0.06	0.06	0.05	0.05	0.05	0.04	0.04	0.04
Plug valve straightway	18	0.49	0.45	0.41	0.40	0.38	0.34	0.32	0.31	0.27	0.25	0.23	0.22
Plug valve 3-way through-flow	30	0.81	0.75	0.69	0.66	0.63	0.57	0.54	0.51	0.45	0.42	0.39	0.36
Plug valve branch flow	90	2.43	2.25	2.07	1.98	1.89	1.71	1.62	1.53	1.35	1.26	1.17	1.08
Swing check valve	50	1.40	1.30	1.20	1.10	1.10	1.00	0.90	0.90	0.75	0.70	0.65	0.60
Lift check valve	600	16.20	15.00	13.80	13.20	12.60	11.40	10.80	10.20	9.00	8.40	7.80	7.22
Standard elbow													
90 degrees	30	0.81	0.75	0.69	0.66	0.63	0.57	0.54	0.51	0.45	0.42	0.39	0.36
45 degrees	16	0.43	0.40	0.37	0.35	0.34	0.30	0.29	0.27	0.24	0.22	0.21	0.19
Long radius 90 degrees	16	0.43	0.40	0.37	0.35	0.34	0.30	0.29	0.27	0.24	0.22	0.21	0.19
Standard tee													
Through-flow	20	0.54	0.50	0.46	0.44	0.42	0.38	0.36	0.34	0.30	0.28	0.26	0.24
Through-branch	60	1.62	1.50	1.38	1.32	1.26	1.14	1.08	1.02	0.90	0.84	0.78	0.72
Mitre bends													
$\alpha = 0$	2	0.05	0.05	0.05	0.04	0.04	0.04	0.04	0.03	0.03	0.03	0.03	0.02
$\alpha = 30$	8	0.22	0.20	0.18	0.18	0.17	0.15	0.14	0.14	0.12	0.11	0.10	0.10
$\alpha = 60$	25	0.68	0.63	0.58	0.55	0.53	0.48	0.45	0.43	0.38	0.35	0.33	0.30
$\alpha = 90$	60	1.62	1.50	1.38	1.32	1.26	1.14	1.08	1.02	0.90	0.84	0.78	0.72

Source: E. Shashi Menon, Piping Calculations Manual, first ed., McGraw-Hill Education, New York, 2005 [46].

Table 3.5 Surface roughness for some pipe materials.

Pipe material	Surface roughness (mm)
Commercial steel	0.046
Galvanized iron	0.15
Cast iron	0.26
Concrete	0.3–3
Riveted steel	0.9–9
Wrought iron	0.046
Drawn tubing	0.0015
Internally plastic coated	0.005
Cement lining	0.033
Fiberglass lining	0.038
Bare carbon steel	0.036

At this stage, it is worth checking if the direct coupled solar pumping system is economically feasible by calculating the hydraulic load

$$\text{Hydraulic Load} = TDH \cdot V_{\text{day}} \tag{3.6}$$

where V_{day} is the daily water requirements. The hydraulic load (m^4/day) is a function of the TDH which is because the cost is expected to increase linearly with the head. Typically, if the value is less than 1500 than the system is feasible, from 1500 to 2000, the system is borderline, and larger than 2000 means that other system configurations must be considered such as those with batteries. Nevertheless, these figures should be taken with caution as it will also depend on other factors such as solar resources. For instance, solar WPSs were reported in some studies to still be more economical than a diesel generator-powered WPS for hydraulic loads in the range of 3000 [20] to below 5750 m^4/day [50].

f. The power required by the motor-pump unit can be calculated as follows:

$$P_e = \rho \cdot g \cdot Q \cdot TDH / (\eta_p \cdot \eta_m) \tag{3.7}$$

P_e is the required electrical power in W, ρ is the water density in kg/m^3, η_p is the pump efficiency (typically between 25% and 50%), and η_m is the motor efficiency. Both efficiencies can be determined from the performance curves.

g. Select a suitable pump that can meet the required pump head and flow rate at its maximum efficiency point. Note that if a direct coupled system is used and the suction head is larger than 7 m, then the pump needs to be submersible.

h. From the torque and speed characteristics of the pump, select a suitable motor. The motor also needs to be designed to operate at its maximum efficiency point. It should be noted that there are commercial integrated motor-pump units which alleviate the need to size the motor.

i. Find the operating voltage and current of the motor.

j. From the voltage, the number of modules connected in series, that is, number of modules in one string can be determined such that

$$N_{\text{series}} = 1.02 \frac{V_{\text{motor}}}{V_{mpp,min,mod}} \tag{3.8}$$

where V_{motor} is the motor voltage. The 1.02 factor is to account for DC cable voltage drop. $V_{mpp,min,mod}$ is the MPP voltage of a single module at the max expected temperature which can be calculated as follows assuming the temperature rise of the cell is 25°C above ambient:

$$V_{mpp,min} = V_{mpp}.\alpha_{Pm}.\left[T_{amb,max}\right] \quad (3.9)$$

V_{mpp} is the module MPP at standard test conditions,[7] α_{Pm} is the MPP temperature coefficient and $T_{amb,max}$ is the maximum expected ambient temperature.

k. From the motor current, the number of parallel strings can be determined such that

$$N_{\text{Parallel}} = \frac{I_{\text{motor}}}{I_{mpp,mod}.\ 0.8\ .\ DF} \quad (3.10)$$

I_{motor} is the motor current, $I_{mpp,mod}$ is the MPP current of the module, and DF is the derating factor. $I_{mpp,mod}$ is obtained from the manufacturer specifications; however, the value is given at standard test conditions, that is, at 1000 W/m^2 irradiance. However, the irradiance varies throughout the day and months from zero and peaks at noon time and is not constant at 1000 W/m^2. Now as a direct coupled system is expected to operate mostly around the noon time, a conservative approach is to adjust $I_{mpp,mod}$ to an average irradiance of 800 W/m^2, and hence the 0.8 factor in Eq. (3.10). Ideally, however, because the irradiance may not necessarily reach 1000 W/m^2 at noon, $I_{mpp,mod}$ should be adjusted to 80% of the average irradiance during the design pumping period $I_{sun,G,tilt,av}$ and its calculation method is shown in the Appendix.

The derating factor accounts for various losses in the solar system. It can typically be assumed to be 75% for simplicity. The various losses are as follows:

1. Temperature derating: module output decreases with increasing ambient temperature. This decrease is particularly in the voltage which typically decreases with a rate of 0.35%–0.4% for every 1°C increase in module temperature above 25°C.

   The temperature derating factor can be calculated as follows:

$$DF_{TEMP} = 1 - \alpha_{Pm}\left[T_{amb,max}\right] \quad (3.11)$$

[7] 25°C, 1000 W/m^2, AM = 1.5.

2. Soiling (DF_{dirt}): occurs because of the accumulation of dirt on the surface of the module. This value typically ranges from 2% to 10% (i.e., 0.9–0.92), where the latter is for heavily polluted/dusty environments. However, in desert areas, this value can reach 50% [51] if no regular module cleaning takes place. With regular cleaning, this value can be assumed to be 3%.
3. Mismatch losses (DF_{mm}): occurs because not all modules have the exact same voltages and currents. This value typically is in the range of 1% with new generation of modules (i.e., 0.99).
4. Module power tolerance (DF_{tol}): this value is given by the manufacturer, for example, ± 3%. However, with new generation modules which mostly have positive power tolerance, this derating factor can be assumed to be 1.
5. Output degradation (DF_{dgrd}): the natural degradation in the energy output of the module with time which is typically guaranteed by manufacturer to be 20% after 25 years. In other words, the design value will depend on the designed lifetime of the system. The value therefore will highly depend on the expected lifetime of the system and how acceptable is to have lower water production in the future. A conservative value of 10%, that is, 0.9 derating factor is typically used.

The total derating factor can then be calculated as follows:

$$DF = DF_{temp} \times DF_{dirt} \times DF_{mm} \times DF_{tol} \times DF_{dgrd} \tag{3.12}$$

Finally, the exercise of designing a solar WPS should be iterative. In other words, the water production needs to be recalculated to check if it matches the water requirements.

3.2.4 Advances in solar pumping technologies

While solar WPS is a mature technology that is widely used, there is still research conducted to improve its performance.

Most research investigated the optimization of the solar WPS to increase its overall efficiency tackling each component separately. For example, many studies focused on optimizing the flow rate and head of a given pump type [38,52–58]. Other studies focused on optimizing the PV system size and predict its performance through mathematical modeling and simulation such as the work by [59–63]. Optimization work also includes building sizing tools [64] to minimize the life cycle costs (LCC) and the loss of power supply probability which takes place when the solar WPS cannot meet the load demand.

Other studies investigated unconventional methods to increase the efficiency of the solar WPS. One technique that was investigated is to spray water or run cool air over the surface of the PV modules to cool the solar cells, which increases their efficiency such as the work by [65–68]. Such a concept has shown significant reduction in the module temperature, and hence an increase in operating efficiency. For example, Abdolzadeh and Ameri [65] reported up to 30% increase in the flow rate when water was intermittently sprayed on the surface of the module. Other innovative concepts explored the use of booster reflector to increase the amount of incident solar irradiance on the solar modules [66].

Some studies considered a more wholistic approach in designing a solar WPS considering not just the water demand but other factors such as the local climate, soil characteristics (depth and moisture), borehole properties, crop water requirement, and the type of irrigation system. These studies reported that such design approach required smaller PV system size than the usual method [59].

Many studies also investigated innovative control algorithms such as fuzzy logic with various targets such as:

1. to match the solar supply with the water demand, avoid pumping excessive water, and ensure a sustainable drawing of underground water [37];
2. maximize discharge rate and system efficiency [69]; and
3. extend the solar WPS operating hours [70].

Other authors looked into innovative ways to use the excess energy from a solar pumping system by storing it using a pumped hydro-storage system [71] and argued that such system ensures higher self-sufficiency after correctly sizing the PV plant and the hydraulic turbine.

The new trend of studies focuses more on the state-of-the-art internet of things, data acquisition, and remote sensing, including the use of various sensors to control and analyze the operation of the solar WPS such as soil humidity, pH, flame (potassium and sodium ions concentration), and temperature values [72–74].

Another similar and interesting concept also relying on data capturing and analysis is improving solar WPS maintenance in terms of reducing human error and improving performance. One particular interesting research looked into using augmented reality (AR) where the repairer can see the maintenance tasks required in AR superimposed on the actual faulty equipment. The AR show the diagnosis results, such as the type of failure and component state, (which were analyzed based on data collected on site from sensors), in addition to the relevant maintenance procedure all through their mobile device [75].

3.2.5 Economics

To properly assess the cost of a solar WPS and properly compare it with a similar system driven by a diesel generator or grid-connected, LCC need to be considered.

LCCs consider the capital cost and all running costs within a certain period which is usually the lifetime of the PV modules (20–25 years). Besides equipment cost, the initial cost also includes the costs associated with engineering and design costs, and installation. For cost breakdown, data from 150 solar WPS's in Bangladesh showed that most of the equipment cost was dedicated to the solar modules [76] as shown in Fig. 3.7. Running costs include all maintenance and equipment replacement cost, for example, the pump needs to be replaced two to three times during 20 years' span [20], in addition to labor wages.

The LCC can be calculated as follows [77–81]:

$$LCC = CC_{tot} + RC_{tot} + \sum_{j=1}^{LP} \frac{OC_{tot}}{(1+IR)^j} \tag{3.13}$$

where LCC is the life cycle cost, LP is the evaluation period which is usually taken to be the lifetime of the PV modules, CC_{tot} and RC_{tot} are the total capital and replacement costs of the equipment, respectively, OC_{tot} is

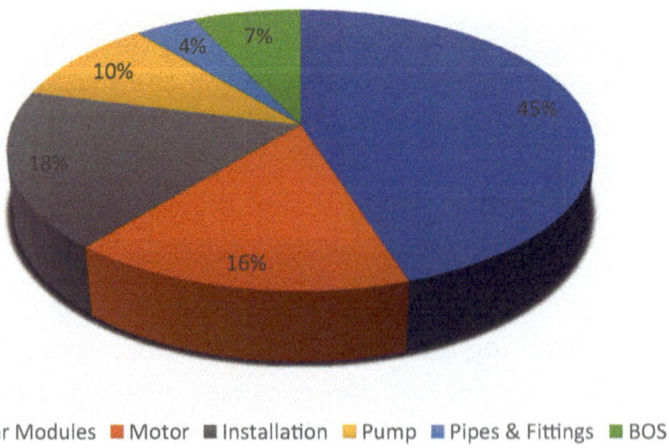

Figure 3.7 Solar WPS cost breakdown.

the total yearly maintenance and operation cost, and IR_r is the real interest rate and is calculated using Eq. (4.6) obtained from Koner et al. [82].

$$IR_r = \frac{IR_n - f}{1 + f} \tag{3.14}$$

where IR_n is the nominal interest rate and f is the inflation rate.

The replacement cost of each asset RC_i is calculated as follows [77,83]:

$$RC_i = CC_i \sum_{j=1}^{Y_i = \frac{LP}{LT_i} - 1} \frac{1}{(1 + IR_r)^{j*LT_i}} \tag{3.15}$$

where Y_i is the number of times the ith asset will be replaced during the lifetime of the project and LT_i is the lifetime of the ith asset.

There are various cost estimation studies in the literature, and they all confirm that a solar WPS is more economical than a diesel generator-powered system. Some studies even reported that a solar WPS as old as 25 years old was still more economical than a diesel generator-powered system with a 4 years payback time in spite of the very high cost of solar modules at that time [18]. In addition, data from 418 wells in Tanzania showed that diesel-powered WPSs have an initial cost of 13,000 USD and annual fuel cost of 5000 USD with an LCC of 93,000 USD. In contrast, while the solar WPS costed more than 30,000 USD, in addition to the high replacement cost of pumps and motors after 10 years, its LCC was estimated to be only 59,000 USD [18,20].

Similarly, a modeling study comparing solar water pumping to diesel-powered water pumping over heads and flow rates ranging from 3 to 50 m^3/day and 10 to 200 m, respectively, in Namibia, showed that the LCC over 20 years for the former was lower even for hydraulic loads in the range of 3000 m^4/day. The result was echoed in many studies such as [18,20,23,50] showing payback times for a solar WPS ranging from 2.3 to 6 years.

3.3 Solar-driven reverse osmosis for water desalination

3.3.1 Historical background

While the osmosis phenomenon was discovered in 1748 with first semipermeable membrane developed in 1867 [84], RO is a relatively new process, compared with other well-established desalination processes such as MSF and MED, that only started to be commercialized in 1960s. It then increasingly gained market share [85] to become the most dominant water desalination technology covering all the spectrum of desalination plant scales with capacities from 1 m^3 to over 600,000 m^3 per day [86].

3.3.2 Application and global status

Similar to solar WPSs, there is an increasing interest in solar PV-driven RO, especially in remote/rural areas. For example, many remote coastal resorts started to use solar PV-RO rather than paying the large operating expenses of a diesel generator or to transport water.

The RO process is currently the cheapest and least energy intensive (Table 3.6) desalination process and together with the low price of solar PV energy, it is no surprise that solar PV-driven RO will be more widely adopted. The use of solar-driven RO process in brackish water desalination is an even more attractive application due to its lower energy use and chemical treatment requirements, discussed in Section 3.3.3.2 [98].

The cost advantages of the RO process over other conventional processes are reflected in the former's dominance in the global desalination market, where in 2013, a positive trend was observed for newly constructed RO plant compared with a decline for MSF and MED plants [86]. Moreover, the fact that RO only requires one form of energy, that is, electricity, makes it ideal for PV coupling. For this reason, based on the review carried out by Ali et al. [97], almost half of 82 solar-driven desalination plants[8] that were installed worldwide from 1977 to 2009 used an RO process. The RO process is also more suitable for decentralized applications due to its modularity. It is no surprise, therefore, that the process is dominating decentralized applications in remote areas [99].

Conversely, solar-driven MSF and MED had little applications because they are generally not suitable for small-scale applications as they become uneconomical or very expensive. For large-scale units, it would be more rational to use cogeneration configuration, that is, thermal desalination in conjunction with a fossil fuel-based plant, which is already applied in many areas in the world, especially the MENA region. It should be also noted that even when MED and MSF are used in a cogeneration setting, the most economical case for such processes, the unit water costs of RO were still lower [93,99,100,101]. For this reason, the cost advantage of the RO process is expected to be more evident for standalone solar-driven desalination plants.

[8] Including hybrid PV-wind energy and solar pond-driven systems.

Table 3.6 Specific energy consumption of the major desalination processes.

	Brackish water	Seawater	
	Electricity (kWh/m^3)	Electricity (kWh/m^3)	Heat (kWh/m^3)
ED	*0.5–10 [a]* *≈ 0.5 for every 1 g/L of salts removed [b]*	–	–
RO	*1–3 [a]* *0.5–3 [b]* *2–3 [c]* *0.5–2.5 [d]*	*10–15 (No ERD) [a]* *9.5 (No ERD) [e]* *2–5 (with ERD) [a]* *2.5–7 (with ERD) [b]* *3–4 (with ERD) [d]* *2.3–3 (with ERD) [f]* *3–5.5 (with ERD) [g]*	None
MED	–	*MED/TVC* *1.5–2.5 [b],[g]* *3.3 [h]* *2.7–2.9 [e]* *3.3–5 [l]* *MED only* *3.5 [j] (MED only)* *1.5–2.5 [g]*	*MED/TVC* *40–108 [b],[g]* *76.1 [h]* *53.8–80.7 [e]* *57.5–70.4* *MED only* *84.9 [j]* *63.9–108 [g]*
MSF	–	*3.6 [a]* *3–5 [b]* *3.9 [e]* *4–6 [g]* *4 [h][i]* *2.5– 5 [j]*	*80.5 [a]* *69.4–91.7 [b]* *80.7 [e]* *52.8–108 [g]* *76.1 [h]* *92.5 [i]* *52.8–78.3 [j]*
MVC	–	*8.5–16 [a]* *8–15 [b]* *7–12 [g]*	*None*
TVC	–	2 [k]	108.5 [k]

[a] [2]; [b] [87]; [c] [88]; [d] [89]; [e] [90]; [f] [91]; [g] [92]; [h] [93]; [i] [94]; [j] [95]; [k] [96]; [l] [97].

For MVC, Garcia-Rodríguez [102,103], and Fiorenza et al. [99] argued that the process could be a better alternative for remote areas than RO due to its robustness, ease of operation and lower operational cost given that the process needs less chemical requirements and because no membrane replacement is needed. However, this conclusion is questionable given that in practice despite its technological maturity, the MVC process has a negligible global market share [86] due to its higher energy consumption. Even among renewable energy-driven desalination plants, which are usually small-scale plants [104] used in remote/rural areas, only 5% used an MVC process [105].

In addition, while the RO process produces water with less quality than distillation processes, that is, MSF, MED, and MVC, this advantage is not crucial when water is used for drinking or irrigation given that water with salinities ranging from 500 to 1000 mg/L is suitable for both purposes [106,107]. Higher quality product water is only needed for industrial purposes and power plants. As a matter of fact, very low salinity water needs re-mineralization to prevent corrosion problems and it may also be unsuitable for irrigation affecting the soil water infiltration rate. Nevertheless, care should be taken with some dissolved solids that are not well rejected by the RO process and may cause health issues, especially Boron [107,108].

Finally, the RO process has various other applications where solar PV energy can play an important role. For example, RO membranes have been used in artificial kidney filtration and urine reclamation (mostly in space applications), a process that might be required in hopsitals in remote/rural areas. In that case, it requires a simple setup and ensures the removal of most of waste metabolites (e.g., >99% of proteins, >95% of salts, and ~70%–96% of urea with polyamide membranes [5,34–36]. However, due to its high driving pressure (minimum 2 Bar, typically) which can damage cells and biological components in the bloodstream, RO is not used directly in the filtration of blood or plasma but only in filtering the substitute solutions (e.g., dialysate) to reclaim clean water [5,37]. However, in the multistep filtration system that includes some prefiltration/separation that allows no cells or proteins to enter the membrane, RO can be used directly to reclaim water from the protein-filtered plasma. In addition, the recent development of microfluidic technologies [38], materials [39,40], 3D printing [41], and advances in RO membrane research [1,6] can be pulled together for a miniaturized RO membrane system [42].

3.3.3 Design and operation

3.3.3.1 Operating principles

The osmotic phenomenon is the natural tendency for a solution with low solute concentration to diffuse through a semipermeable membrane to dilute a solution with a higher solute concentration. Semipermeable

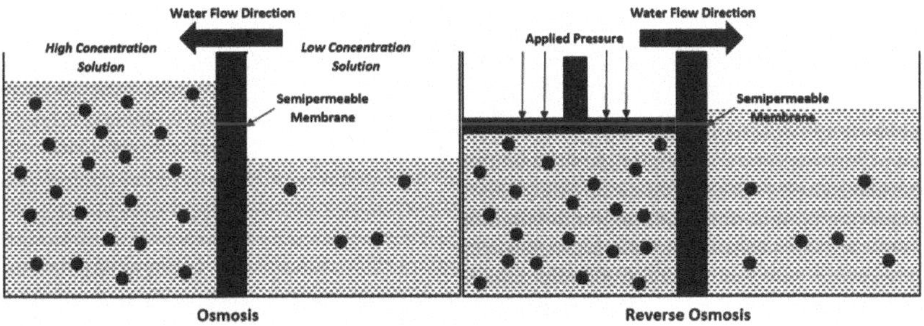

Figure 3.8 Osmosis and reverse osmosis process [109].

means highly permeable to water molecules while nonpermeable or has low permeability to other molecules such as various salt molecules.

As shown in Fig. 3.8, the height of the column of the more concentrated solution, therefore, increases as more solution from the low concentrated side crosses the membrane until equilibrium is reached. This pressure increase in the higher concentrated solution at this point is the osmotic pressure. Accordingly, to desalinate a water solution, this process is reversed, hence the name reverse osmosis, by applying an external pressure on the high-concentration solution to overcome the osmotic pressure. Reverse osmosis is therefore a pressure-driven process where a semipermeable membrane is used to block most of the dissolved solids in the water solution. The separation mechanism in RO is complex and occurs through three different mechanisms:

1. Electrostatic repulsion: the RO membrane surface has a negative polarity repelling the anions in water. Conversely, cations usually follow the anions to maintain the electro-neutrality of the feed solution. For this reason, RO membranes have more effectively reject ions with higher valence such as calcium and magnesium compared to low-valence ions such as sodium [110–113].
2. Sieving effect: due to its narrow pore size, that is, 0.1–1 nm, RO membranes are impenetrable to particles, as they mostly have large sizes. This mechanism is crucial as it permits the rejection of non-charged substances such as viruses, bacteria, and dissolved organic matter in water [110–113].
3. Solubility and diffusion: due to the polarity and small molecular size of water molecules, they dissolve in the membrane material and diffuse through it. This in turn further increases their permeability compared to other solute molecules [22–24].

The solute transport across the membrane is, therefore, determined by pore flow and diffusion and can be described as follows [18]:

$$N_{Ax} = \frac{\rho_A k}{\mu}\frac{dp}{dx} - D_{AB}\frac{dp_A}{dx} \quad (3.16)$$

where N_{Ax} is the mass flux of element A in the x-direction (perpendicular to the membrane surface), ρ_A is the mass density of A, k is the permeability, μ is the viscosity, dp/dx is the pressure gradient in the x-direction, and D_{AB} is the diffusion coefficient for the diffusion of element A in membrane B.

Mass transport through RO membranes can be calculated as follows [6,18,21]:

$$N_A = L(TMP - \Delta\pi) \quad (3.17)$$

where N_A is the water flux through the membrane, L is the permeability coefficient, TMP is the transmembrane pressure, and $\Delta\pi$ is the osmotic pressure difference between the feed and the permeate membrane surface.

For a thermodynamically ideal solution, osmotic pressure can be calculated based on ion concentration (e.g., salt) and the solution temperature as follows [6,25]:

$$\pi = CRT \quad (3.18)$$

where C is the ion concentration (mol/L), R is the ideal gas constant (0.08206 L atm/mol K), and T is the operating temperature (K).

Membrane salt rejection (R) is a measure of overall membrane system performance and is given by the equation below (for membrane with crossflow operation) [28]:

$$R = \left(1 - \frac{C_{\text{permeate}}}{C_{\text{feed}}}\right) \times 100\% \quad (3.19)$$

Normally, RO membranes achieve a salt (NaCl) rejection rate of 98%–99.8% [28]. Nanofiltration (NF), which is a very similar process to RO but uses membrane with larger pore sizes, exhibits rejection values greater than 90% for multivalent ions and between 60% and 70% for monovalent ions [29,30]. NF, therefore, uses less energy than RO but produces water with lower quality.

3.3.3.2 Reverse osmosis plant design

3.3.3.2.1 Plant design

The RO desalination system has four main stages: pretreatment system; high-pressure pump(s) (HPP); the RO stage(s); and the posttreatment system (Fig. 3.9).

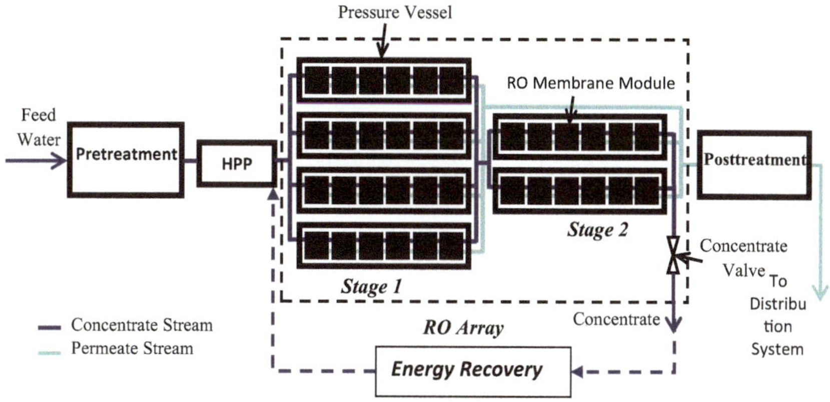

Figure 3.9 Schematic showing the main components of an RO desalination plant.

3.3.3.2.1.1 Pretreatment In the pretreatment stage, the water is initially treated to prevent fouling, scaling, and chemical attacks from damaging or blocking the RO membranes. Fouling, scaling, and chemical attacks are a major limitation to the RO process which can reduce the performance, shorten the lifetime, or damage the RO membrane [111]. Membrane fouling is the accumulation of solids on the membrane surface and/or within its pores, eventually resulting in clogging of these pores or reducing their effective area which in turn reduces the permeate flow.[9] Fouling also increases the pressure drop along the membrane feed/concentrate channel which can eventually mechanically damage the membrane [114].[10] There are four main forms of fouling: fouling caused by microorganism in water, referred to as bio-fouling; fouling by the adsorption of organic matter to the membrane surface, referred to as organic fouling; fouling by metal oxides; fouling due to the precipitation of inorganic dissolved solids, referred to as scaling; and fouling by suspended solids and colloids[11] in water, referred to as particle or colloidal fouling [115]. Cartridge, sand and media filters, and clarification in addition to coagulation and flocculation are typically used to remove large particles and suspended solids, including organic matter in the feed water [116,117].

[9] In practice, RO plants operate at constant permeate flow by adjusting the feed pressure. Accordingly, membrane fouling will result into an increase in feed pressure and therefore the energy consumption.

[10] Membrane telescoping.

[11] Colloids are a smaller type of suspended solids that are distributed evenly in the water due to their surface charge and therefore are harder to precipitate and cause water turbidity.

Particulate and suspended solids fouling potential is usually assessed by the silt density index (SDI) [118]. The SDI measures the rate at which a certain amount of the feed water will plug a 0.45-μm filter at a fixed pressure; the higher the SDI, the more the colloidal fouling potential of the feed water [116]. Typically, SDIs less than five are recommended [119]. The type of water, whether surface or from the ground, greatly impacts the quality of the feed water. For instance, surface water has high colloidal fouling potential as it contains high amounts of suspended solids [87] and natural organic matter [120][12]. In contrast, groundwater has usually a low SDI (<3) because the ground itself filters the water [121]. For this reason, filtering groundwater typically does not require media filters, clarification, coagulation, and flocculation and 5-μm cartridge filters are typically sufficient[13] [122]. Bio-fouling is controlled by disinfecting the water using oxidants such as chlorine, that is, pre-chlorination, and non-oxidant biocides to inactivate bacteria and other microorganisms [123]. Groundwater has also low bio-fouling potential [122] compared with surface waters, and for this reason, disinfecting the water may not be needed as observed in some existing groundwater RO desalination plants [118]. Metal oxides fouling occurs when the feedwater has high concentrations of dissolved iron and manganese, which is typical for groundwater with low dissolved oxygen. When the water is then exposed to air, these metals can oxidize and precipitate [160]. The precipitated metal oxides accumulate on the membrane surface clogging the pores. Fouling caused by iron and manganese oxides is prevented by deliberately oxidizing both metals so they can be easily filtered. For this reason, filters with a manganese dioxide media are typically used, and are referred to as iron filters, and with some types,[14] additional oxidizing chemicals[15] such as chlorine and potassium permanganate are required to help with the oxidation process [123]. There are also other methods to handle metal oxide fouling such as the use of oxidizing chemicals in conjunction with MF to remove iron oxides [124]; using lime softening in conjunction with settling tanks to remove iron (P. Roginson, 2014, pers. comm., 5 Feb.); and the use of an ion exchange process [125]. It should be noted that other pressure-driven processes such as microfiltration and ultrafiltration can also be used to

[12] Organic matter from natural sources

[13] RO membranes manufacturers usually recommend a feed water with SDI less than five to meet the warranty requirements.

[14] That is, manganese greensand media filter.

[15] For example, chlorine, chlorine dioxide, potassium permanganate, and ozone.

replace conventional pretreatment methods. Both are also pressure-driven membrane treatment technologies but with larger pore sizes than RO ranging from 0.1 to 5 μm and 0.005 to 0.1 μm, respectively. Their separation mechanisms rely solely on sieving. The advantages of using such membranes are their ability to produce higher feed quality for the RO plant removing the suspended solids and microorganisms preventing the RO modules from excessive fouling, hence reducing their cleaning requirements [126,127] and increasing their reliability [128]. Microfiltration is also reliable in pretreating water with iron content in contrast with cartridge filters which cannot be used for such water types because they will be rapidly fouled and may need to be replaced every day (P. Roginson, 2014, pers. comm., 5 Feb.). Scaling is the precipitation of salts on the membrane surface as they exceed their solubility limit in water, especially because of CP [129]. In addition to fouling the membrane and increasing the pressure drop, scaling increases the salt passage through the membrane because the high concentration of salts on the membrane surface causes more salts to diffuse, reducing the permeate quality [116].

The most common types of scales are calcium carbonate, calcium sulphate, reactive silica and magnesium hydroxide [98,130] due to their low solubility in pure water [111]. Other less common types of scales are barium sulphate, calcium phosphate, and strontium sulphate. Calcium carbonate, magnesium hydroxide, and calcium phosphate scaling can be controlled by lowering the pH of water, usually to values from 5.5 to 6.2 [109,116,131], by adding acids such as sulphuric or hydrochloric acids to increase their solubility. Calcium sulphate, reactive silica, barium sulphate and strontium sulphate scaling is more difficult to control and require the use of scale-inhibitors which prevent crystals from forming and growing [116]. Reactive silica, in particular may limit the recovery rate [117]. Scaling can be also controlled through the use of an ion exchange process which is more effective than using chemicals as it removes most of the ions causing scaling [130] or through lime softening to precipitate calcium and magnesium ions in addition to silica, barium and strontium by adding chemicals such as lime and caustic soda to one or more clarifier tanks where the precipitation occurs [132]. However, the ion exchange process is expensive due to the high costs of the regeneration process [88] and is not economically feasible to treat high salinity water (Dow Chemical Company, 2013, pers. Comm., 8 Jun.). Conversely, lime softening requires additional components such as clarifier tanks and re-carbonation reactors where carbon dioxide is added to the water to decrease its pH [132]. Chemical attack refers to the degradation of the membrane active layer[16], particularly with the more widely used polyamide membranes [133], because of the presence of oxidizing agents [134] in water such as chlorine, used during pre-treatment. Oxidizing agents reduce the membrane's salt rejection and therefore the permeate water quality. All oxidizing agents, therefore, need to be removed before reached the RO membrane by adding a reducing agent, such as sodium metabisulphite, to the water or through using activated carbon filters [114]. It should be noted that in addition to pre-treatment, the RO membranes require periodic cleaning where the number of cleanings per year will vary according to the manufacturer and the feed water type. DOW Chemical Company for example recommends four cleanings per year as long as the RO membranes operate within the specified operating conditions [135]. The cleaning involves the use of acid to remove scaling from the membranes surface then exposing the membrane to a high pH solution ($pH > 12$), e.g. by using caustic soda, which acts as a biocide [136] removing microorganisms while organic material dissolve in the water at this high pH.

3.3.3.2.1.2 Water Pressurizing Right after the pre-treatment stage, a high-pressure pump is used to pressurize the feed water to overcome the osmotic pressure and obtain the desired permeate flow. The pressure is controlled using a valve located at the concentrate side at the last RO plant stage. The feed water pressurized against the membrane partially crosses the membrane leaving most of the salts behind and is referred to as the permeate water. The remaining and now more concentrated feed water is referred to as the concentrate or the brine.[17] The more pressure is applied, the more permeate water is produced for the same water flow rate. This in turn increases the water recovery rate hence the concentrate salinity and osmotic pressure [116].

3.3.3.2.1.3 The RO Array The RO array consists of one or more stages, each has one or more pressure vessels connected in parallel (Fig. 3.9). The pressure vessels house the RO modules and usually can fit from six to eight modules connected in series where the concentrate of one module is fed to the subsequent module. The number of RO modules in series

[16] The active layer is a very thin layer which is responsible for retaining the undesired substances in water.

[17] Both terminologies are used interchangeably in this study.

determines the recovery rate, such that the more modules in a pressure vessel and the more stages used, the higher the recovery rate. This is because it allows the concentrate to be treated several times. The number of pressure vessels in parallel is set based on the feed flow rate. RO modules have typically a maximum recommended feed flow rate that should not be exceeded to avoid damaging the membranes. Therefore, higher feed flow rate requires more pressure vessels in parallel to evenly distribute the feed flow rate. Conversely, the number of pressure vessels used in the subsequent stage should be less than the previous stage. This allows enough concentrate to flow in each RO module ensuring enough tangential flow to reduce the fouling potential and concentration polarization (CP). CP is the concentration of salts at the membrane surface as the water crosses the membrane. This high concentration at the membrane surface further increases the osmotic pressure; hence, for the same feed pressure, the permeate flux[18] decreases [110]. The minimum concentrate flow, therefore, rapidly removes the accumulating salts away from the membrane surface, thereby reducing CP.

3.3.3.2.1.4 Energy recovery device Energy recovery devices (ERDs) are typically used in seawater desalination applications to recover the pressure energy of the concentrate. The discharged concentrate is usually one to four bars less than the feed pressure [112] and is otherwise wasted by friction in the concentrate valve [137]. The ERD then allows the reuse of the pressure energy of the concentrate to pressurize the feed water such that only an amount equal to the RO plant permeate flow will need to be pressurized by the pump. The use of ERDs therefore results in a significant reduction of specific energy consumption of RO plants. Several types of ERDs are used such as Pelton wheels, positive displacement pumps, and isobaric pressure exchangers. For brackish water desalination, defined as water with salinity less than 10,000 kg/L, ERD devices are likely to be uneconomical. This is because, with brackish water, higher recovery rates are possible compared to seawater desalination, and therefore concentrate pressure energy that could be recovered may not be significant to justify the high costs of the ERDs. It is no surprise then that most PV-RO plants for brackish water desalination reported in the literature [138–147] did not use ERDs. This is in contrast with seawater PV-RO plants reported in the literature which almost all used an ERD [121,148,149].

[18] The permeate flow rate per unit area of membrane surface.

3.3.3.2.1.5 Posttreatment Posttreatment is required to ensure that the RO permeate water is suitable for drinking and that it is noncorrosive. For example, all acids that were added to the feed water in the pretreatment stage would severely corrode any metal in the distribution system. Accordingly, the pH of the permeate water needs to be increased typically to values ranging from 6.5 to 8.5 [126]. This is achieved by using degassers to remove carbon dioxide,[19] formed from the reaction of the added acid with calcium bicarbonate [111], which would otherwise redissolve in water forming carbonic acid, further reducing the pH; and by remineralizing the water by increasing the carbonate and bicarbonate content, for example, adding sodium bicarbonate [98,124]. Remineralizing the water could be also achieved by blending the raw feedwater with the permeate water [98]. Degasifiers are also used to remove hydrogen sulfide gases which may occasionally exist in groundwater [150]. Hydrogen sulfide produces offensive odor that reduces the acceptability of water and increases the corrosiveness of water, and can also increase the turbidity of the water if oxidized [150]. The last step of posttreatment is disinfection, usually through chlorination with a typically recommended residual chlorine of 0.5 mg/L [137]. There are also other methods such as the use of ozone and UV radiation [132], to ensure that the water is free from viruses and germs in the permeate tank and the distribution system [134].

3.3.3.3 Solar photovoltaic reverse osmosis plant configuration

Fig. 3.10 shows a typical solar PV-RO system configuration. It shares the same components and configurations as a solar WPS in addition to the RO plant itself. While surface water can be used with solar PV-RO systems, typically water is extracted from the ground even if it is seawater to make use of the natural filtering property of the soil to block harmful substances that can plug the RO membrane or at least increase the pretreatment requirements as discussed in Section 3.3.3.2.

The main difference, however, is that high pressure pumps are typically driven by AC motors and therefore DC–AC inverters are always required. Moreover, typically, batteries are required not just to ensure longer operation of the plant but because RO plants cannot handle the variation in operating conditions. For example, RO modules have operating constraints, mainly, maximum feed and permeate flow rates, maximum feed pressure, and minimum concentrate flow rate to avoid damaging the membranes or shortening their lifetime. Managing these constraints with intermittent solar

[19] Carbon dioxide is not rejected by the RO membrane.

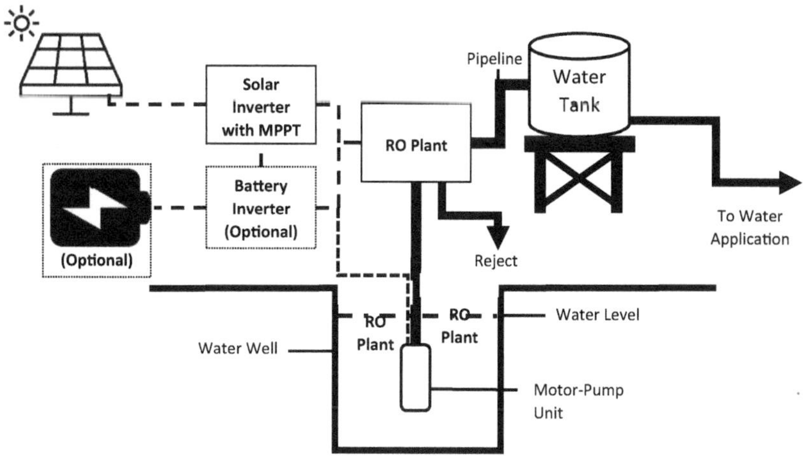

Figure 3.10 Typical solar PV-RO plant.

resources becomes then a daunting task. For this reason, batteries are used at least for power conditioning.

There is, however, numerous research work aiming to develop control strategies to avoid using batteries which will be discussed in Section 3.3.4. These systems are referred to as battery-less systems.

3.3.3.4 Solar photovoltaic reverse osmosis plant design considerations

Solar PV-RO systems share similar design considerations as those of a solar WPS discussed in Section 3.2, such as the water demand, water source, and solar resources. However, there are few more aspects to be considered in the design.

3.3.3.4.1 Feed water properties

Studying the feed water properties is critical for the RO system design. This includes water salinity, chemical composition, and temperature. Water salinity, as discussed in Section 3.3.3.1, determines the pressure requirements of the RO plant and the allowed recovery rate. Along with the water chemical composition, water salinity also determines the pre-treatment requirements including antiscalant and acid dosing.

The chemical composition also impacts the pressure requirements. It is observed that in contrast with seawater, which is mainly composed of sodium and chloride [151], treating brackish water as a sodium chloride solution results into inaccurate estimation of the osmotic pressure, and hence the energy consumption. For example, using the Reverse Osmosis

System Analysis (ROSA) software [152], the average osmotic pressure[20] and the RO plant specific energy consumption with 4000 mg/l feed water salinity and a typical groundwater chemical composition obtained from [153] was found by the authors to be 25% and 7.6%, respectively, lower than that of a sodium chloride solution with the same salinity. The permeate salinity when treating the groundwater as a sodium chloride solution is also found to be 20% higher than the case where the actual chemical composition is used. The reduction in the osmotic pressure is because for the same water salinity, groundwater typically contains a higher percentage of the higher molar mass bicarbonate, and sulfate ions. Accordingly, lower number of moles is present in the groundwater compared with a sodium chloride solution with the same salinity. Moreover, the higher rejection of RO membrane to multivalent ions [116] results in a lower permeate salinity when the actual water composition is used.

Finally, feed water temperature is also essential for the design of RO plants as it affects the membrane permeability.

3.3.3.4.2 Produced water salinity

The required acceptable water salinity for domestic purposes in typically 500 mg/L [10,154–157], although values up to 1000 mg/L are acceptable [107]. However, if the water will be mainly used for livestock or irrigation, higher water salinity is allowed depending on the animal species and crop type. For example, FAO guidelines recommend water salinities less than 3200 mg/L for livestock and poultry, as long as specific element concentrations do not exceed the specified limits. Similarly, for some salt-tolerant crops such as cotton, water salinities up to 5200 mg/L can be used without any impact on crop yield[21] [106].

3.3.3.4.3 Battery size and battery-less systems

As discussed in the previous section, solar PV-RO plants can be battery-less or with batteries. Systems with batteries are more widely used and are designed similarly to any standalone PV system. While batteries are expensive, a solar PV-RO plant can still be more economical with a diesel generator-powered system due to the impact of capacity factor.

[20] Osmotic pressure based on the average salinity between the feed water and the RO concentrate.

[21] It should be noted that the type of soil and irrigation method also impact the water salinity of each crop.

In other words, the use of batteries allows the RO plant to produce more water, and hence more profit for the same amount of capital. This is then determined by comparing the LCC for PV plant and RO plant separately. If the LCC balance shifts more towards the PV plant, i.e., the PV system has higher CAPEX, which is the case with treating high-salinity water increasing the energy demand, then the system may be designed to operate only during sunshine hours. Conversely, if the LCC of the RO plant is higher than that of the PV, than larger batteries, hence longer operating hours, should be considered.

3.3.4 Advances in solar pumping technologies

Most work in the literature focused on assessing the feasibility of PV-RO plants [139] given that they are still in the early commercialization phase. One particular important area that raised great deal of research interest is battery-less systems. Battery-less PV-RO plants necessitate complex control algorithms to maximize the use of solar energy while protecting the RO membrane from extreme variations in operating conditions. These algorithms also include weather prediction models and artificial intelligence (AI). Numerous studies tackled the challenges associated with battery-less systems. The studies mostly focused on studying the impact of the intermittency of solar supply on the performance of the RO plant, in particular, the impact of varying feed pressure and flow rate on the energy consumption, permeate flow rate, and quality [140,143,146,147,158,159]. However, with the dropping in battery prices, solar PV-RO plants with batteries are likely to be more widely used compared with battery-less systems. In that case, research on battery-less systems may decline and the research focus will be more in developing more robust RO membranes and cheaper battery units.

3.3.5 Economics

In addition to the LCC, with solar PV-RO plants, an important measure to compare different water treatment systems and configuration is the levelized cost of water (LCOW). The LCOW can be calculated as follows:

$$\mathrm{LCOW} - \frac{LCC}{\sum_{j=1}^{LP} \frac{\dot{V}_{p,tot}}{(1+IR_r)^j}} \tag{3.20}$$

where LCC is the life cycle cost calculated using Eq. (3.13), LP is the evaluation period which is taken to be the lifetime of the PV modules, $V^{\cdot}_{prod}$ is the overall annual water production of the desalination plant, and IR_r is the real interest rate and is calculated using Eq. (3.14).

Regarding the LCOW of solar PV-RO systems, it will depend on the type of PV-RO system. Due to the low LCOE of the PV plants compared with conventional fossil fuel-based electricity, any RO plant connected to the grid, but having portion of its electricity supplied from a PV system whether behind the meter or through a power purchase agreement, will have an LCOW lower than an RO plant fully buying electricity from the grid (unless the electricity is heavily subsidized by the government). In remote areas, the LCOW of the PV-RO plant, hence its feasibility compared to a diesel generator-driven RO plant, will greatly depend on the following factors: (1) cost of the battery bank; (2) the operating hours, hence the capacity of the battery bank; and (3) CAPEX of the RO compared to the PV plant. For example, in a study carried out by the authors [161], it was found that under Egyptian climate conditions, a PV-RO plant designed to operate only daytime has a similar LCOW to a plant driven by a diesel generator, also operating daytime, over a range of feed water salinities of 2000–20,000 mg/L. Nevertheless, due to the low-capacity factor of the RO plant in that case, the LCOW was found to be up to 36% higher than a similar RO plant designed to operate 24 hours especially at low feed water salinities. At high feed water salinities, however, the LCOW of day-only PV-RO starts to approach that of a PV-RO designed to operate 24-hours. This is because the benefit achieved by increasing the capacity factor of the RO plant is offset by the significant increase in the cost of the PV equipment given that a much larger battery bank is required. In that case, the life cycle cost of the PV equipment dominated that of the RO equipment. The high cost of the PV equipment also caused a 24-hour operating PV-RO plant to have higher LCOW compared to one using a diesel generator. This is further exacerbated at higher feed water salinities.

Likewise, in battery-less systems, despite the very low LCOE, the low-capacity factor of the RO plant will increase the LCOW compared to a conventional RO plant. Therefore it can be concluded that in contrast to solar WPS's, there is not yet a clear cut to conclude that the PV-RO will be more economical than diesel generator-driven plants in remote areas due to the high prices of batteries and a detailed case-by-case basis feasibility study will be required.

3.4 Conclusion

In this chapter, the technologically mature renewable energy-driven water applications, namely water pumping and water desalination, were discussed. The chapter particularly focused on solar-driven systems and the use of reverse osmosis process in desalination due to their commercial widespread use compared to other renewable energy water applications. A historical overview followed by the application status of each technology were first discussed. A thorough system design and operation was then covered elaborating on the main system components, and system design selection and considerations. An overview of the economics of each system with a focus on the economic feasibility was discussed. Finally, the main advances for each technology were reviewed to help the readers finding research opportunities and improvement, and understand the direction both technologies are heading.

References

[1] A. Omar, A. Nashed, Q. Li, G. Leslie, R.A. Taylor, Pathways for integrated concentrated solar power – Desalination: A critical review, Renewable & Sustainable Energy Reviews 119 (2020) 109609.

[2] M. Abu-Arabi, Status and prospects for solar desalination in the MENA region, in: L. Rizzuti, H. Ettouney, A. Cipollina (Eds.), Solar Desalination for the 21st Century, Springer Netherlands, 2007, pp. 163–178.

[3] M.A. Eltawil, Z. Zhengming, L. Yuan, A review of renewable energy technologies integrated with desalination systems, Renewable & Sustainable Energy Reviews 13 (2009) 2245–2262.

[4] IRENA, Renewable Power Generation Costs in 2019, IRENA, 2020.

[5] C. Kost, T. Schlegl, J. Thomsen, S. Nold, J. Mayer, Study Levelized Cost of Electricity Renewable Energies, Fraunhofer ISE, 2012.

[6] M.Z.A. Ab Kadir, Y. Rafeeu, N.M. Adam, Prospective scenarios for the full solar energy development in Malaysia, Renewable & Sustainable Energy Reviews 14 (2010) 3023–3031.

[7] M. Schachinger, Module Price Index, <www.pv-magazine.com>, 2021.

[8] M. Aliyu, G. Hassan, S.A. Said, M.U. Siddiqui, A.T. Alawami, I.M. Elamin, A review of solar-powered water pumping systems, Renewable & Sustainable Energy Reviews 87 (2018) 61–76.

[9] G.E. Ahmad, J. Schmid, Feasibility study of brackish water desalination in the Egyptian deserts and rural regions using PV systems, Energy Conversion and Management 43 (2002) 2641–2649.

[10] B.S. Richards, A.I. Schafer, Design considerations for a solar-powered desalination system for remote communities in Australia, Desalination 144 (2002) 193–199.

[11] E. Tzen, D. Theofilloyianakos, M. Sigalas, K. Karamanis, Design and development of a hybrid autonomous system for seawater desalination, Desalination 166 (2004) 267–274.

[12] M. Retnanestri, The I3A Framework: Enhancing the Sustainability of Off-grid Photovoltaic Energy Service Delivery in Indonesia, University of New South Wales, Sydney, 2007.
[13] J. Rheinländer, De-central water and power supply integrating renewable energy – Technical and economic performance prediction, in: L. Rizzuti, H. Ettouney, A. Cipollina (Eds.), Solar Desalination for the 21st Century, Springer Netherlands, 2007, pp. 111–126.
[14] S.I. Yannopoulos, G. Lyberatos, N. Theodossiou, W. Li, M. Valipour, A. Tamburrino, et al., Evolution of water lifting devices (pumps) over the centuries worldwide, Water 7 (2015) 5031–5060.
[15] J. Chigerwe, N. Manjengwa, P. van der Zaag, W. Zhakata, J. Rockström, Low head drip irrigation kits and treadle pumps for smallholder farmers in Zimbabwe: A technical evaluation based on laboratory tests, Physics and Chemistry of the Earth, Parts A/B/C 29 (2004) 1049–1059.
[16] K. Meah, S. Fletcher, S. Ula, Solar photovoltaic water pumping for remote locations, Renewable & Sustainable Energy Reviews. 12 (2008) 472–487.
[17] K. Welsien, B. Scalise, Solar Pumping 101: The What, Why, and the How, World Bank, 2018.
[18] R. Foster, A. Cota, Solar water pumping advances and comparative economics, Energy Procedia 57 (2014) 1431–1436.
[19] T. Jenkins, Designing Solar Water Pumping Systems for Livestock, New Mexico State University, 2014.
[20] World Bank Group, Solar Pumping: The Basics (English), World Bank Group, 2018. Available from: www.worldbank.org.
[21] C. Gopal, M. Mohanraj, P. Chandramohan, P. Chandrasekar, Renewable energy source water pumping systems—A literature review, Renewable & Sustainable Energy Reviews 25 (2013) 351–370.
[22] S. Verma, S. Mishra, S. Chowdhury, A. Gaur, S. Mohapatra, A. Soni, et al., Solar PV powered water pumping system – A review, Materials Today: Proceedings (2020).
[23] J. Burney, L. Woltering, M. Burke, R. Naylor, D. Pasternak, Solar-powered drip irrigation enhances food security in the Sudano–Sahel, Proceedings of National Academy of Sciences 107 (2010) 1848–1853.
[24] P. Periasamy, N.K. Jain, I.P. Singh, A review on development of photovoltaic water pumping system, Renewable & Sustainable Energy Reviews 43 (2015) 918–925.
[25] J.K. Kaldellis, G.C. Spyropoulos, K.A. Kavadias, I.P. Koronaki, Experimental validation of autonomous PV-based water pumping system optimum sizing, Renewable Energy 34 (2009) 1106–1113.
[26] D.B. Singh, A. Mahajan, D. Devli, K. Bharti, S. Kandari, G. Mittal, A mini review on solar energy based pumping system for irrigation, Materials Today: Proceedings (2021).
[27] C. Honsberg, S. Bowden, PVCDROM, <www.pveducation.org>, 2013.
[28] O.C. Vilela, N. Fraidenraich, C. Tiba, Photovoltaic pumping systems driven by tracking collectors. Experiments and simulation, Solar Energy 74 (2003) 45–52.
[29] S.R. Wenham, M.A. Green, M.E. Watt, R. Corkish, Applied Photovoltaics, second ed, Centre for Photovoltaic Engineering, UNSW, 2006.
[30] M. Kolhe, J.C. Joshi, D.P. Kothari, Performance analysis of a directly coupled photovoltaic water-pumping system, IEEE Transactions on Energy Conversion 19 (2004) 613–618.
[31] P. Arun, R. Banerjee, S. Bandyopadhyay, Optimum sizing of battery-integrated diesel generator for remote electrification through design-space approach, Energy 33 (2008) 1155–1168.

[32] M. Muselli, G. Notton, A. Louche, Design of hybrid-photovoltaic power generator, with optimization of energy management, Solar Energy 65 (1999) 143–157.
[33] S. Rehman, L.M. Al-Hadhrami, Study of a solar PV–diesel–battery hybrid power system for a remotely located population near Rafha, Saudi Arabia, Energy 35 (2010) 4986–4995.
[34] Merkasol, Enersol OPzS Classic Solar Battery 4600Ah 2V, Merkasol, 2014.
[35] R. Dufo-López, T. Cortés-Arcos, J.S. Artal-Sevil, J.L. Bernal-Agustín, Comparison of lead-acid and li-ion batteries lifetime prediction models in stand-alone photovoltaic systems, Applied Sciences 11 (2021) 1099.
[36] GSES, Stand Alone Power Systems: Design and Installation, eighth ed., GSES, 2017.
[37] P.E. Campana, Y. Zhu, E. Brugiati, H. Li, J. P.V. Yan, Water pumping for irrigation equipped with a novel control system for water savings, Energy Procedia 61 (2014) 949–952.
[38] A.K. Tiwari, V.R. Kalamkar, Effects of total head and solar radiation on the performance of solar water pumping system, Renewable Energy 118 (2018) 919–927.
[39] A. Allouhi, M.S. Buker, H. El-houari, A. Boharb, M. Benzakour Amine, T. Kousksou, et al., PV water pumping systems for domestic uses in remote areas: Sizing process, simulation and economic evaluation, Renewable Energy 132 (2019) 798–812.
[40] M. Forstmeier, W. Feichter, O. Mayer, Photovoltaic powered water purification - Challenges and opportunities, Desalination 221 (2008) 23–28.
[41] USDA, Design of Small Photovoltaic (PV) Solar-Powered Water Pump Systems, Programme Solidarité Eau, 2010.
[42] C. Brouwer, M. Heibloem, Irrigation water management: Irrigation water needs, Irrigation and Water Management Training Manual No. 3, FAO, 1986.
[43] Michael Smith Engineer Ltd. Useful information on centrifugal pumps. Michael Smith Engineer Ltd, 2021.
[44] G. Li, Y. Jin, M.W. Akram, X. Chen, Research and current status of the solar photovoltaic water pumping system – A review, Renewable & Sustainable Energy Reviews 79 (2017) 440–458.
[45] F. Trieb, S. Kronshage, V. Quaschning, J. Dersch, H. Lerchenmüller, G. Morin, et al., SOKRATES-Projekt Solarthermische Kraftwerkstechnologie für den, Schutz Des Erdklimas (2004).
[46] E. Shashi Menon, Piping Calculations Manual, first ed., McGraw-Hill Education, New York, 2005.
[47] Neutrium, Pressure Loss From Fittings – Excess Head (K) Method, Neutrium, 2012.
[48] B.S. Massey, Mechanics of Fluids, Springer United States, 1989.
[49] F.F. Farshad, H.H. Rieke, Surface roughness design values for modern pipes, SPE Drilling & Completion 21 (2006) 212–215.
[50] I. Odeh, Y.G. Yohanis, B. Norton, Economic viability of photovoltaic water pumping systems, Solar Energy 80 (2006) 850–860.
[51] S. Sulaiman, H. Hussain, N. Leh, M. Razali, Effects of dust on the performance of PV panels, International Journal of Mechanical and Mechatronics Engineering. 5 (2011) 2021–2026.
[52] M. Abdolzadeh, M. Ameri, M.A. Mehrabian, The effects of an operating head on the performance of photovoltaic water pumping systems: An experimental investigation, Proceedings of the Institution of Mechanical Engineers, Part A: A Journal of Power Energy 223 (2009) 341–347.
[53] M. Benghanem, K.O. Daffallah, S.N. Alamri, A.A. Joraid, Effect of pumping head on solar water pumping system, Energy Conversion and Management 77 (2014) 334–339.

[54] Z.A. Firatoglu, B. Yesilata, New approaches on the optimization of directly coupled PV pumping systems, Solar Energy 77 (2004) 81–93.

[55] D. Fiaschi, R. Graniglia, G. Manfrida, Improving the effectiveness of solar pumping systems by using modular centrifugal pumps with variable rotational speed, Solar Energy 79 (2005) 234–244.

[56] Boutelhig A, A. Hadjarab, Bakelli Y. et al. Study to select an optimum photovoltaic pumping system (PVPS) configuration upon experimental performances data of two different DC pumps tested at Ghardaïa site. Energy Procedia. 2011;6:769–776.

[57] M. Benghanem, K.O. Daffallah, A.A. Joraid, S.N. Alamri, A. Jaber, Performances of solar water pumping system using helical pump for a deep well: A case study for Madinah, Saudi Arabia, Energy Conversion and Management 65 (2013) 50–56.

[58] A. Hamidat, B. Benyoucef, T. Hartani, Small-scale irrigation with photovoltaic water pumping system in Sahara regions, Renewable Energy 28 (2003) 1081–1096.

[59] Z. Glasnovic, J. Margeta, A model for optimal sizing of photovoltaic irrigation water pumping systems, Solar Energy 81 (2007) 904–916.

[60] V. Badescu, Time dependent model of a complex PV water pumping system, Renewable Energy 28 (2003) 543–560.

[61] I. Odeh, Y.G. Yohanis, B. Norton, Influence of pumping head, insolation and PV array size on PV water pumping system performance, Solar Energy 80 (2006) 51–64.

[62] T. Martiré, C. Glaize, C. Joubert, B. Rouvière, A simplified but accurate prevision method for along the sun PV pumping systems, Solar Energy 82 (2008) 1009–1020.

[63] C. Olcan, Multi-objective analytical model for optimal sizing of stand-alone photovoltaic water pumping systems, Energy Conversion and Management 100 (2015) 358–369.

[64] Y. Bakelli, A. Hadj Arab, B. Azoui, Optimal sizing of photovoltaic pumping system with water tank storage using LPSP concept, Solar Energy 85 (2011) 288–294.

[65] M. Abdolzadeh, M. Ameri, Improving the effectiveness of a photovoltaic water pumping system by spraying water over the front of photovoltaic cells, Renewable Energy 34 (2009) 91–96.

[66] H. Tabaei, M. Ameri, Improving the effectiveness of a photovoltaic water pumping system by using booster reflector and cooling array surface by a film of water, Iranian Journal of Science and Technology: Transactions of Mechanical Engineering 39 (2015) 51–60.

[67] M. Mohanraj, P. Chandramohan, M. Sakthivel, K. Sopian, Performance of photovoltaic water pumping systems under the influence of panel cooling, Renewable Energy Focus 31 (2019) 31–44.

[68] A. Kordzadeh, The effects of nominal power of array and system head on the operation of photovoltaic water pumping set with array surface covered by a film of water, Renewable Energy 35 (2010) 1098–1102.

[69] K. Benlarbi, L. Mokrani, M.S. Nait-Said, A fuzzy global efficiency optimization of a photovoltaic water pumping system, Solar Energy 77 (2004) 203–216.

[70] S. Sallem, M. Chaabene, M.B.A. Kamoun, Energy management algorithm for an optimum control of a photovoltaic water pumping system, Applied Energy 86 (2009) 2671–2680.

[71] M. Petrollese, P. Seche, D. Cocco, Analysis and optimization of solar-pumped hydro storage systems integrated in water supply networks, Energy 189 (2019) 116176.

[72] A. Reghukumar, V. Vijayakumar, Smart plant watering system with cloud analysis and plant health prediction, Procedia Computer Science 165 (2019) 126–135.

[73] A.R. Al-Ali, A. Al Nabulsi, S. Mukhopadhyay, M.S. Awal, S. Fernandes, K. Ailabouni, IoT-solar energy powered smart farm irrigation system, Journal of Electronic Science and Technology 17 (2019) 100017.

[74] A. Mahjoubi, R.F. Mechlouch, A.B. Brahim, Data acquisition system for photovoltaic water pumping system in the desert of Tunisia, Procedia Engineering 33 (2012) 268–277.
[75] S. Benbelkacem, M. Belhocine, A. Bellarbi, N. Zenati-Henda, M. Tadjine, Augmented reality for photovoltaic pumping systems maintenance tasks, Renewable Energy 55 (2013) 428–437.
[76] M.A. Hossain, M.S. Hassan, M.A. Mottalib, M. Hossain, Feasibility of solar pump for sustainable irrigation in Bangladesh, International Journal of Energy and Environmental Engineering 6 (2015) 147–155.
[77] M. Kolhe, S. Kolhe, J.C. Joshi, Economic viability of stand-alone solar photovoltaic system in comparison with diesel-powered system for India, Energy Economics 24 (2002) 155–165.
[78] P.C. Ghosh, B. Emonts, D. Stolten, Comparison of hydrogen storage with diesel-generator system in a PV–WEC hybrid system, Solar Energy 75 (2003) 187–198.
[79] S. Al-Hallaj, S. Parekh, M.M. Farid, J.R. Selman, Solar desalination with humidification–dehumidification cycle: Review of economics, Desalination 195 (2006) 169–186.
[80] K. Branker, M.J.M. Pathak, J.M. Pearce, A review of solar photovoltaic levelized cost of electricity, Renewable & Sustainable Energy Reviews 15 (2011) 4470–4482.
[81] C. Kost, J. Mayer, J. Thomsen, N.M. Hartmann, C. Senkpiel, S. Philipps, et al., Levelized Cost of Electricity Renewable Energy Technologies, Fraunhofer ISE, 2013.
[82] P.K. Koner, V. Dutta, K.L. Chopra, A comparative life cycle energy cost analysis of photovoltaic and fuel generator for load shedding application, Solar Energy Materials and Solar Cells 60 (2000) 309–322.
[83] M.M.H. Bhuiyan, M.A. Asgar, R.K. Mazumder, M. Hussain, Economic evaluation of a stand-alone residential photovoltaic power system in Bangladesh, Renewable Energy 21 (2000) 403–410.
[84] J. Glater, The early history of reverse osmosis membrane development, Desalination. 117 (1998) 297–309.
[85] E.R. Reahl, Half A Century of Desalination with Electrodialysis, The McIlvaine Company, 2006.
[86] IDA, IDA Desalination Plants Inventory, Global Water Intelligence and Water Desalination Report, 2013.
[87] National Research Council, Desalination: A National Perspective, The National Academic Press, 2008.
[88] J.E. Miller, Review of Water Resources and Desalination Technologies, Sandia National Laboratories, 2003.
[89] N. Ghaffour, T.M. Missimer, G.L. Amy, Technical review and evaluation of the economics of water desalination: Current and future challenges for better water supply sustainability, Desalination 309 (2013) 197–207.
[90] N.M. Wade, Technical and economic evaluation of distillation and reverse osmosis desalination processes, Desalination 93 (1993) 343–363.
[91] Low energy desal study funded, <Desalination.com>, 2008.
[92] The Encyclopedia of Desalination and Water Resources, Energy Requirements of Desalination Processes, 2013.
[93] R. Borsani, S. Rebagliati, Fundamentals and costing of MSF desalination plants and comparison with other technologies, Desalination 182 (2005) 29–37.
[94] R.G. Raluy, L. Serra, J. Uche, Life cycle assessment of desalination technologies integrated with renewable energies, Desalination 183 (2005) 81–93.
[95] A. Al-Karaghouli, L.L. Kazmerski, Energy consumption and water production cost of conventional and renewable-energy-powered desalination processes, Renewable & Sustainable Energy Reviews 24 (2013) 343–356.

[96] M.A. Darwish, N.M. Al-Najem, Energy consumptions and costs of different desalting systems, Desalination 64 (1987) 83–96.
[97] M.T. Ali, H.E.S. Fath, P.R. Armstrong, A comprehensive techno-economical review of indirect solar desalination, Renewable & Sustainable Energy Reviews 15 (2011) 4187–4199.
[98] I.C. Watson, O.J. Morin, L. Henthome, Desalting Handbook for Planners, third ed., In Reclamation USDotIBo, 2003.
[99] G. Fiorenza, V.K. Sharma, G. Braccio, Techno-economic evaluation of a solar powered water desalination plant, Energy Conversion and Management. 44 (2003) 2217–2240.
[100] N.M. Wade, Distillation plant development and cost update, Desalination 136 (2001) 3–12.
[101] K.V. Reddy, N. Ghaffour, Overview of the cost of desalinated water and costing methodologies, Desalination 205 (2007) 340–353, doi:10.1016/j.desal.2006.03.558.
[102] L. García-Rodríguez, Seawater desalination driven by renewable energies: A review, Desalination 143 (2002) 103–113.
[103] L. Garcia-Rodriguez, Renewable energy applications in desalination: State of the art, Solar Energy 75 (2003) 381–393.
[104] S.A. Kalogirou, Seawater desalination using renewable energy sources, Progress in Energy and Combustion Science 31 (2005) 242–281.
[105] Tzen E., Desalination Units Powered by RES: Opportunities and Challenges, in: ADU RES Seminar, Center for Renewable Energy Sources (CRES), Hammamet 2005.
[106] R.S. Ayers, D.W. Westcot, Water Quality for Agriculture, Food and Agriculture Organization (FAO), Rome, 1994.
[107] WHO, Guidelines for Drinking-Water Quality, fourth ed., World Health Organization, Geneva, 2011.
[108] J.-S. Choi, J.-S. Cho, S. Lee, T.-M. Hwang, H. Oh, D.R. Yang, et al., Understanding boron rejection by reverse osmosis membranes, Desalination and Water Treatment 15 (2010) 129–133.
[109] Water Quality Association. Osmosis Process, HM Digital, 2005.
[110] (E.D.) Howe, Fundamentals of Water Desalination, Marcel Dekker, Inc., New York, 1974.
[111] A. Porteous, Desalination Technology Developments and Practice, Applied Science Publishers, 1984.
[112] O.K. Buros, The ABCs of Desalting, International Desalination Association, 2000.
[113] J. Pinnekamp, H. Friedrich, Membrane Technology for Waste Water Treatment, second ed., FiW-Verlag, 2006.
[114] DOW Chemical Company, Water Chemistry and Pretreatment: Colloidal and Particulate Fouling Prevention, DOW Chemical Company, 2014.
[115] J.C. Crittenden, R.R. Trussell, D.W. Hand, K.J. Howe, G. Tchobanoglous, Membrane Filtration, MWH's Water Treatment: Principles and Design, third ed., Wiley, 2012.
[116] J.C. Crittenden, R.R. Trussell, D.W. Hand, K.J. Howe, G. Tchobanoglous, Reverse osmosis, MWH's Water Treatment: Principles and Design, third ed., Wiley, 2012.
[117] J. Kucera, Techniques and Technologies Reverse Osmosis – Design, Processes, and Applications for Engineers, Wiley-Scrivener, 2010.
[118] AWWA Research Foundation Staff, Lyonnaise des Eaux-Dumez, Water Research Commission of South Africa, Water Treatment Membrane Processes, McGraw-Hill, New York, 1996.
[119] DOW Chemical Company, Water Chemistry and Pretreatment: Guidelines for Feedwater Quality, DOW Chemical Company, 2014.

[120] C. Liu, S. Caothien, J. Hayes, T. Caothuy, Membrane Chemical Cleaning: From Art to Science. Water Quality Technology Conference, American Water Works Association, Denver, CO, 2001.

[121] M. Thomson, D. Infield, A photovoltaic-powered seawater reverse-osmosis system without batteries, Desalination 153 (2003) 1–8.

[122] DOW Chemical Company, Water Chemistry and Pretreatment: Scale Control, Chemical Company, 2013.

[123] J. Kucera, Water Quality Guidelines. Reverse Osmosis – Design, Processes, and Applications for Engineers, Wiley-Scrivener, 2010.

[124] M.A. Al-Sayed, M.K. Mahmoud, A. Hamdy, Z.M. Khaled, M.A. Abbas, Reverse osmosis pretreatment: removal of iron in groundwater desalination plant in Shupramant-Giza – A case study, Current World Environment 7 (2012) 23–32.

[125] DOW Chemical Company, Water Chemistry and Pretreatment: Colloidal and Particulate Fouling Prevention, DOW Chemical Company, 2014.

[126] L.K. Wang, J.P. Chen, Y. Hung, N.K. Shammas, Membrane Systems Planning and Design. Membrane and Desalination Technologies, Springer, Dordrecht, 2010.

[127] A. Schafer, A. Broeckmann, B. Richards, Membranes and renewable energy – A new era of sustainable development for developing countries, Membrane Technology (2005) 6–10.

[128] E. Drioli, A. Criscuoli, E. Curcio, Integrated membrane operations for seawater desalination, Desalination 147 (2002) 77–81.

[129] J. Kucera, Basic Terms and Definitions. Reverse Osmosis – Design, Processes, and Applications for Engineers, Wiley-Scrivener, 2010.

[130] Dow Chemical Company, Water Chemistry and Pretreatment: Scale Control, Dow Chemical Company, 2013.

[131] C.J. Gabelich, M.D. Williams, A. Rahardianto, J.C. Franklin, Y. Cohen, High-recovery reverse osmosis desalination using intermediate chemical demineralization, Journal of Membrane Science 301 (2007) 131–141.

[132] J.C. Crittenden, R.R. Trussell, D.W. Hand, K.J. Howe, G. Tchobanoglous, Removal of selected constituents, MWH's Water Treatment: Principles and Design, third ed., Wiley, 2012.

[133] B. Nicolaisen, Developments in membrane technology for water treatment, Desalination 153 (2003) 355–360.

[134] G. Micale, A. Cipollina, L. Rizzuti, in: G. Micale, L. Rizzuti, A. Cipollina (Eds.), Seawater Desalination for Freshwater Production, Springer, Berlin, Heidelberg, 2009, pp. 1–15.

[135] DOW Chemical Company, FILMTEC™ Membranes System Design: Membrane System Design Guidelines for Commercial Elements, DOW Chemical Company, 2013.

[136] J. Thomson, M. Slunjski, M. Fabig, G. Crisp, Potable water HERO process at Yalgoo, Water Practice & Technology (2009) 4.

[137] F. Banat, H. Qiblawey, Membrane desalination driven by solar energy, in: L. Rizzuti, H. Ettouney, A. Cipollina (Eds.), Solar Desalination for the 21st Century, Springer Netherlands, 2007, pp. 271–291.

[138] S. Alawaji, M.S. Smiai, S. Rafique, B. Stafford, PV-powered water pumping and desalination plant for remote areas in Saudi Arabia, Applied Energy 52 (1995) 283–289.

[139] Z. Al Suleimani, V.R. Nair, Desalination by solar-powered reverse osmosis in a remote area of the Sultanate of Oman, Applied Energy 65 (2000) 367–380.

[140] A. Joyce, D. Loureiro, C. Rodrigues, S. Castro, Small reverse osmosis units using PV systems for water purification in rural places, Desalination 137 (2001) 39–44.

[141] D. Weiner, D. Fisher, E.J. Moses, B. Katz, G. Meron, Operation experience of a solar- and wind-powered desalination demonstration plant, Desalination 137 (2001) 7–13.

[142] P.C.M. de Carvalho, D.B. Riffel, C. Freire, F.F.D. Montenegro, The Brazilian experience with a photovoltaic powered reverse osmosis plant, Progress in Photovoltaics: Research and Applications 12 (2004) 373–385.
[143] A.I. Schafer, B.S. Richards, Testing of a hybrid membrane system for groundwater desalination in an Australian National Park, Desalination 183 (2005) 55–62.
[144] A.I. Schafer, B.S. Richards, From concept to commercialisation: Student learning in a sustainable engineering innovation project, European Journal of Engineering Education 32 (2007) 143–165.
[145] E.S. Hrayshat, Brackish water desalination by a stand alone reverse osmosis desalination unit powered by photovoltaic solar energy, Renewable Energy 33 (2008) 1784–1790.
[146] B.S. Richards, D.P.S. Capao, A.I. Schafer, Renewable energy powered membrane technology. 2. The effect of energy fluctuations on performance of a photovoltaic hybrid membrane system, Environmental Science and Technology 42 (2008) 4563–4569.
[147] A. De Munari, D.P.S. Capao, B.S. Richards, A.I. Schafer, Application of solar-powered desalination in a remote town in South Australia, Desalination 248 (2009) 72–82.
[148] E.S. Mohamed, G. Papadakis, Design, simulation and economic analysis of a stand-alone reverse osmosis desalination unit powered by wind turbines and photovoltaics, Desalination 164 (2004) 87–97.
[149] S.A. Kershman, J. Rheinländer, T. Neumann, O. Goebel, Hybrid wind/PV and conventional power for desalination in Libya—GECOL's facility for medium and small scale research at Ras Ejder, Desalination 183 (2005) 1–12.
[150] J.C. Crittenden, R.R. Trussell, D.W. Hand, K.J. Howe, G. Tchobanoglous, Chemical oxidation and reduction, MWH's Water Treatment: Principles and Design, third ed., Wiley, 2012.
[151] H.E. Hömig, Seawater and Seawater Distillation, Vulkan-Verlag, 1978.
[152] DOW Chemical Company, Reverse Osmosis System Analysis (ROSA). 9.1 ed., 2013.
[153] S.A. Atta, A.M. Sharaky, A.S. El Hassanein, K. Khallaf, Salinization of the Groundwater in the Coastal Shallow Aquifer, Northwestern Nile Delta, Egypt, in: First International Conference on the Geology of Tethys. Cairo, Egypt, 2005.
[154] D. Herold, V. Horstmann, A. Neskakis, J. Plettner-Marliani, G. Piernavieja, R. Calero, Small scale photovoltaic desalination for rural water supply – Demonstration plant in Gran Canaria, Renewable Energy 14 (1998) 293–298.
[155] M.E. Soltan, Characterisation, classification, and evaluation of some ground water samples in upper Egypt, Chemosphere 37 (1998) 735–745.
[156] C.R. Martinetti, A.E. Childress, T.Y. Cath, High recovery of concentrated RO brines using forward osmosis and membrane distillation, Journal of Membrane Science 331 (2009) 31–39.
[157] D.B. Riffel, P.C.M. Carvalho, Small-scale photovoltaic-powered reverse osmosis plant without batteries: Design and simulation, Desalination 247 (2009) 378–389.
[158] B.S. Richards, A.I. Schafer, Photovoltaic-powered desalination system for remote Australian communities, Renewable Energy 28 (2003) 2013–2022.
[159] A.I. Schafer, A. Broeckmann, B.S. Richards, Renewable energy powered membrane technology. 1. Development and characterization of a photovoltaic hybrid membrane system, Environmental Science and Technology 41 (2007) 998–1003.
[160] J.C. Crittenden, R.R. Trussell, D.W. Hand, K.J. Howe, G. Tchobanoglous, Water quality management strategies, MWH's Water Treatment: Principles and Design, third ed., Wiley, 2012.
[161] A. Nashed, High Recovery Rate Solar Driven Reverse Osmosis and Membrane Distillation Plants for Brackish Groundwater Desalination in Egypt, University of New South Wales, Sydney, 2014.

Appendix

This appendix shows an alternative more detailed method to estimate the number of operating hours of a directly coupled solar pumping system.

1. Instead of doing the calculation based on the global horizontal irradiation, the more accurate tilted irradiation is to be calculated. This can be found using a solar design tool such as SAM or PVSYST or can be calculated as follows:
   - **a.** The global horizontal, beam/direct horizontal, and diffuse irradiation must be known.
   - **b.** For simplicity, the tilted irradiation can be estimated at solar noon as it simplifies the calculations such that (Fig. 3.A.1):

$$S_{b,tilt} = S_{b,Hor} \cdot \frac{\sin(\alpha + \beta)}{\sin(\alpha)} \quad (3.A.1)$$

$S_{b,tilt}$ is the beam irradiation incident on the module surface, $S_{b,Hor}$ is the beam horizontal irradiation, α is the sun altitude, and β is the module tilt angle.

The sun altitude at noon can be calculated as follows:

$$\alpha_{\text{north hemisphere}} = 90 - \varphi + \delta \quad (3.A.2)$$

$$\alpha_{\text{south hemisphere}} = 90 - \varphi - \delta \quad (3.A.3)$$

φ is the latitude (absolute value) of the location and δ is the solar declination angle which is calculated as follows:

$$\delta = 23.45 \sin\left[(d - 81).\frac{360}{365}\right] \quad (3.A.4)$$

d is the day of the year, for example 15th of January means $d = 15$ and 31st of December $d = 365$.

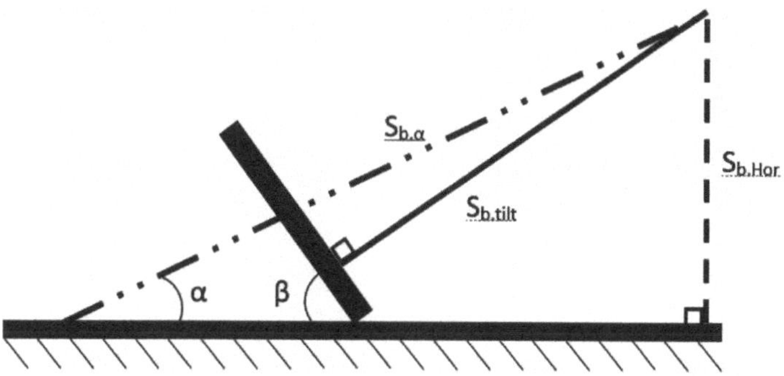

Figure 3.A.1 Solar irradiance components.

Now as the diffuse irradiation is not impacted by the module tilt angle, it is added directly to the titled beam irradiation to get the global tilted irradiation such that:

$$S_{G,\text{tilt}} = S_{B,\text{tilt}} + S_D \tag{3.A.5}$$

2. Both the peak sunny and cloudy day irradiance need to be estimated for an average day, for example, mid-month, on each month/season as follows:

$$I_{\text{sun},b,\alpha} = 1.353 \times 0.7^{(AM^{0.678})} \tag{3.A.6}$$

$$I_{\text{sun},G,\text{tilt}} = 1.1 \times I_{\text{sun},b,\alpha} \times \sin(\alpha + \beta) \tag{3.A.7}$$

$$I_{\text{cloud}} = 0.2 \times \quad I_{\text{sun},b,\alpha} \tag{3.A.8}$$

$I_{\text{sun},b,\alpha}$ is the beam irradiance falling on a plane perpendicular to sun rays and $I_{\text{sun},G,\text{tilt}}$ is the global tilted irradiance, both on sunny days. The 1.1 factor in Eq. (3.A.6) is to account for diffuse irradiance. The cloudy day irradiance is assumed to be 20% of $I_{\text{sun},b,\alpha}$. AM is the air mass and can be calculated as follows:

$$AM = 1/\sin(\alpha) \tag{3.A.9}$$

3. Now that both the global tilted irradiation and based on the sunny and cloudy days irradiance calculated in the previous step, the percentage of sunny and cloudy days in each month or season can be determined as follows:

$$\text{Days}_{prd} \times S_{G,\text{tilt}} = \text{Days}_{prd} \times 6.76 \times N_i \times \left[X_i \times I_{\text{sun},G,\text{tilt},i} + Y_i \times I_{\text{cloud},i}\right] \tag{3.A.10}$$

X_i and Y_i are the percentages of sunny and cloudy days respectively in the ith month/period $I_{\text{sun},G,\text{tilt},i}$ and $I_{\text{cloud},i}$ are the peak sun global tilted irradiance on a typical sunny and cloudy day during the ith month or period. The $6.76 \times N_i$ comes from integrating the typical solar profile shown in Fig. 3.A.2.

4. During cloudy days, it can be assumed that the solar energy available is not sufficient for a direct coupled solar water-pumping system to operate. For this reason, only the sunny day irradiance is considered to be the useful irradiance.
5. Now that the sunny day irradiance and the percentage of sunny days in each month are known, the equivalent number of sunshine hours, hence operating hours can be estimated. This should be fixed at $6.76 \times N_i \times X_i$, that is, the equivalent number of sunny hours in the

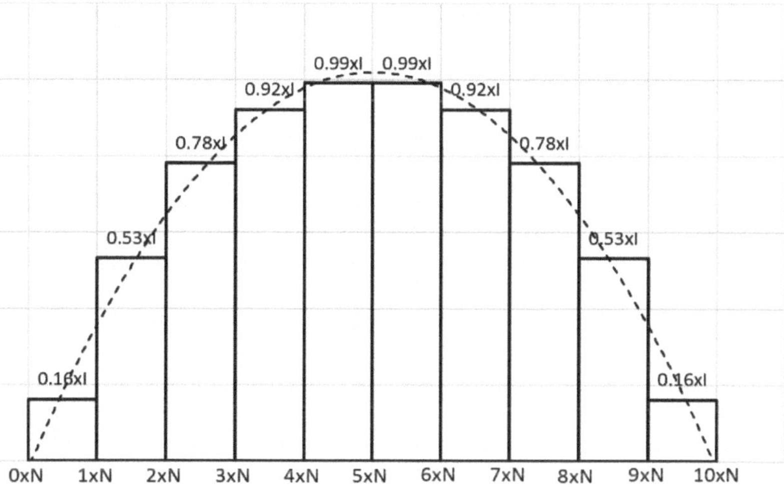

Figure 3.A.2 Typical sunny day solar irradiance profile as a function pf the peak day irradiance. *N* represents an approximation of the sunlight hours, that is, width of the profile for each month and can be obtained from table below.

Hemisphere	JAN	FEB	MAR	APR	MAY	JUN	JUL	AUG	SEP	OCT	NOV	DEC
South	1.4	1.3	1.2	1.1	1.0	1.0	1.0	1.1	1.2	1.3	1.4	1.4
North	1.0	1.1	1.2	1.3	1.4	1.4	1.4	1.3	1.2	1.1	1.0	1.0

*i*th month of the solar irradiance is $I_{\text{sun},G,\text{tilt},i}$. However, to be more conservative, typically a weighted average value of the sunny day peak irradiance based on the number of days M_i and the percentage of sunny day X_i in each month is used, which is calculated as follows:

$$I_{sun,G,tilt,av} = \frac{\sum_i \left(X_i M_i I_{\text{sun},G,\text{tilt},i}\right)}{\sum_i (X_i M_i)} \tag{3.A.11}$$

Accordingly, the equivalent number of sunny hours in each month can be calculated as follows:

$$Hr_{eq}, i = \frac{6.76 \times N_i \times I_{\text{sun},G,\text{tilt},i} \times X_i \times M_i}{I_{\text{sun},G,\text{tilt},av}} \tag{3.A.12}$$

It should be noted that if the solar water-pumping system is coupled with a MPPT than the same design concept would apply. However, it is expected that it will result in overestimating the size of the system given that the MPPT will extend the operating period of the plant.

CHAPTER 4

Solar-driven water treatment: generation II technologies

Maziar Dehghan[1], Mahsa Ghasemizadeh[1] and Saman Rashidi[2]
[1]Department of Energy, Materials and Energy Research Center (MERC), Tehran, Iran
[2]Department of Energy, Faculty of New Science and Technologies, Semnan University, Semnan, Iran

4.1 An introduction to desalination

Increasing water demand along with population growth and economic growth, and at the same time reducing water resources owing to climate change and pollution of water resources, exacerbate water shortages in most parts of the world [1]. According to a research, 40% of the world's population is facing severe water shortages and by 2025 this amount will increase to 60% [2]. In addition, 66% of the world's population experience drastic water scarcities for at least one month per year [3]. This indicates that conventional water resources such as rain, snowmelt, rivers, and aquifers no longer meet the water needs of humans in low-water areas. This has a conflict with sustainable development, which aims to make clean water available to present and future generations [4].

Countries and societies facing water crises need to rethink water resources planning and management, which includes the creative use of feasible but unconventional water resources [5]. While methods of reducing water demand, such as consumption savings and improved productivity, are able to decrease the gap between water supply and demand, these methods should be combined with supply-side strategies to struggle water shortage [6]. Such strategies of conserving water resources and improving supply have been used in some water-scarce areas [4]. However, these strategies need to be extended to areas where water scarcity and decline are intensifying [7].

Among the strategies to increase water supply, the desalination processes have received much attention and are increasingly used as a sustainable option in meeting water needs [4]. Desalination is the process of eliminating salt from water in order to generate freshwater in accordance with the requirements of drinking water quality required by humans [8].

Solar-Driven Water Treatment
DOI: https://doi.org/10.1016/B978-0-323-90991-4.00006-2

Today, there are more than 15,906 desalination units with a total desalination capacity of 95.37×10^6 m^3/day (34.81×10^9 m^3/year) in 177 countries worldwide. About half of the world's desalination capacity can be observed in the Middle East and North Africa (48%) [4].

Sixty percent of the total capacity of online desalination is used for urban water and more than 30% of it provides water required by industry. Increasing awareness of the strategic value of water for national security has placed desalination at the heart of the food–water–energy nexus. Water supply, energy production, and food production are recognized as integral elements in any sustainable current and future solution to the water crisis. Water can be used to extract, process, and purify fossil fuels and generate electricity. At the same time, energy reserves are used to pump, move, distribute, and purify water. Water is also used to produce, process, and transport food. The largest consumer of water is the agricultural sector. The intensity of water–energy–food relationship in each region depends on the composition of local energy, demand characteristics, availability and access to resources, etc. [9]. In the desalination process, there is a close relationship between water and energy. In order to reduce the effects of desalination on climate change, sustainable (renewable) energy sources have been proposed to reduce carbon emissions. It is estimated that 25 kg of carbon dioxide is emitted to produce 1 m^3 of freshwater. As a result, water–energy–carbon nexus should be considered in providing a sustainable solution. In this regard, renewable energy is a good alternative to conventional energy sources. Until a few decades ago, renewable energy desalination was only practiced in countries with high levels of solar radiation or wind energy sources, such as the countries surrounding the Persian Gulf. However, the recent developments in the costs of renewable technologies have led to an increase in the tendency to use renewable energy in the desalination industry [10]. On the other hand, in remote areas, the lack of infrastructure for drinking water and electricity is a challenge. In these areas, low population density has made it economically disadvantageous to set up local water and electricity supply systems. In this situation, there is a unique opportunity to implement small-scale, autonomous, and decentralized renewable energy–based (RE) desalination systems [9].

4.1.1 Desalination history

The origin of desalination goes back to 1943 and as a verb, "desalt" was registered in 1909 [11]. The primary purpose of desalination was not the production of freshwater, but the extraction and use of salt in salty water by natural evaporation [12]. According to the evidence, Aristotle was one of the first philosophers to describe the desalination technique. According to his observations, when saline water is converted to steam, the condensed steam does not contain any salt [13]. In ancient paintings, there is a picture of sailors boiling seawater, placing a large sponge in the opening of a brass vessel to absorb anything that evaporates [14]. As a result, it can be said that the history of desalination backs to the 4th century BCE [15].

Early desalination techniques were employed on naval ships from the 17th to 19th century. Eventually, the first desalination plants were designed for ships employed to supply fresh boiler water, so there was no need to travel by water cargo [15]. Years later, an English patent was granted in 1852 [16], and after that in 1872, a Swedish engineer named Carlo Wilson built the first distillation plant with a production capacity of 22.70 m^3/day in Las Salinas, Chile [17]. In 1912 a desalination unit with a capacity of 75 m^3/day was designed in Egypt [18], and later in 1938 a larger one in Saudi Arabia [12]. During the period 1929–1937, the total capacity of water desalination increased owing to the emergence of the oil industries [18].

In the 1940s, during World War II, owing to the shortage of drinking water for soldiers, desalination research was conducted to find suitable solutions for drinking water supply [12]. Telkes [19], for instance, made a plastic still employed by the US Air Force and Navy during World War II. After World War II, the US Congress passed the Saline Water Conversion Act (PL82-448) in 1952. Subsequently, the Office of Saline Water was established and funded in the Department of the Interior's Bureau of Reclamation [12].

In the 1960s the knowledge of desalination has advanced a lot. Owing to the rapid population growth and water scarcity around the world, it was a historical time to commercialize the industry. New desalination methods used fossil sources, because many oil-producing countries in the Middle East and North African region faced water scarcities and as a result, they devoted a portion of their energy resources, oil or gas, to water desalination instead of export [15].

The first desalination plants were installed in Kuwait and Channel Island in 1960 [18]. In the late 1960s, desalination units with a generation capacity of 8×10^3 m^3/day were employed in different regions around the world. Most of the installed units were based on thermal processes, and this technology was expensive and required a lot of energy [20]. However, this technology was suitable for oil-rich countries in the

Middle East [15]. Since the 1970s, membrane processes are widely employed and commercialized [18]. Eventually, since the 1980s, desalination is recognized as a purely commercialized technology [20].

4.1.2 Desalination methods

Desalination methods are categorized into two general groups: thermal and membrane based [21] (see Fig. 4.1). In thermal processes, by imitating the natural cycle of evaporation and condensation, heat is used to distill feedwater and produce freshwater [21]. Solar still desalination [22,23], humidification–dehumidification (HDH) desalination [24,25], multieffect distillation (MED) [26,27], multistage flash (MSF) distillation [28,29], vapor compression distillation, in which steam is produced mechanically [30,31] or thermally [32], freezing desalination [33,34], and hydrate formation [35,36] are among the thermal desalination processes. Membrane-based processes are: reverse osmosis (RO) operated by hydraulic power (pressure difference) [37,38], electrodialysis (ED) operated by a potential difference (direct current, DC) [39], and membrane distillation (MD) operated by temperature difference [40]. Other methods include adsorption desalination [41], hydrogel desalination [42], and ionic exchange desalination [43].

Among the mentioned membrane-based methods, RO is recognized as the most commercialized one, which can be used for seawater and ED

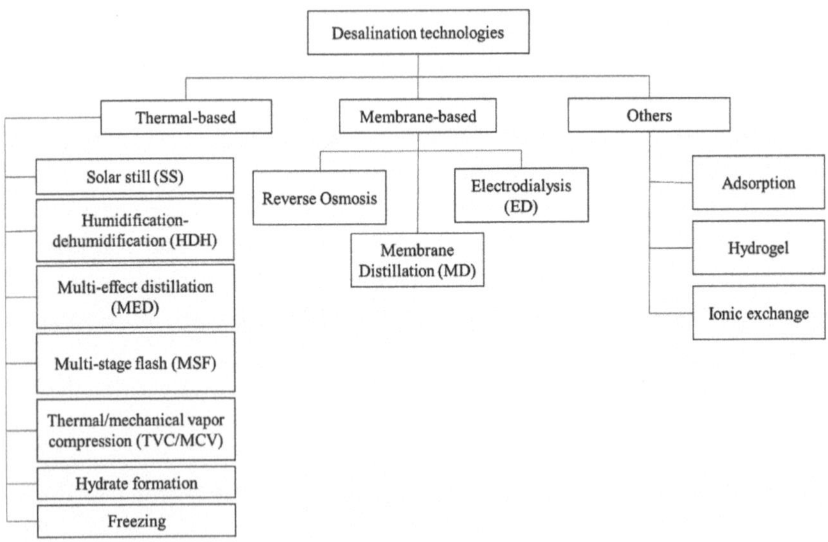

Figure 4.1 Classification of desalination technologies.

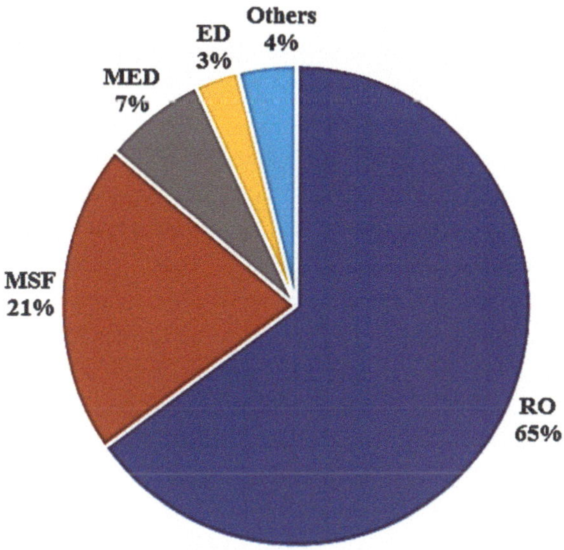

Figure 4.2 Contribution of desalination processes to global production [46].

process and is very suitable for brackish water desalination [44,45]. MED and MSF processes are also the most popular thermal processes [18]. In 2000, the volume of freshwater generated by thermal methods (mostly MSF) and RO was 11.6×10^6 and 11.4×10^6 m^3/day, respectively, which account for 93% of the total freshwater generated. Since 2000, the number of RO units has increased at an exponential rate, while thermal methods have increased marginally. The current production capacity of the RO process is 65.5×10^6 m^3/day, which is 69% of the volume of freshwater generated [4]. Fig. 4.2 displays the share of various desalination technologies installed in the world. Unlike MED and MSF, RO does not have the ability to integrate with a power plant to use the plant's by-product heat, as an advantage that low-grade heat is available and cheap [21]. The choice of appropriate technology is related to different factors, including the purpose of desalination, the economy, the physical condition of the site, the quality of feedwater and desalinated water, local capacity, and technical knowledge [21].

4.2 Renewable energy–based desalination

Existing desalination units are often large-scale and require a lot of materials and energy. On the other hand, large-scale units are reserved for rich countries and are not suitable for developing countries and remote areas. Today, many desalination plants use large amounts of fossil fuels to produce thermal and electrical energies in the thermal units and electrical energy needed for membrane units [47]. Total energy demand (thermal and electrical) of thermal processes, especially MSF and MED processes is between 14 and 30 kWh/m^3 [48]. In contrast, the energy consumption of membrane processes, especially the RO process, is less and between 2 and 5 kWh/m^3 of electrical energy [49]. As a result, the emission of carbon dioxide and other greenhouse gases from the desalination plants will have environmental consequences. Decarbonization of desalination technology is a fundamental solution to decrease carbon dioxide pollutants and mitigate climate change, while meet global water demand [50]. In recent years, renewable resources have attracted considerable attention. These resources require little maintenance, are a free and durable source of energy, and reduce environmental pollution [51].

Utilization of RE desalination technology is a proper option for producing freshwater and is quite suitable for remote regions where access to drinking water and electricity is difficult [23]. Renewable resources such as solar, wind, hydroelectric, biomass, and geothermal can be integrated with desalination technologies [52]. Since solar energy is the most abundant sustainable source of energy on earth, it has received much attention. Earth receives 1.361 kW/m^2 of solar radiation annually at the top of the atmosphere [53]. Thirty percent of it is scattered and 70% is exploited [54]. About 70% of RE desalination units worldwide are based on solar energy, because of having the ability to produce both thermal and electrical energies [55]. Fig. 4.3 shows the share of renewable energy employed in various desalination processes. In the forthcoming sections of this chapter, the photovoltaic RO (PV-RO), solar MED, and solar MSF processes are fully discussed.

4.2.1 Solar-driven desalination processes

Solar-based desalination methods are categorized into two groups, including direct and indirect techniques (see Fig. 4.4). In direct solar desalination, the solar energy received by the solar collector is used directly to produce freshwater, like what happens in solar stills. In the indirect method, solar energy is collected by solar thermal collectors and/or PV modules and converted into thermal and/or electrical energy and employed in desalination methods such as MED, MSF, MD, and RO.

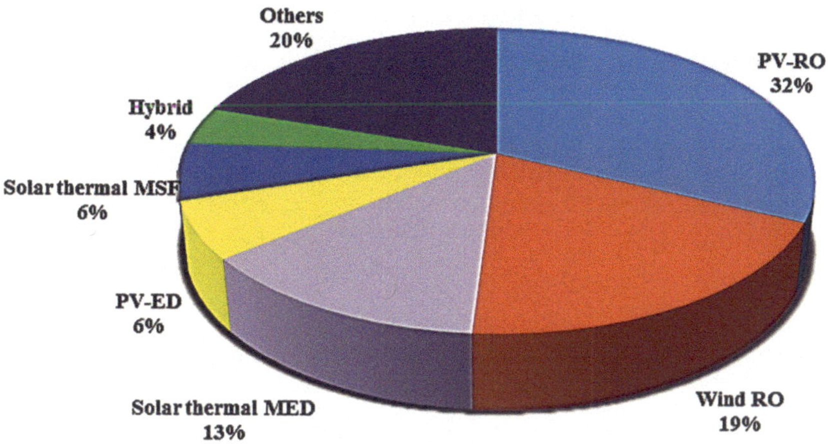

Figure 4.3 Desalination methods integrated with renewable resources at plants worldwide [56].

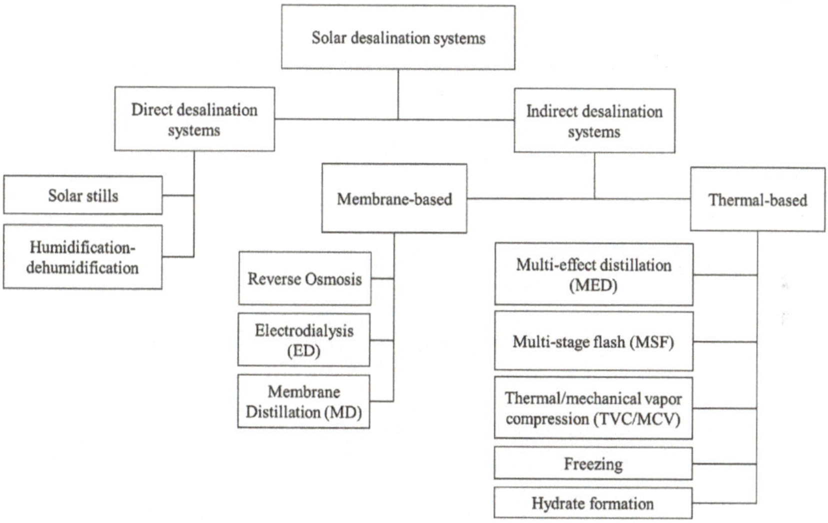

Figure 4.4 Categorization of direct and indirect solar desalination methods.

Table 4.1 summarizes the benefits and disadvantages of the solar desalination processes.

As mentioned earlier, in solar desalination processes, solar energy is used in both direct and indirect methods. In the indirect method, in which solar energy is first collected and then exploited, solar desalination system consists of two subsystems: the solar collector and the desalination

Table 4.1 Benefits and disadvantages of various solar desalination methods [57].

Desalination type	Benefits	Disadvantages
Solar stills	• Environmental friendly • Low Operations and Maintenance (O&M) cost • Produced water has high quality • High construction material availability • Suitable for household and communities living on islands • Removal of fluoride, arsenic, bacteria, etc. from the water	• Large area occupation • Low efficiency • Is not appropriate for high capacities of water production
Humidification—dehumidification	• Low installation and O&M cost • High flexibility • Suitable for decentralized operation • Simpler brine pretreatment • Operates with any type of energy • Requirements availability	• High capital investment cost • High overall costs of produced water
Multistage flash (MSF)	• Producing high-quality distilled water • Reliable device operation • Appropriate for large-scale distillation plants • Water with any quality can be treated • Minimum or no feedwater pretreatment is needed	• High energy consumption rates • High operation temperature causes corrosion in devices • Heavy structure • High capital cost
Multieffect distillation	• Lower thermal energy consumption levels • Reliable device operation • Producing high-quality distilled water	• Costly and heavy structure • Electricity consumption for vacuum pump

(Continued)

Table 4.1 (Continued)

Desalination type	Benefits	Disadvantages
	• High operation temperature is not required • No feedwater pretreatment is required • Lower carbon dioxide emission compared to the MSF desalination	
Solar PV-powered reverse osmosis (RO)	• Smooth operation • Operation in ambient temperature • Flexibility in capacity expansion • Can be constructed as a compact or portable device • Low energy consumption • Highly suitable for the treatment of groundwater and brackish water	• Membranes have a short lifetime • A high-pressure pump is required • Possibility of biological fouling of membranes • Requires pretreatment of feedwater • Using a battery is not recommended owing to the high capital cost and the need of battery replacement
Solar thermal-powered RO	• Device is safe • Flexible operation • Environmental friendly • Batteries are not required • A low-temperature source is sufficient • Solar collectors could cover a wide temperature range • No efficiency losses • Low O&M cost • Nonskilled labor would suffice • Suitable for large capacity operation • Consumes less energy for posttreatment	

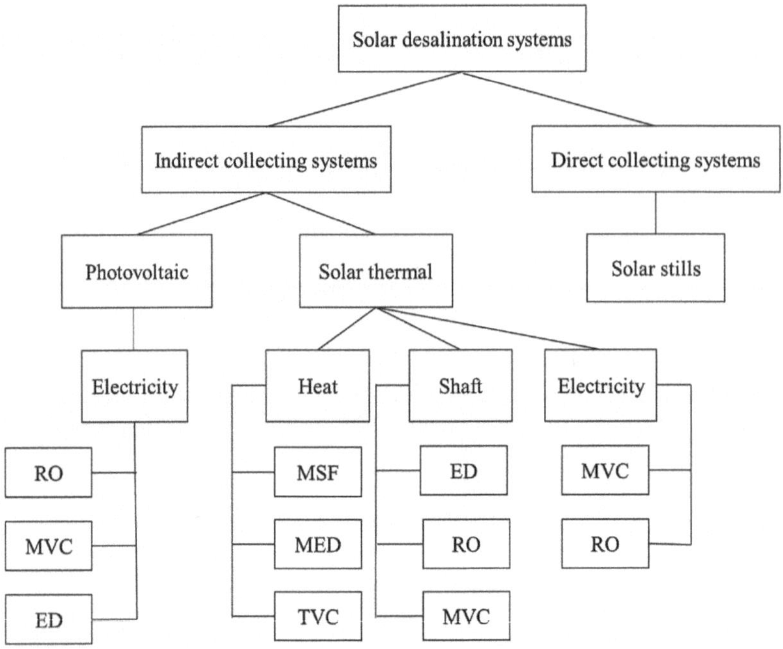

Figure 4.5 Combination of solar power systems with conventional desalination technologies [59].

unit. In the solar collector, solar energy is converted into useful forms of energy (electricity and heat) either by the PV-based systems or through thermal collectors like concentrator solar power (CSP) systems [58]. Fig. 4.5 shows the combination of solar systems with desalination units to supply electricity or heat. In this section, PV and solar heating systems used in desalination processes are discussed.

4.2.2 Solar PV-based desalination

PV power generation systems are suitable for remote areas with relatively low power demand. PV cells are semiconductors that produce DC. The set of cells forms a PV module with a clear glass cover on the surface and a waterproof substrate on the back surface. The set of modules also consists of strings and arrays. PV modules have two main configurations: on-grid and off-grid. The off-grid system does not inject generating power into the network. The on-grid systems, on the other hand, transmit the output power from the PV modules to an inverter and then to the distribution network. PV-based desalination technology is an excellent option for small- to medium-sized communities in remote regions with high access to solar energy and saline water [58].

In recent years, solar PV has become an inexpensive renewable technology compared to hydropower and wind energy sources [60]. Rapid progress in technology has decreased the cost of PV modules by 80% in the last 10 years [61]. Compared to 2015, PV costs can be reduced by 59% more, reducing the global average from 0.05 to 0.06 $/kWh [62]. By 2050, PV is predicted to account for 20% of the global energy supply, with a 50% decrease in carbon dioxide emissions by 2100 [63]. Among the desalination technologies combined with solar PV, PV-RO technology is a mature and commercially available one [64].

4.2.3 Solar thermal-based desalination

Solar thermal systems use the thermal energy of solar radiation. Solar collectors are the simplest solar thermal device. They can absorb sunlight and transfer its heat to a fluid. Flat plate collectors and evacuated tubes are the most popular types of collectors. The flat plate collector consists of a black absorber plate in which the heat received by the plate is transferred to the fluid in the tube. The evacuated tube collector consists of double-walled borosilicate glass tubes under vacuum. The black coating on the inner tube can absorb solar heat and transfer it to the liquid in the tube. The vacuum between the two pipes minimizes heat loss. Water, air, and oil are used as operating fluids in solar collectors [52].

CSP systems are another type of solar thermal collectors working at higher temperatures and usually used to generate electricity. This type of collectors uses mirror(s) to concentrate solar radiation and generate heat. The two main kinds of CSP collectors are power towers and parabolic trough collectors (PTC). The PTC consists of a reflective parabolic material and a receiving tube located in the parabolic focal line. Concentrated radiation is received by the heat exchange fluid in the receiver tube and converted into heat. The fluid temperature in the PTC reaches up to 350°C–400°C, which is a very considerable amount as compared to the fluid temperature in the flat plate and evacuated collectors. The performance of collectors in the solar thermal units is related to the type of technology employed, the operating temperature of the working fluid, the environment temperature, and the solar radiation. At present, solar collectors can provide thermal efficiency varied from 60% to 75% [52].

In desalination systems with photovoltaic-thermal (PV-T) or CPV-T, the generated electricity and heat energy wasted in PV panels are used simultaneously in the desalination process. HDH-PVT, MED-PVT, solar still/-PVT, and RO-PVT processes are among the favorite combined desalination systems in the literature [65].

4.3 Direct solar desalination

Solar still and HDH process can be considered the direct solar water desalination technologies. In these technologies, freshwater is continuously produced during the evaporation/condensation cycle. Thus initially, solar energy directly causes saline water to evaporate and produce steam. Then, due to the condensation of the resulting vapors, freshwater is produced. In this section, these two technologies are discussed in detail.

4.3.1 Solar stills

A solar distiller can convert saline water into freshwater employing direct solar energy. The basis of the process is the same as the rain production cycle in nature [66]. Solar energy causes water to evaporate from the surface of the ocean, lakes, and reservoirs on the earth's surface. The vapors produced in the atmosphere remain until they condense and turn into water droplets. The vapors are in the form of clouds and are spread by the wind all over the earth. As water droplets merge in the clouds and grow, they fall from the sky like rain. Solar stills work in a similar way. The saline water is evaporated in the distillation basin by solar energy. The resulting vapors move upward toward the surface of the condenser glass cover, such as the formation and movement of clouds. On the glass surface, vapors condense and the produced water droplets flow under the influence of gravity and are stored in the collection chamber [67].

Solar stills have a simple configuration. They are inexpensive and easy to operate. They also need low maintenance. A typical solar still has a black bed to absorb heat, and the saline water in this bed is exposed to sunlight. The bed is covered with transparent materials such as glass. In solar stills, the phenomenon of mass and heat exchange occurs simultaneously. Heat exchange in solar stills can be observed in the forms of convection and radiation and is absorbed by the water in the bed. The evaporated water condenses on the inner surface of the glass. Freshwater is collected through a canal and stored in a bottle or tank [68]. The only drawback of solar stills is the low productivity rate [69]. Therefore this

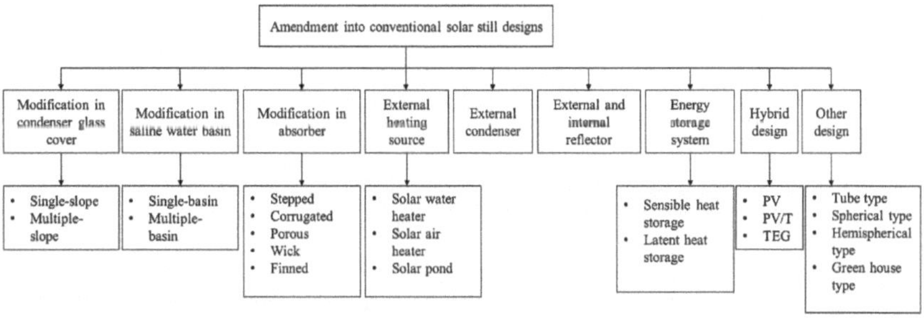

Figure 4.6 Amendment in solar still designs [70].

desalination method is not fully commercialized compared to other desalination technologies. Numerous studies have been conducted in the field of design control and operational parameters to maximize the energy efficiency and solar efficiency of stills [70]. Fig. 4.6 shows the modifications made in studies to improve the productivity of solar stills.

4.3.1.1 Energy efficiency of solar stills

The yield of solar stills is defined as the ratio of evaporative heat exchange to solar radiation received by the absorber plate and is calculated according to the following equation [71]:

$$\eta_{th} = \frac{\sum \dot{m}_{ew} h_{fg}}{3600 A_p \sum I(t)} \tag{4.1}$$

where h_{fg} denotes the latent heat of evaporation of water and $\dot{m}_{ev}$ is the productivity of freshwater that is calculated by:

$$\dot{m}_{ew} = \frac{h_{ew} A_w \left(T_w - T_g\right) \times 3600}{h_{fg}} \tag{4.2}$$

T_w and T_g denote the temperatures of water and glass cover, respectively.

4.3.1.2 Types of solar stills

In general, solar stills are categorized into two groups: active and passive systems [72,73]. In passive stills, water evaporates directly by solar energy and by natural heat exchange [67]. In active solar stills, an extra source, including solar collectors, solar concentrators, solar ponds, waste heat from industries as well as power plants, is used to preheat and boost the saline water temperature in the basin [70]. The capital cost of active solar stills is high, so they are not suitable for domestic applications and are mostly used in industrial applications [70]. Distilled water is needed for many industries, including wet batteries, pharmacology, hospitals, and laboratories, which can be easily produced by active solar stills [74]. On the other hand, passive solar stills are proper for home applications as they are affordable and have a simple design [70]. In this section, different types of passive and active solar stills with their shapes will be discussed in more detail.

4.3.1.2.1 Passive solar stills

Passive solar stills in terms of shape (single slope, double slope, pyramidal, spherical, etc.), number of basins (single basin and multibasin), type of absorber (wick, cotton, etc.), its structure (stepped, finned, and corrugated), and equipment used to increase maximum solar radiation collection and distillation efficiency (both reflector and condenser) can be classified. In this section, first the simplest type of passive solar still, that is, single-slope single-basin solar still is introduced, and more advanced types of solar stills are discussed.

1. *Single-basin single-slope passive solar still*

   The single-basin single-slope (SBSS) passive solar still is the basic design of a solar still. Typically, the water depth in the basin is about 2 cm, system efficiency is about 35%, and productivity is about 3–4 $kg/m^2/day$ [75]. A picture of SBSS passive solar still is illustrated in Fig. 4.7. Modified models of this type of still are presented as follows:

   a. *SBSS passive solar still with wick*

      Wick is a band made of porous materials, such as jute, cotton [70], and sponge [88], that transfers fluids according to the capillary effect. In wick-type solar stills, the saline water moves slowly on the perforated surface and evaporates [68]. Because the liquid flow on the wick is small, the water heats up quickly and reaches a high temperature. Wick-type solar stills are usually inclined; because tilted still is well exposed to sunlight [68]. According to studies, using jute wick, wool wick, and sponge wick can increase the productivity of solar stills by 85%, 19.76%, and 29.6%, respectively as compared to conventional stills [70]. A wick-type solar still is disclosed in Fig. 4.7.

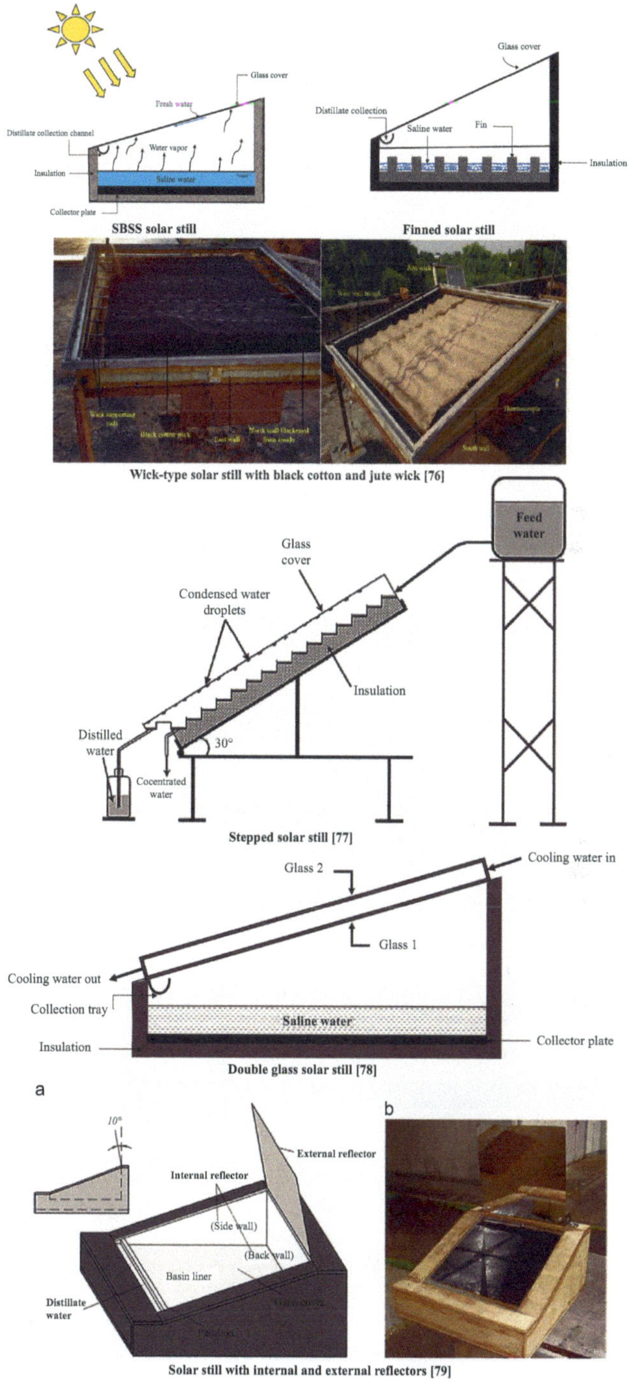

Figure 4.7 (Continued).

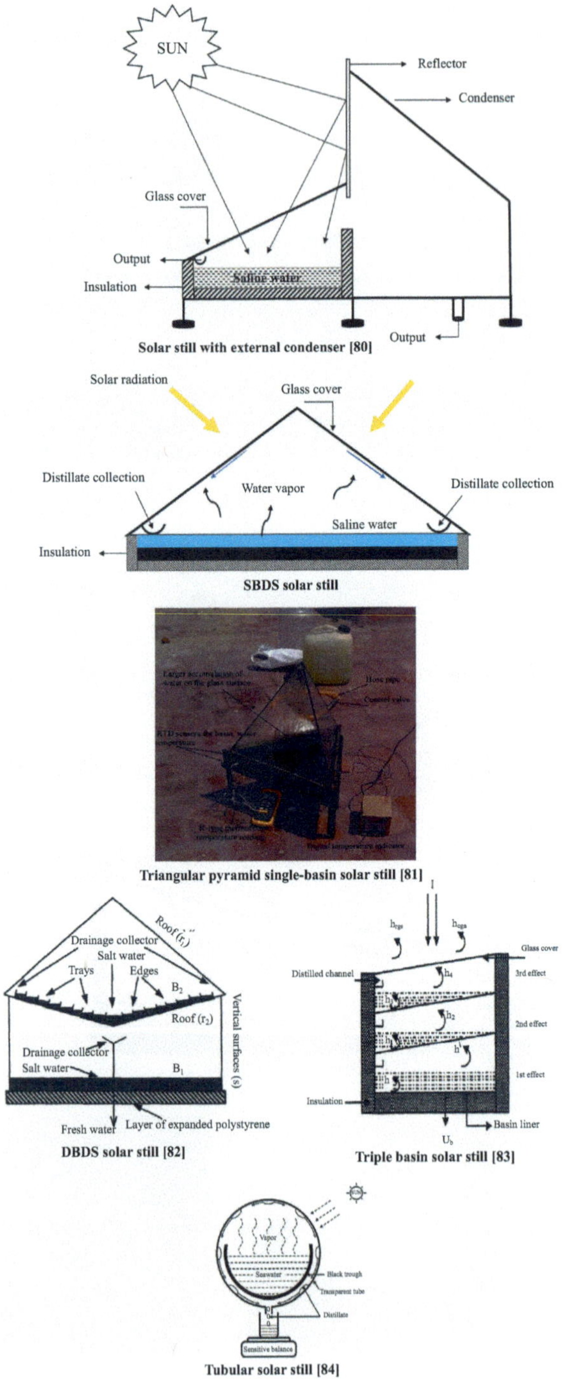

Figure 4.7 (Continued).

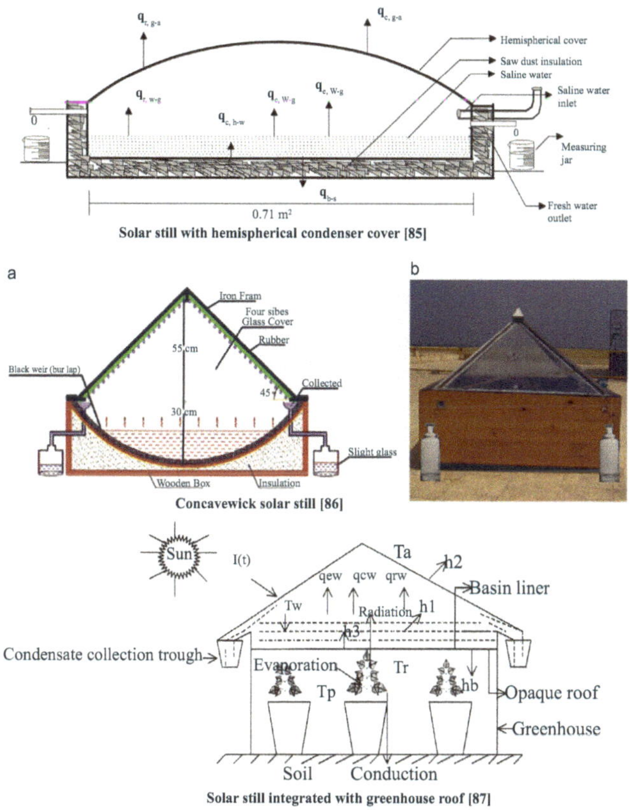

Figure 4.7 Passive solar stills [87].

b. *SBSS passive solar still with finned and corrugated absorber*

The absorber is a necessary part of a solar still. The radiation received by the absorber determines the efficiency of the solar still. As a result, the larger area of the adsorbent surface leads to higher solar radiation received and consequently, the efficiency is improved [89]. However, the adsorbent surface cannot be increased directly, because it increases the size of the system. In this case, using corrugation and fin is the best approach. This type of design, by increasing the exposure area, absorbs more solar radiation and thus reduces the preheating time of saline water [70]. The finned solar still is displayed in Fig. 4.7. Studies show that the efficiencies of finned, corrugated, and conventional stills are 47.5%, 41%, and 35%, respectively [90]. In addition, the costs of production of 1 L of freshwater in these stills are \$0.041, \$0.047, and \$0.049, respectively [90].

c. *SBSS passive solar still with stepped absorber*

In stepped solar still or weir-type cascade solar still, the absorber plate is stepped. In this type of solar still, the angle of inclination is equal to the latitude of the location, and hence the received irradiance would be the maximum. Consequently, the rate of water evaporation and distillation is high in this type of solar still with an energy efficiency of between 72% and 83% [70]. A stepped solar still is displayed in Fig. 4.7.

d. *SBSS passive solar still with double glazing*

Boosting the temperature difference between saline water and glass increases the condensation rate. This can be done by boosting the saline water temperature or cooling the glass cover by air or water [72]. Water cooling of the glass surface has been shown to double daily distilled water production [91]. Abu-Arabi et al. [78] used the double-walled glass coating in a single-basin solar still. The cooling fluid (water) moves between two walls and cools the glass coating (see Fig. 4.7). According to the results, the efficiency of this system is 34% higher than a simple solar still.

e. *SBSS passive solar still with reflectors*

Reflectors are used in solar stills to concentrate solar radiation on the absorber plate and thus to increase the temperature and evaporation rate. This method is an economical approach to enhance the performance of stills [92]. Based on the installation configuration, reflectors are classified into two groups: internal reflectors (IR) and external reflectors (ER) [68]. A single-slope solar still with IR and ER is disclosed in Fig. 4.7. The use of ER and IR increases the productivity about 72.8% in winter and 40.33% in spring [93]. Parabolic, dish, and compound parabolic reflectors are various kinds of reflectors employed in single-basin solar still [70]. Comparison of studies shows that the combination of parabolic reflector and paraffin wax phase-change material (PCM) in the single-basin still causes the highest productivity of 5.243 $kg/m^2/day$ [94].

In general, indoor or outdoor reflectors are used in relatively cold climates with low solar irradiance. It should be noted that, to achieve the desired efficiency throughout the year, the installation angle of ER should be changed during different seasons and solar radiation should be tracked during the day [92,95].

f. *SBSS passive solar still with external condenser*

Many attempts have been made to increase the distilled water by integration of solar stills with condensers. A condenser is a heat exchange device in which the phase of a heat exchange fluid changes owing to latent heat exchange (gas phase to liquid phase). In solar stills, using a condenser increases the glass water temperature difference and thus increases productivity compared to conventional setups. Condensers are divided into three categories according to the installation location: internal, external, and built-in [68]. A picture of the solar still with the external condenser is provided in Fig. 4.7.

2. *Single-basin double-slope passive solar stills*

Increasing the condensation area decreases the thickness of the condensate film and ultimately boosts the heat and mass exchange rates from the basin that results in higher productivity [68]. A double-slope solar still consists of two glass covers with two different slopes for receiving more solar radiation (see Fig. 4.6). The optimal tilt angle of the glass is related to the direction of each surface relative to the south [70]. Different enhancement techniques namely adopting of finned bed, black granite, wick, ER, and all the mentioned modifications can be employed in a single-basin double-slope (SBDS) solar still. It was concluded that the increase in productivity by these techniques compared to conventional still is equal to 58.47%, 69.84%, 42.33%, 93.39%, and 171.43%, respectively [96].

3. *Single-basin multiple-slop passive solar stills*

In recent years, various designs of solar stills are proposed to overcome the limitations of conventional SBSS solar still. Pyramid solar still is one of these designs. The pyramid solar stills are more efficient than ordinary ones. With pyramid solar still, there is no need for a solar tracking system and the effect of shadows on the side walls is reduced. To achieve the maximum efficiency in a pyramid solar still, the inclination angle of the glass cover must be equal to the latitude of the site [97]. Comparing pyramidal and rectangular (single-slope) solar stills with the same size and materials shows that rectangular still with a large value of water temperature and a small value of glass cover temperature has an efficiency of 36% [98]. In contrast, in pyramidal still, productivity efficiency is equal to 28.9% owing to low water temperature and high glass cover temperature [98]. Using PCM in triangular pyramid solar still increases productivity by 20%. Thus the daily efficiency increases from 45% without PCM to 53% with PCM [99]. A picture of a triangular pyramid single-basin solar still is provided in Fig. 4.7.

4. *Multiple-basin single-slope passive solar stills*

   In a single-effect solar still, the heat released from condensation is unused and only increases the temperature of the glass. To overcome this problem and use the latent heat released, a multieffect design is proposed. In such a scheme, the reuse of latent heat of the upper basin is provided, which will ultimately increase the efficiency and productivity of the device [68]. Owing to the existence of several layers of glass, a small amount of thermal energy needed to evaporate the upper basin water is provided by the lower effect and the rest can be provided by direct sunlight [22]. Numerous studies have been performed on solar double, triple, and quadruple basin stills. Most studies believe that double-basin still has higher efficiency as compared to single-basin one. Triple and quadruple basins increase efficiency at night [68]. The schematics of multiple-basin solar stills are illustrated in Fig. 4.7.

   Comparing the performance of SBDS and double-basin double-slope (DBDS) solar stills illustrates that the double-basin efficiency is 85% higher than the single-basin still [100]. The amount of freshwater production from DBDS still could be 1.7–1.8 $L/m^2/day$ [82].

5. *Tubular solar stills*

   Tubular stills have been introduced with the aim of simplifying the structure and to increase the efficiency up to 40% higher than conventional solar still [101]. Experiments on tubular stills with a glass cover cooling system show that the productivity and yield of the system could be boosted by 31.4% and 32.6%, respectively [102]. In addition, combining a tubular still with a parabolic concentrator results in a 676% increase in the productivity (see Fig. 4.7) [84].

6. *Hemispherical solar stills*

   Hemispherical covers are designed for maximum collection of solar radiation. The energy efficiency of solar stills with hemispherical covers is about 34% and if the cooling system is used, the energy efficiency increases to 42% [85]. A solar still with a hemispherical cover is shown in Fig. 4.7.

7. *Concave basin solar still*

   A laboratory-scale concave wick still prototype is shown in Fig. 4.7. The idea is to increase the basin area using a concave surface, which results in the daily distilled production of 4.1 L/m^2 and the efficiency of 30% [86].

Unusual types of solar stills, such as pyramidal, tubular, hemispherical, and concave solar stills, have been studied but no clear advantages have been reported [73].

8. *Combination of solar still and greenhouse*

The schematic of the combination of greenhouse and solar still is shown in Fig. 4.7. All solar radiation reaching the ceiling is received by the solar still basin. Some of this received radiation enters the greenhouse through the evaporation of water and the rest enters the greenhouse through conduction, convection, and radiation. The produced freshwater owing to condensation is used to irrigate greenhouse plants. Part of the incoming radiation to the greenhouse is received by the plants and the floor of the greenhouse and after reflection leads to heating of the greenhouse air. The other part enters the soil and improves grain growth [72]. In this configuration, solar radiation with wavelengths between 400 and 700 nm is used for photosynthesis and the rest, which includes infrared radiation, is used for desalination. Since water is a good absorber for infrared radiation, the combination of passive solar stills with the greenhouse leads to the creation of a selective optical filter. In general, the efficiency of the greenhouse solar stills decreases to 1.6 L/m^2/day, which is two or three times less than conventional solar stills. As a result, it can be said that greenhouse solar stills are not suitable for commercialization, unless the system costs are offset by the benefits of the greenhouse. Otherwise, it is more economical to use solar stills separately to supply the water needed to irrigate the plants in the greenhouse [73].

4.3.1.2.2 Active solar stills

As previously described, active solar stills employ external equipment to preheat the saline water. These types of stills work at higher temperatures, so the rate of evaporation and finally the rate of distilled production in active stills are higher than the passive ones [70]. The efficiency of active solar stills depends on the number of passes of heat transfer pipes, water mass flow rate (for effective heat transfer), natural or forced circulation, collector slope, heat transfer pipe material, collector heat loss coefficient, etc. [103]. In the following sections, the types of external equipment used in active solar stills are described.

- *Solar still coupled to flat plate collector*

  The image of an SBSS active solar still integrated with a flat plate collector is illustrated in Fig. 4.8. The saline water heats by the collector and enters the basin. High water temperatures lead to higher evaporation rates, resulting in increased productivity. Using a flat plate collector increases the productivity and efficiency as compared to the passive mode by 44.63% and 24.91%, respectively [110]. A combination of internal collector and IR with a solar tracking system in single-slope still could increase the thermal performance of this system in the range of 30%–380% [111].
- *Solar still coupled to evacuated tube collectors*

  The evacuated tube collector consists of several evacuated glass tubes and internal absorber surfaces (see Fig. 4.8) [73]. Using pulsating heat pipe (PHP) leads to the maximum hourly efficiency in single-slope solar still with an optimal basin water depth of 1 cm and a PHP filling ratio of 40% is 875 mL/m^2 [112].
- *Solar still coupled to a concentrator*

  Another configuration is single-slope solar still integrated with a parabolic trough concentrator and a heat exchanger. The system includes a solar still panel, a parabolic concentrator that conducts heat exchange through the oil fluid, a circulating oil pump, and a heat exchanger to transfer heat from the oil to the basin water (see Fig. 4.8). The efficiency of such systems can be up to 18% higher as compared to passive solar stills [104]. The parabolic collectors will be more efficient with the tracking system. Distillation efficiency will increase up to 137% if the concentrated solar collectors with PCM and glass cover cooling are used. In general, the integration of solar still with solar collectors will increase the efficiency of freshwater production by 36% [103].
- *Solar still coupled to solar air heater*

  Solar air heater has a high efficiency compared to a collector, evacuated tube, and concentrator. According to studies, stills with an air heater reach the highest efficiency and productivity when it is combined with a heat storage unit and water spraying [73]. Comparing the performance of a stepped solar still with an air heater and aluminum coating as a heat energy storage with ordinary (SBSS) still with aluminum coating reveals that the efficiency in the first system is increased by 112%, while in the second one is increased by 53% [113]. In SBSS solar stills, the solar air heater has higher distillation efficiency than the flat plate collector [70].

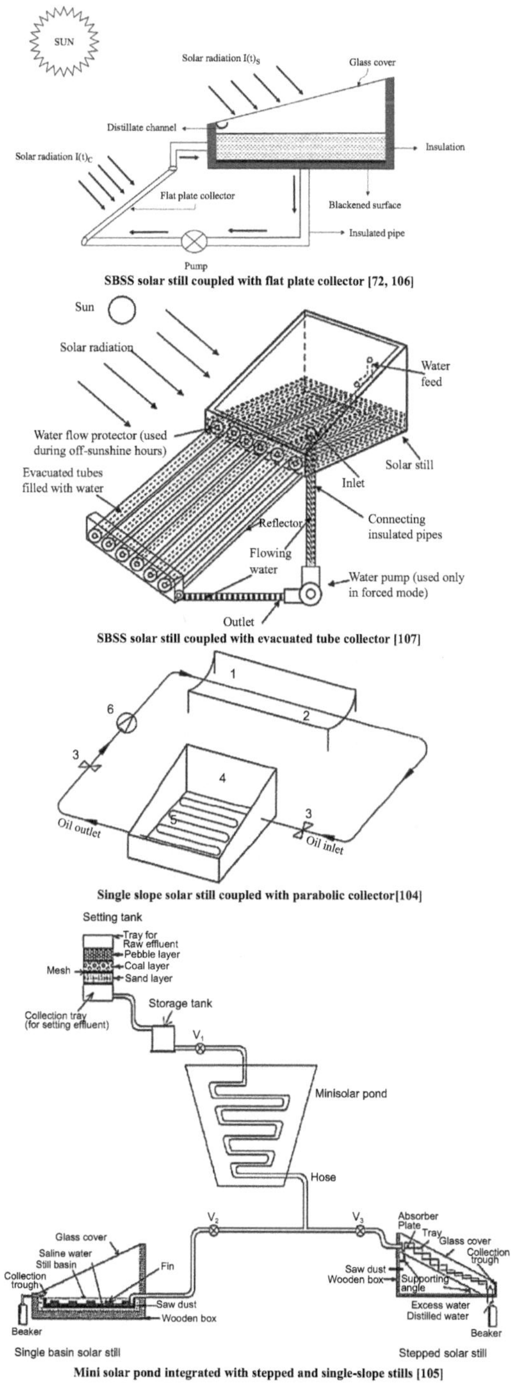

Figure 4.8 (Continued).

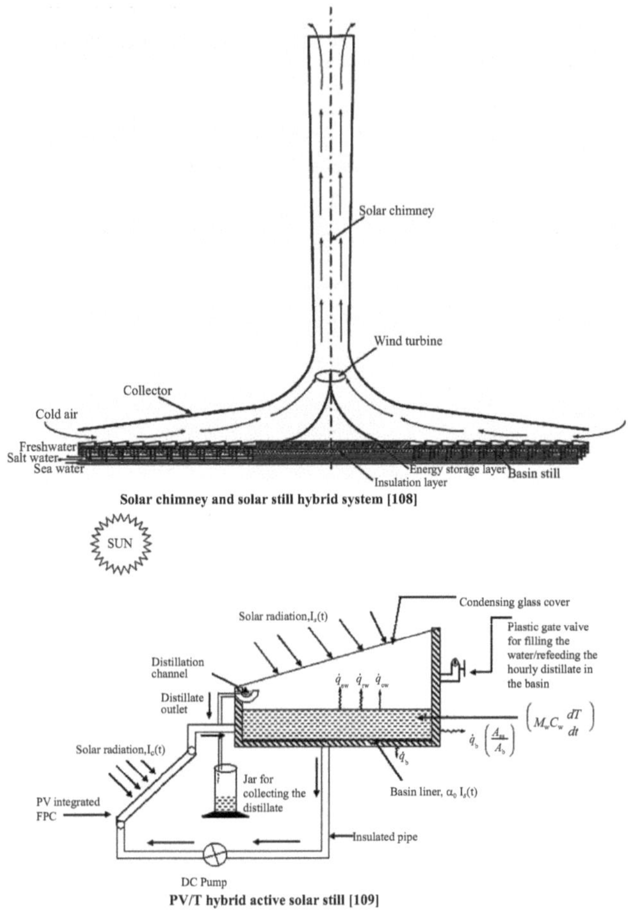

Figure 4.8 Active solar stills [109].

- *Solar still coupled to a solar pond*

  A solar pond has three zones: upper convective, nonconvective, and lower convective zones, and is employed for the storage of heat and could be integrated with solar stills to preheat saline water leading to increased productivity [73]. Combining a minisolar pond with a finned single-slope solar still and a stepped solar still (see Fig. 4.8) is one of the proposed ways to reach higher efficiencies [105]. Compared to flat plate collectors and vacuum tubes, the use of solar ponds results in an approximately 80% increase in the efficiency of active solar stills over conventional solar still. In this active configuration, a fan and/or pump is required, which leads to complexity and increased system costs [73].

- *Solar still coupled to solar chimney*

  A solar chimney was first employed for the conversion of solar energy into electrical energy. In this system, the air is heated by solar energy and its internal energy, especially kinetic energy, increases, and resultantly, air density decreases [114]. Hot and energetic floating air moves to the turbine installed at the foot of the chimney and generates electricity by operating the turbine [115]. Combining solar chimneys with solar desalination systems has been proven to be an effective way to acquire the major portion of the received solar energy leading to the simultaneous generation of electricity and freshwater [68]. The solar chimney takes up a large space and therefore the combination of this system with the desalination plant causes the optimal use of space and solar energy [68]. A schematic of a combined solar chimney and solar still system is shown in Fig. 4.8.

- *Solar still integration with PV-T hybrid system*

The hybrid PV-T system, in addition to the conversion of solar energy into electrical energy, also generates beneficial heat that can be used for preheating water in a solar still [72]. This active hybrid solar still system consists of a solar still, a PV integrated collector, and a DC pump motor powered by a PV system for circulating water between solar still and PV-T. The schematic of PV-T hybrid active solar still is displayed in Fig. 4.8. The daily production of freshwater from a passive still ranges between 2 and 5 kg/m^2, while the daily production of freshwater from a PV-T hybrid active solar still would fall in the range of 6–12 kg/m^2 [116].

4.3.1.3 Important parameters in solar stills

The main problem in employing solar still technology in desalination is the low productivity rate of distilled water. Compared to other desalination technologies, such as thermal and membrane technologies, solar stills have a relatively low efficiency rate. The two main parameters involved in the low productivity of solar still are environmental and design parameters [69]. As a result, the parameters affecting their efficiency and productivity should be taken into consideration.

4.3.1.3.1 Environmental parameters

- *The intensity of solar radiation*

The most important parameter, which affects the productivity of solar stills, is the intensity of solar radiation. There is a direct relationship between

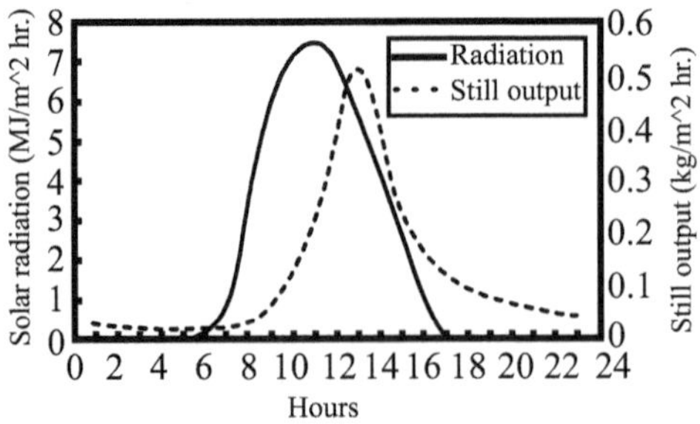

Figure 4.9 Changes of the efficiency of a solar still with the intensity of sunlight [117].

this parameter and the efficiency of a solar still. In other words, at higher intensities of solar radiation, the temperature of the water mass boosts and the evaporation process intensifies [69,88]. The changes of the efficiency of a solar still with the intensity of sunlight are illustrated in Fig. 4.9.

- *Wind speed*

  Increasing the wind speed leads to an increase in the convective heat exchange between the glass and the environment and thus the distillation rate can be improved [88]. In other words, wind augments the condensation rate by reducing the temperature on the glass and thus improves the yield of solar still [69]. Experiments show that the daily production improves with increasing wind speed to a specific speed, and more than that will have no effect on increasing water production. This specific wind speed is estimated to be 10 m/s in summer and 8 m/s in winter [88].

- *Dust accumulation*

If the condensing cover is not kept clean, the transmittance of solar radiation is reduced annually by 70% [69]. The accumulation of dust on the glass surface is directly related to its angle of inclination. Thus it has been observed that at an angle of inclination of 30 degrees, the agglomeration of dust on the glass cover diminishes the transmission of solar radiation by an average of 1% [69].

4.3.1.3.2 Design parameters

- *Basin water depth*

  Basin water depth is one of the important design parameters affecting the efficiency of solar stills [69]. Evaporation rate in solar still has an inverse relation with water depth [88]. In other words, the productivity of a solar still decreases with increasing water depth. In fact, as the water depth in the basin is increased, the amount and volumetric heat capacity of the water boost. Increasing the volumetric heat capacity leads to a decrease in water temperature in exchange for incoming solar radiation. Therefore in shallow basins, the water mass has lower heat capacity and higher temperature. As a result, the evaporation rate and efficiency of the solar still are higher. During the day, decreasing water depth leads to increased productivity, while during the night the opposite trend is seen. In the shallow bed, owing to the low amount of water, heat is not stored in the system, and the production rate decreases during the night [69]. By increasing the water depth from 20 to 70 mm, the production rate decreases by 14% [118]. In some studies, the optimal depth is mentioned to be 40 mm and in others between 20 and 60 mm [69].

- *The angle of inclination of the cover plate*

  Materials with high transmissivity, including clear glass plates, are used to cover solar stills. Depending on the latitude of the place, the covering plate of solar stills is placed at a certain angle and in a sloping manner [88]. It is said that when the slope of a glass plate is equal to the latitude, the annual efficiency of solar still is maximized [119]. The slope of the covering plate equal to latitude $\pm 10°$ has been used in many studies [88]. In areas with a latitude of more than 20°, the use of single-slope still is recommended [120]. If a double-slope still is used in these areas, it receives the solar irradiance from only one side, while the other side is in the shade [121].

- *The thickness of the glass cover*

  Glass has the feature that it selectively allows only more energetic radiation to pass through and prevents the passage of longer wavelengths. This feature helps solar stills to absorb most of the high energy radiation and prevent it from being reflected. Since glass is in contact with the atmosphere, it acts as a condensing plate and has a lower temperature than water [88]. By exploring different glass thicknesses in solar stills, in the range of 3–6 mm, a higher generation rate of 15.5% was obtained in the solar still with 3 mm glass thickness [122].

- *Insulation*

  In a solar still, the proper insulation reduces heat loss from the floor and walls of the basin. Thick insulation is usually done to diminish heat loss from the solar still to the surrounding area [69]. Materials such as thermocol, sawdust, wood, polyurethane, and gypsum are employed as insulation in solar stills [88]. In Section 4.3.1.3, the parameters affecting the solar productivity of solar stills were investigated. These parameters include environmental and design parameters. Environmental parameters are radiation, wind speed, and dust. Radiation and wind speed, also called meteorological parameters, are highly unpredictable and uncontrollable. Increasing the radiation, increasing the wind speed to a certain extent, and reducing the accumulation of dust will increase the efficiency of solar stills. Maximizing radiation, that is maximizing the flux of radiation reaching the glass coating, will be possible through solar detectors and reflectors. Increasing the wind speed, up to a certain speed, increases the distillation efficiency by decreasing the glass surface temperature and increasing the temperature difference between water and glass. Proper maintenance must be done to prevent heat transfer reduction due to the accumulation of dust on the glass surface and achieve constant efficiency. On the other hand, the design parameters such as water depth, slope angle, and thickness of glass cover and insulation are completely controllable. Reducing water depth will increase efficiency. In this case, the phase change from liquid to gas occurs rapidly. The slope of the glass cover depends on the installation location, operating time, and glass material. In general, the optimal slope equal to the latitude of the place is recommended. However, in order to avoid wastage in condensate collection, it is recommended that the glass cover slope be equal to latitude $\pm 10°$. It will also be effective to choose the right thickness of glass in order to make the best use of the solar radiation reaching the glass surface. The use of insulation with an optimal thickness also has a great effect on reducing heat loss from still floor and walls.

4.3.1.4 Energy storage system in solar stills

A high amount of thermal energy is wasted from the basin and side walls of a solar still. According to studies, 38.4% of solar energy is employed to evaporate the saline water and the remainder of this amount is wasted [67]. Up to 25.7% of heat is wasted through the basin and side walls that decreases the efficiency of solar still [123]. Fig. 4.10 illustrates the portion of solar energy used and dissipated in a typical solar still.

The thermal energy storage units can store energy through cooling or heating storage medium for cooling or heating or power generation applications. Two kinds of energy storage systems can be employed by solar stills. They are thermal and chemical systems (see Fig. 4.11) [67].

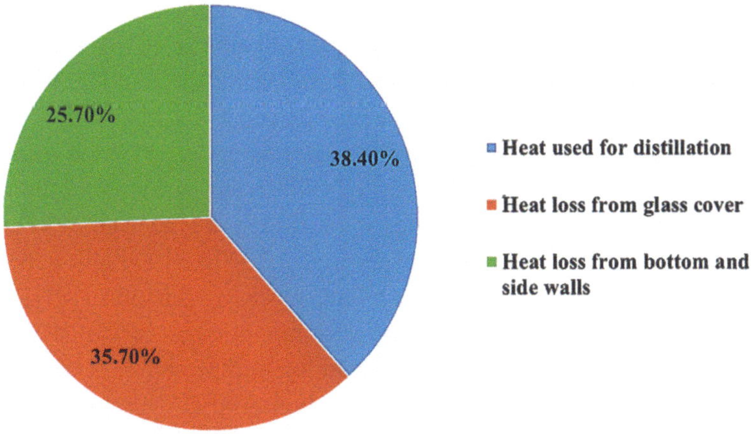

Figure 4.10 Portion of solar energy used and dissipated in a typical solar still [67].

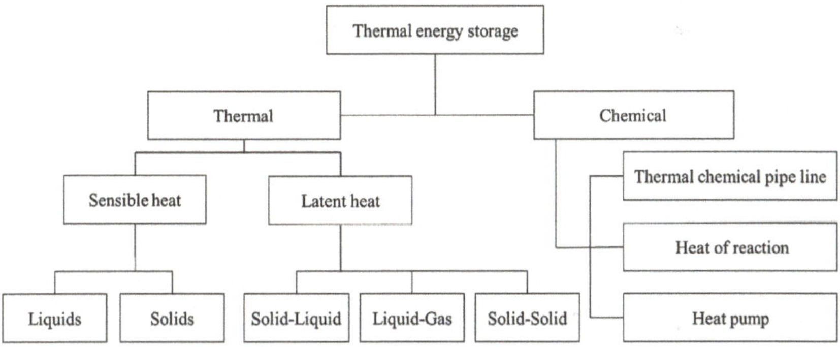

Figure 4.11 Thermal energy storage systems used for solar stills [124].

For applications such as solar stills, the thermal energy storage system is employed owing to economic considerations [125]. Solar heat storage in a still can be either sensible or latent [67]. A sensible heat storage material stores thermal energy by changing the temperature of the material. Basalt and black stones and steel wool fibers can be used as sensible energy storage materials [67]. Other materials such as rubber, gravel, pebbles, and quartzite are also used to store heat energy [88].

A latent heat storage material absorbs/releases thermal energy during the phase change process. The sensible energy storage systems are often large and take up a lot of space. PCMs are categorized as a latent energy storage system, which has the potential to store 5–14 times more heat than the energy stored in a counterpart sensible energy storage system.

Hence, they are proper for solar stills [126]. During the charge duration, as the temperature boosts, the chemical bonds of the PCM are broken, and a phase change between solid and liquid states occurs. The temperature remains constant until the phase change process is completed. When the temperature around the liquid PCM decreases, the PCM is solidified and the stored latent heat is released [67]. When the working temperature is low, PCM with a low melting point should be selected. However, when the thermal power is relatively high, PCMs with higher melting points are a good option to provide a longer heating duration [127].

Organic compounds with low melting points, 10°C–60°C, are suitable for applications with lower working temperatures. The eutectic compound is effective at a temperature of about 100°C. Meanwhile, eutectic mixtures of inorganic compounds are often used at temperatures between 100°C and 1250°C [67]. Mixtures of organic compounds, such as sodium and potassium that melt at about 170°C, are suitable for applications with higher working temperatures [128]. Metal hydrides can be used for long-term energy storage [129]. In a solar still system, PCM releases thermal energy during the evening and night. An ideal PCM for a solar still should have large values of latent heat and thermal conductivity and low melting point [67]. The absorber plate and the PCM substrate should have complete contact. Otherwise, the empty space in the PCM causes thermal contact resistance and decreases the heat exchange between the absorber plate and the PCM [130]. The amount of thermal energy stored in the PCMs during the day affects the efficiency of the device during the evening and night.

Fig. 4.12 shows the effect of PCM on evaporative heat exchange coefficient in a solar still. As can be seen, the evaporative heat exchange occurs even during nonsunny hours [95]. Organic PCM has a low thermal conductivity, 0.1–0.7 W/m/K, which leads to increased melting and charging time [128]. Therefore to boost the heat exchange between the PCM and the absorber plate, the thermal conductivity of the PCM should be improved. Increased thermal conductivity is achieved by placing a stationary metallic structure in the PCM [132]. Aluminum, copper, and nickel metal foams have good thermal conductivity and can enhance the total thermal conductivity of PCM [133]. Another technique is to employ hollow pin fins on the absorber plate toward the PCMs. The thermal conductivity of PCMs increases 40 times employing aluminum blades and 33 times employing carbon blades [134].

Numerous PCMs have been used in solar stills [67]. Paraffin wax is known as the most suitable PCM for both inactive and active solar stills. This material has the highest latent heat of melting in comparison with

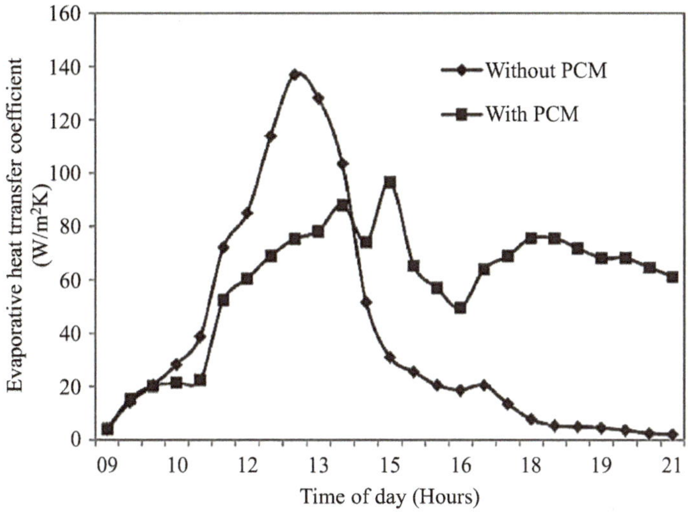

Figure 4.12 Influence of PCM on the evaporative heat exchange coefficient in a solar still [131].

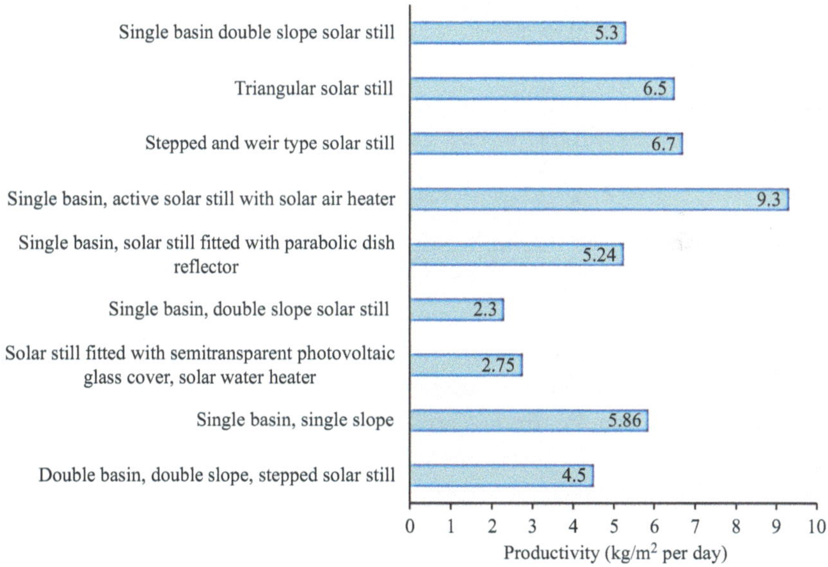

Figure 4.13 Daily productivities of various solar stills with paraffin wax PCM [67].

other PCMs employed in solar stills and has all the favorable features for energy storage. Paraffin wax is a cheap and available material, safe, nonreactive, and compressible with metal containers [67]. Fig. 4.13 compares the productivity of various types of solar stills integrated with paraffin wax as PCM. The maximum daily productivity is 9.3 kg/m^2/day for a single-basin integrated with solar air heater active solar still. Among passive stills, weir and stepped types have maximum daily efficiency. SBDS solar still has the lowest daily productivity (2.3 kg/m^2/day).

4.3.2 Solar-powered humidification–dehumidification process

One of the decentralizing water desalination methods on a small scale is HDH technology. This technology has many economic and environmental benefits, including the ability to integrate with sustainable energy sources, low operating temperature, low maintenance, and simple structure [47]. Many efforts have been made to combine the HDH desalination process with renewable energy sources; so that all the energy needs of the system are met from the local renewable energy source. In this regard, the HDH based on solar energy has received considerable attention. Because, by available solar energy even in remote areas facing a shortage of drinking water, the process of diffusion and condensation of saline water and air will occur effectively [135]. Due to its environmental issues, energy and economic aspects, the solar HDH process is considered the most desirable and reliable technology for local freshwater production. In this section, the HDH process and its types will be described along with the types of solar technologies used in the solar HDH system.

4.3.2.1 Humidification–dehumidification process

The HDH is a desalination technique in which a fluid, including air or water, receives heat from a heat source and enters an evaporator (humidifier), leading to the evaporation of saline water. Then, the produced vapors are directed to the condenser (dehumidifier) and owing to the condensation of vapors, freshwater is produced [25,135]. In fact, the HDH process has three parts, including a humidifier, dehumidifier, and heater. In the humidification section, water has contact with unsaturated air and penetrates into the air. The concentration difference between the air–water interface and water vapor in the air can be considered as the driving force for this diffusion process [23].

The humidifier consists of three parts [136]: (1) spray nozzle on the top for spraying hot seawater into the humidifier, (2) humidifier bed which is usually filled with packed bed material to boost heat and mass exchange between seawater and carrier gas, and (3) brine collection chamber at the bottom. Spray, wetted wall and packed bed towers, and bubble columns are the types of humidifiers used in the HDH process [23]. In the dehumidifier section, water vapor in humid air condenses in the form of distilled water. In a typical dehumidifier, a heat exchanger like copper coils is located with or without blades, and condensation is performed by low-temperature saline water. The most common dehumidifiers in the HDH process are flat plate and finned tube heat exchangers. Pumps and blowers can also be used to flow water and air, respectively [136].

4.3.2.2 Classification of humidification–dehumidification systems

HDH units can be categorized into three groups [135]. The first category is based on the energy source employed, including wind, solar, geothermal, electrical energies, or hybrid devices. An attractive and promising aspect of this classification is the use of renewable and low-grade energy resources [47,136]. The second category is related to the type of heated stream (air-heated or water-heated). In HDH, hot air is used to exchange water content from seawater to air owing to its high water-carrying capacity (high absolute humidity). This hot air can be generated from the heating of seawater or air [136].

Cycle configuration is the basis of the third category. The HDH system is divided into two cycles: open-air closed-water (OACW) and closed-air open-water (CAOW) systems [47,136] (see Fig. 4.14). In the OACW cycle, the air warms, humidifies, condenses, and eventually leaves the system. In the CAOW cycle, the airflow circulation occurs in a closed cycle. In both configurations, airflow can circulate naturally or forcibly. Forced flow, however, is preferred owing to the production of more water. Fig. 4.14 shows a schematic of the two cycles, OACW and CAOW, in both air-heated and water-heated modes in the HDH system. Unlike the other three cases, where the heater is placed between the humidifier and the dehumidifier, in OAOW, the heating system is placed at the bottom of the humidifier and both water and air leave the system at the end of the process.

4.3.2.3 Humidification–dehumidification systems driven by solar energy

Compared to other desalination technologies, the HDH process is proper for small- to medium-scale applications. Also, HDH units require little maintenance [57]. Hence, it can be effectively powered by

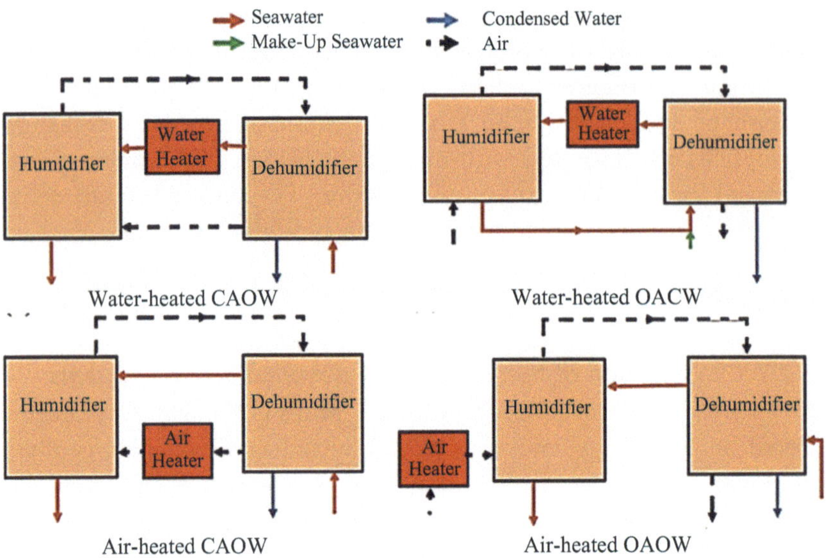

Figure 4.14 Schematic of basic HDH units [136].

renewable energy sources [136]. The thermal energy sources used in the HDH process fall into two groups: renewable energy sources such as solar and geothermal and low-grade energy sources such as PV-T systems, refrigeration systems, and heat pumps [136]. Solar energy is the most widely employed renewable energy in the HDH process [47,136]. Solar-driven HDH systems are very suitable for remote regions with no access to drinking water. In these areas, solar energy meets all the thermal needs of the system and there is no need for electricity or any other power sources [57].

4.3.2.3.1 Solar air heaters

Increasing the temperature of the air increases the water vapor absorption capacity of the air. The HDH system uses a solar air heater to increase the air temperature. Various kinds of solar air heaters are integrated with different cycle types of HDH systems, including OACW, CAOW, and OAOW [136]. Fig. 4.15 illustrates a simple schematic of a single-pass solar air heater integrated with the packed bed OAOW HDH system.

Focusing on the influences of different parameters, such as solar radiation, wind speed, ambient temperature, air mass flow rate, air inlet

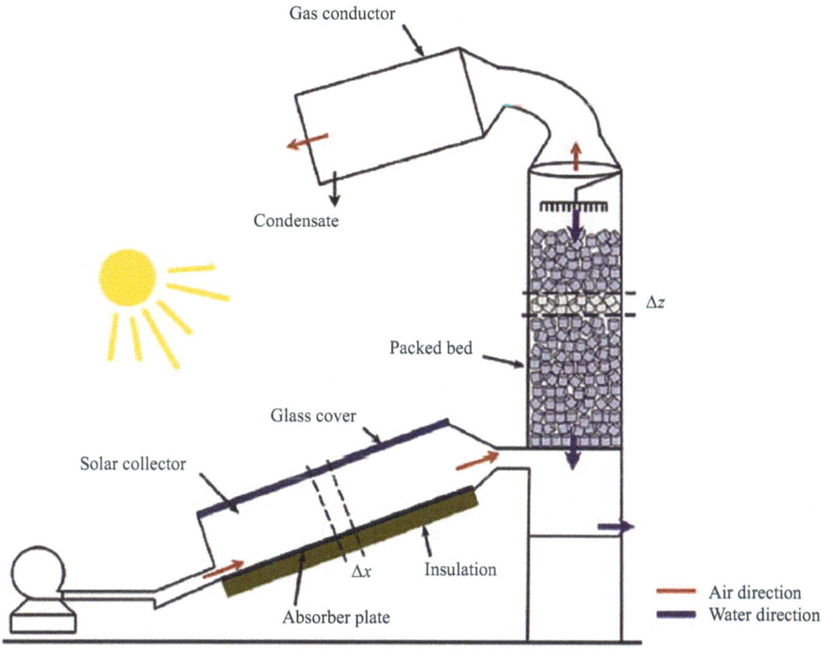

Figure 4.15 Schematic of a solar air heater integrated with the packed bed HDH system [137].

humidity, and air inlet temperature on the HDH system integrated with solar air heater, shows that the system efficiency enhances as the mass flow rate and temperature of the inlet air are boosted [138]. Double-pass solar air heaters increase productivity by about 8% in comparison with a single-pass counterpart [139]. The combination of solar air heater and PCM in the HDH system shows improved overall system performance by maintaining the output temperature at a constant value [140]. Solar air heaters are categorized into two groups: passive and active [57]. In passive operation, air flows owing to the specific gravity difference between the hot and cold sides. In the active systems, the air is forced to flow by an external source, such as a fan.

One of the solar air heaters employed in the HDH system is the PTC. Fig. 4.16 shows that how this collector can be integrated with the HDH system. Using a parabolic trough between the humidifier and the dehumidifier has a greater advantage than placing the collector before the humidifier [141]. An evacuated tube collector can also be employed as a solar air heater in HDH systems [142].

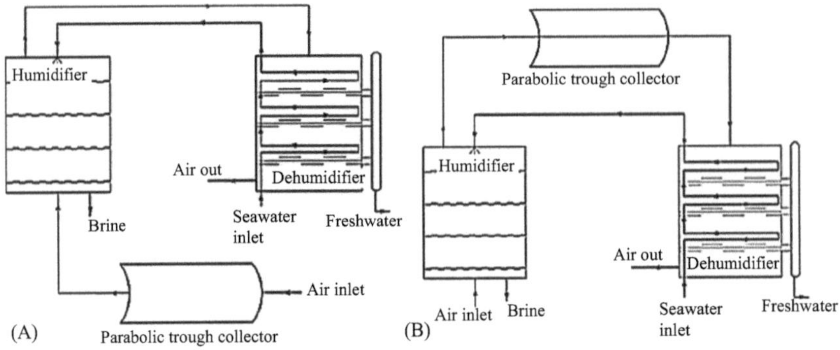

Figure 4.16 Parabolic trough collectors integrated with HDH system: (A) before humidifier; (B) after humidifier [141].

4.3.2.3.2 Solar water heaters

By boosting the temperature of the saline water entering the process, the desalination efficiency increases [143]. For this purpose, solar water heaters are employed to increase the temperature of the water entering the humidifier. In HDH processes, solar water heating systems have received more attention than solar air heating systems because the heat capacity of water is more than air [136]. Fresnel lens [144], flat plate solar collectors [145], PTC [136], and evacuated tube solar collectors [57] are employed as water heaters in the HDH desalination process. An image of the HDH system integrated with the evacuated tube collector is illustrated in Fig. 4.17. In addition, solar dishes are employed as solar water heaters in HDH. Different types of solar dishes including cubical, cylindrical, and hemispherical are used in HDH systems [147]. The thermal efficiencies of flat plate collectors, Fresnel lens, and evacuated tubes are about 80%, 70%, and 65%, respectively [136]. In the water-heated HDH system, the gained output ratio (GOR) boosts with increasing concentration factor of the solar collector [148]. The GOR of a thermal desalination unit is equal to the ratio of the mass flow rate of distilled water produced to the mass flow rate of steam used, also known as the performance ratio of the desalination unit [149].

4.3.2.3.3 Solar dual air and water heaters

In this case, solar heaters are used to heat both air and water. Fig. 4.18 shows a schematic of this type of system in the HDH process. In this system, a dual-purpose flat panel collector is used to heat the air and water. The combination of evacuated tube collector (for heating water) and flat plate collector (for heating air) is also applicable [151,152].

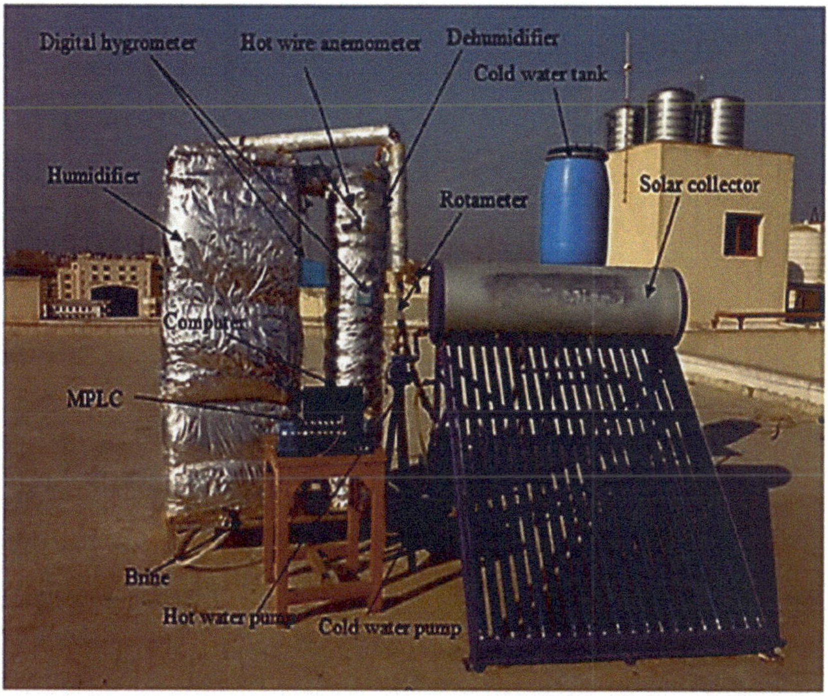

Figure 4.17 HDH system equipped by the evacuated tube water solar collector [146].

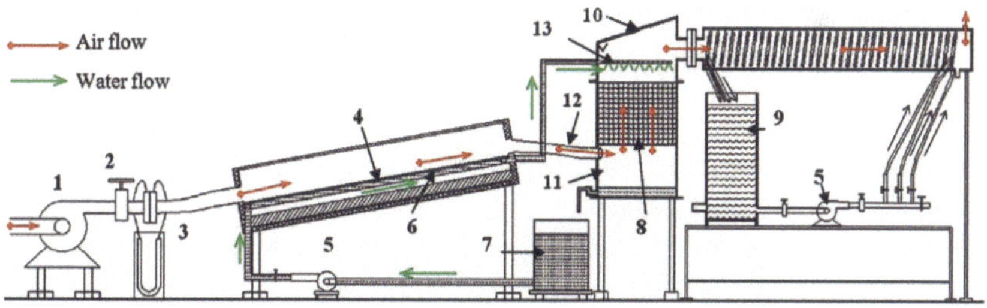

Figure 4.18 HDH system with dual-purpose collector [150].

The performances of solar air-heated, water-heated, and both water- and air-heated-driven HDH systems are compared in Fig. 4.19. It can be seen that among the three heating methods, the highest GOR, the lowest freshwater production, and the lowest water production cost are related to air-heated HDH system. The water-heated

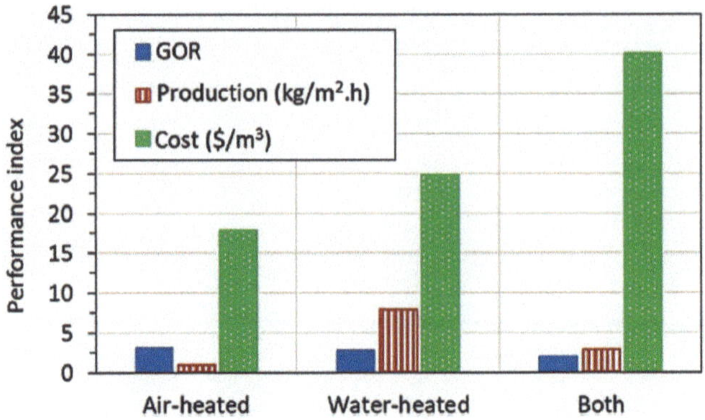

Figure 4.19 Performance of solar air-heated, water-heated, and both water- and air-heated-driven HDH systems [136].

HDH system has an acceptable efficiency and is not desirable in terms of cost and GOR. Dual air and water heated has the highest production cost and the lowest efficiency and GOR. As a result, the use of solar water heaters in the HDH system seems to be a good option. The low density and heat capacity and the problems caused by air leakage have made solar air heaters unfavorable for HDH systems. The cost of freshwater production for all three methods is considerably high, because most of the systems studied are on the pilot scale. The cost of water production on an industrial scale is expected to be less than 5 $\$/m^3$ [136].

4.3.2.3.4 Humidification–dehumidification systems driven by photovoltaic-thermal panels

The heat generated by the cooling of the PV panels can be used in an HDH system. The idea of combining HDH with PV-T was first suggested by Wang et al. [153]. They combined the water-heated HDH system with PV modules and obtained the heat required for the HDH process by cooling the panels. Air is also used to cool the panel and provide the heat required by the air-heated HDH system [154]. A schematic of the air-heated HDH system combined with PV-T is illustrated in Fig. 4.20. Studies show that the efficiency of an HDH system combined with PV-T is equal to 0.57 $kg/m^2/h$ and the cost of freshwater production is 11.5 $\$/m^3$ [136].

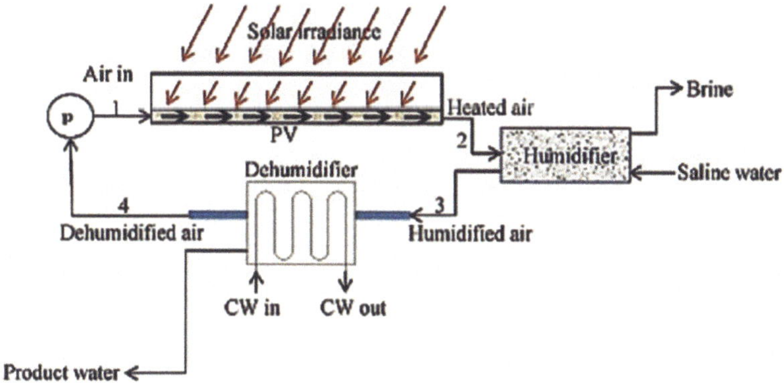

Figure 4.20 HDH integrated with PV-T panels [154].

4.4 Indirect solar desalination

As mentioned earlier, indirect solar desalination technologies fall into two categories: thermal and membrane. Here, thermal processes, such as diffusion-driven desalination (DDD), MSF desalination and MED, and RO membrane process are investigated. In thermal processes, desalination occurs based on the evaporation–condensation cycle and with phase change. In membrane processes, freshwater is produced by the passage of saline water through the membrane without changing the phase. In the following sections, the mentioned processes, their combination with solar energy, and the types of solar equipment used in these desalination units are explained.

4.4.1 Solar-powered diffusion-driven desalination

DDD is a low-cost and low-energy process that requires little maintenance. Hence, it can be driven entirely by solar energy [155]. This method, introduced by Klausner et al. [156], is based on the direct contact evaporation and condensation to desalinate seawater and brackish water. The evaporation and condensation processes are related to the inlet water and air temperature, inlet humidity ratio, and water to airflow ratio [157].

The solar-powered diffusion-driven desalination (SDDD) process consists of a direct contact evaporator and condenser and a solar collector. Initially, saline water circulation occurs in a flat plate solar collector to heat it. The packed bed material is used to fill direct contact between the evaporator and condenser and as a result, a direct connection between air and water can be provided. Heated saline water is sprayed through a nozzle to the top of the packed bed. The airflow is forced as a counter current stream from below by a fan into the evaporator and is in direct contact with the falling liquid film. The airflow is evaporated and humidified by the heated water. Humidified air in a completely saturated state exits the top of the evaporator and is blown into the condenser, where it is in direct contact with the falling liquid layer of freshwater. The air stream dehumidifies and returns to the evaporator. The heat exchanger is by-passed when the system is operated with a solar heater. The saline water discharged from the evaporator is returned to the saline water storage tank for recirculation. Similarly, freshwater is discharged from the condenser into the freshwater tank to be recirculated in the condenser. Since the system works in a closed manner, except for heat lost by the system components, no heat discharge occurs during operation [155,157]. Fig. 4.21 shows the flow diagram of SDDD process.

The amount of water produced in the SDDD process is three times that of a solar still. Since solar stills do not use electricity in comparison with SDDD or any other process, a larger surface area and volume per unit liter

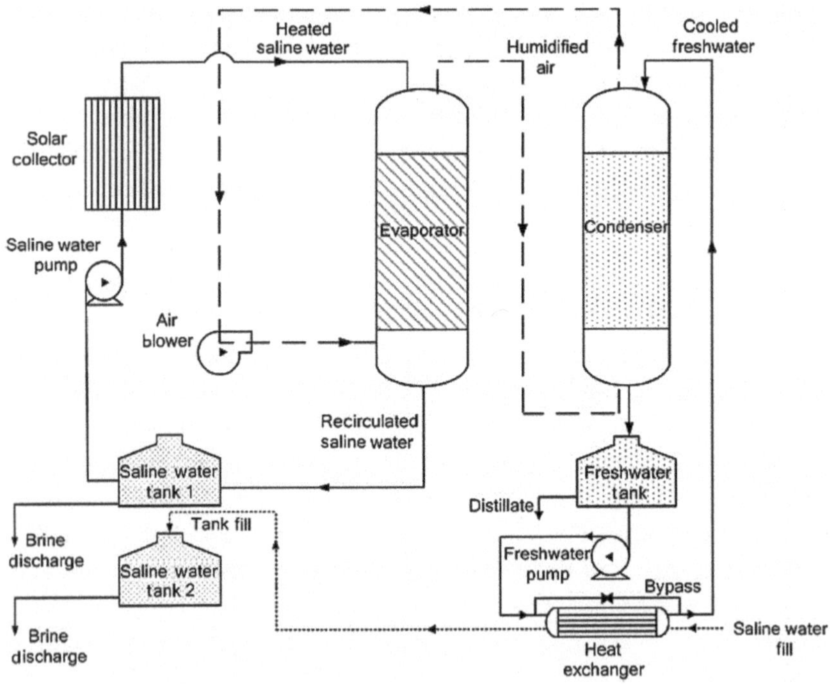

Figure 4.21 Process flow diagram of SDDD [157].

of distilled water are required [155]. The cost of water production in SDDD in the case of low-cost collectors is about 4 $\$/m^3$ [155]. According to simulations, for the specific production of 100 L/day of freshwater by SDDD method and using eight flat plate collectors with an area of 2 m^2, specific energy consumption (SEC) is equal to 3.6 kWh/m^3. Given the low SEC of this process and the low construction cost, it can be said that the SDDD process is competitive with other desalination systems with small scales [155].

4.4.2 Solar-powered multistage flash desalination

MSF desalination was the first large-scale commercial desalination process and came to prominence in the 1970s. Since then it has a significant share of the market in the Middle East [29]. MSF accounts for 21% of the world's installation capacity for desalination, second only to RO [49]. The production capacity of MSF may vary significantly, ranging from units that can produce 23,000 m^3/day to very large units which can produce 528,000 m^3/day. The production cost also varies, typically ranging between 0.52 and 1.75 US $\$/m^3$ [48]. The energy demand of MSF technology is high and varies between 13.5 and 25.5 kWh/m^3. Hence, most MSF units are built next to an existing power plant [158]. This method of thermal desalination is based on a flash distillation of heated brine at reduced temperature and pressure. Seawater/brackish water feed into this process, where the flow goes through successive heating stages, and after flashing, some of the same heat exchangers are used (on the shell side) to condense the freshwater, recycling latent and sensible heat. Typically the seawater is preheated by an external heat source before entering the first stage. The brine temperature then rises to 90°C−110°C. The heated brine flows continuously in stages and at each stage, a small amount of water is vaporized. The resulting steam condenses at each stage and freshwater is produced. Finally, concentrated brine and freshwater are drained from the last stage. Fig. 4.22 shows a schematic of the MSF desalination process. MSF units usually consist of 4−40 stages, and each stage operates at a lower temperature and pressure than the previous stage. Thus the boiling point of the feedwater is reduced during successive stages, and owing to the continuous boiling of the brine, there is no need for an extra heat source in addition to the seawater preheating heat source [158]. The performance and GOR of the MSF process are enhanced by boosting the top brine temperature (TBT), the brine temperature in the first flashing stage, decreasing the intake saline water temperature, boosting the stages, and increasing the specific heat exchange area [74].

In solar-powered MSF systems, various solar technologies, including parabolic collectors, flat plate collectors, central tower receivers, linear Fresnel reflectors, evacuated tubes, solar ponds, and PV panels, are used to combine with MSF desalination system [159]. To integrate MSF with solar energy, the TBT (between 90°C and 110°C) must be adjusted to prevent unstable operation [159]. Fig. 4.23 illustrates the MSF unit integrated with a solar collector. Since the 1980s, several solar MSF units with a capacity of

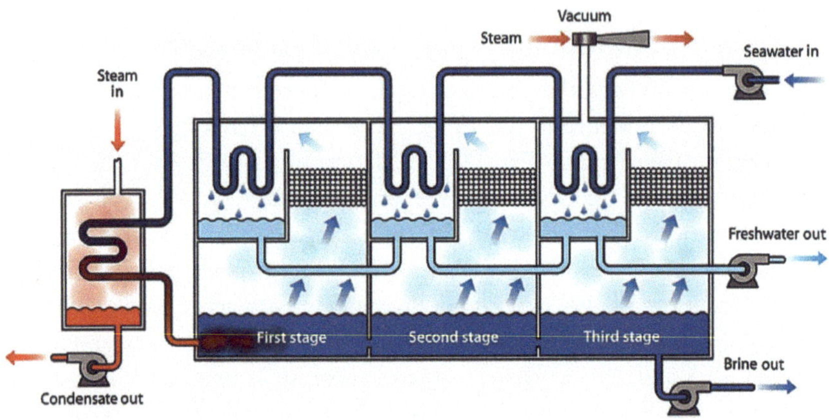

Figure 4.22 Schematic of MSF process [48].

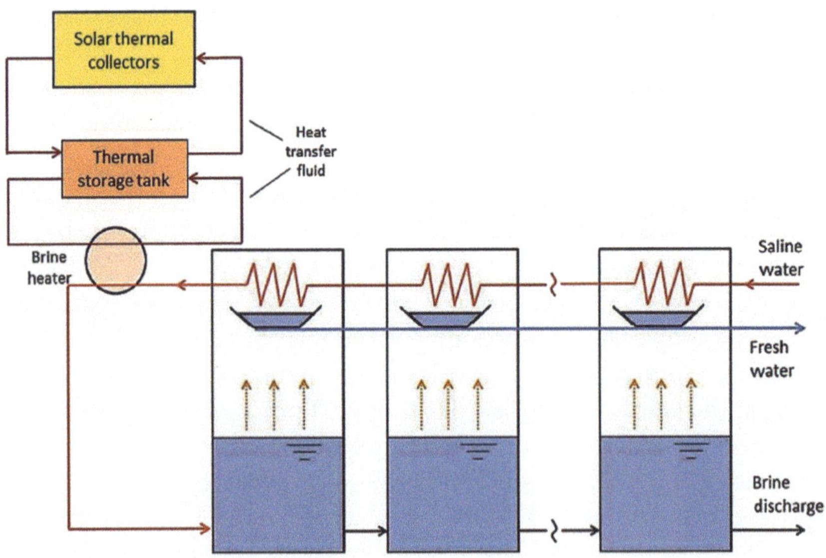

Figure 4.23 A solar-powered MSF desalination system [74].

10–20 m^3/day have been launched [160]. In 1983, in Safat, Kuwait, a solar-powered MSF plant using PTC collectors was installed with 10 m^3/day capacity and the SEC of 81–106 kWh/m^3 [161].

Over the past 20 years, few studies have been conducted on solar-powered MSF units compared to the solar-powered MED process, most of which have focused on pilot and small-scale units. This shows that solar MSF is less technologically and economically competitive than solar MED, owing to some reasons [74]: (1) The need for a relatively high TBT in this process has made it unfavorable to combine with solar energy. (2) Higher TBT means higher fouling and scaling rates. (3) Thermodynamically, MSF is less efficient as compared with MED.

4.4.3 Solar-powered multieffect desalination

Although the multieffect desalination (MED) process is thermodynamically more efficient as compared to the MSF process, it is not a usual process [149]. The MED process accounts for 7% of global installation capacity and consumes two to three times as much energy as the RO process [149]. MED units have a capacity between 600 and 91,000 m^3/day, however, the investment and energy consumption costs of MED units are high [149]. The MED desalination process consists of a set of cells (i.e., effects), in general between 2 and 16 effects, which work at reduced pressure. Hot steam enters the first effect from an external heat source through the tube. As a result of spraying seawater on the tube, heat is transferred from steam to water and hence, the seawater evaporates. The steam from the evaporation of water and the boiling of the brine enters the second effect and the evaporation/condensation process is performed and freshwater is produced. The steam and brine move between the effects at reduced temperature and pressure, and this continues until the last effect. There are several configurations of the MED system, forward feed (FF), backward feed (BF), and parallel cross feed (PCF) [162]. In the FF configuration, seawater and steam enter the first effect and the output brine from each effect is used to condense the steam into the next effect. In this case, the steam and brine stream move in the same direction and forward. In BF, brine and steam move in opposite directions. In this way, seawater enters the last effect and steam enters the first effect. In PCF, seawater enters all effects and steam enters the first effect. In this case, the brine enters each effect in parallel with the steam. This configuration has the highest performance and lowest SEC compared to the two previous

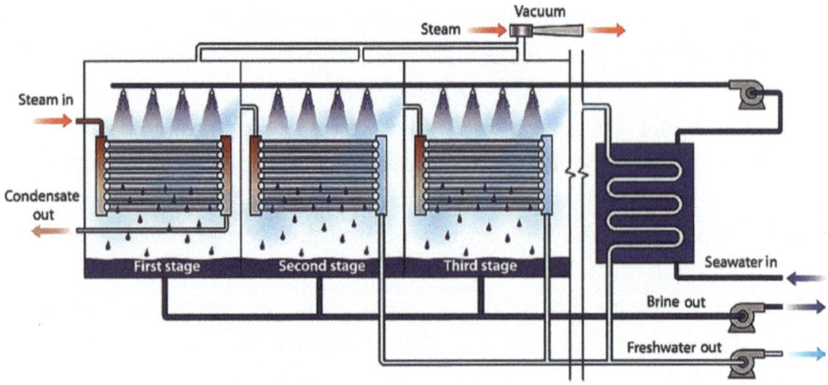

Figure 4.24 Schematic of MED unit [48].

ones, and is the most common configuration in industrial MED units [162]. Fig. 4.24 shows a schematic of the PCF-MED process. The GOR of an MED unit is related to the evaporator temperature at the last effect and is not strongly influenced by the feed stream temperature [163].

A heat pump can be used to recover steam from the last effect and thus increase the efficiency of MED. Four types of heat pumps, a mechanical vapor compression (MVC), a thermal vapor compression (TVC), absorption heat pumps (AHP), and adsorption heat pumps (ADHP), can, theoretically, be combined with MED [74]. TVC is one of the most widely used heat pumps in MED commercial units. TVC is a steam jet ejector which, according to the Venturi effect, increases the temperature and pressure of the steam output from the last effect by means of motive steam. In MED-MVC, the temperature and pressure of steam are increased by a compressor. MVC is used for small- to medium-scale power plants with a maximum capacity of 5000 m^3/day. In MED-AHP, the steam required for the first effect is produced by the latent heat recovery of the last effect. AHP loop consists of an evaporator, absorber, generator, and condenser. The integration of AHP with MED has several advantages, including reduction of thermal energy consumption, which in solar mode leads to reduction of solar field size, reduction of electrical energy consumption, reduction of seawater intake capital cost owing to less need for cooling seawater flow, the feasibility of decreasing operating temperature influence to less than ambient temperature, and add more effects in the same TBT [164]. In the MED-ADHP hybrid system, the output steam is absorbed in the adsorber/desorber bed. With this combination, the last effect works at a temperature lower than the ambient temperature, and more effects in the same TBT can be added to the unit.

In MED, the temperature of TBT is between 55°C and 120°C, and is compatible with solar thermal technologies [74]. There are several MED solar systems, including PTC integrated with MED, MED-TVC system integrated with a solar thermal power plant, MED-AHP driven by PTC or hybrid solar gas energy source [74]. MED integrates with PTC in two ways [162]: (1) The heat energy generated by PTC is consumed directly in MED; (2) the heat energy generated by the PTC is consumed in a Rankine cycle and the MED unit replaces the Rankin cycle condenser unit. The combination of solar organic Rankine cycle (ORC) with MED is more attractive for power generation and freshwater, but requires a larger solar field [163]. CPV-T is one of the solar heat sources that has good potential for integration with large-scale thermal desalination units. Integration of PV-T system with MED leads to lower cost of using the solar system; because the cost of collectors is offset by generating electricity [65]. In an MED-CPV-T unit with nine effects and a TBT of 60°C, the GOR is 6.59 [165].

In solar MED units, the cost of producing freshwater depends on the solar thermal technology, the intensity of the solar radiation, the rate of water production, and the temperature and the salinity of the seawater. The cost of water production in MED units combined with solar low-temperature collectors (60°C–95°C), medium-temperature collectors (165°C–200°C), and high-temperature collectors (370°C–530°C) is equal to 2–3.6, 1.4–3.1, and 1.8–2.2 $\$/m^3$, respectively, and the payback period is between 4 and 16 years [162]. The average GOR of solar MED units is varied between 2.7 and 15 [74].

4.4.4 Solar-powered reverse osmosis

Among the desalination technologies, RO is growing rapidly, and is expected to exceed a market size of 9 billion US dollars by 2022 [52]. RO is one of the most efficient desalination technologies owing to its low SEC (2–5 kWh/m^3) and high rates of adoption, commanding a 65% share of the installed desalination capacity in the world [49]. The superiority of RO over other methods stems from the following [166]:

- RO production capacity varies from standalone units with a generation capacity less than 1 m^3/day to units with larger scales and a generation capacity of more than 5×10^5 m^3/day.
- This process works in a wide range of feedwater salinity, such as brackish water and seawater.

- RO units operate continuously and reliably without extended shutdown periods.
- RO units with low SEC operate in the range of 2–4 kWh/m^3, not too far from the thermodynamic limit of 1 kWh/m^3 for seawater.
- Carbon dioxide emissions from seawater RO (SWRO) units are between 1.7 and 2.8 kgCO_2/m^3, the lowest amount compared to other desalination processes. However, the amount of carbon dioxide emissions from MSF distillation units is between 15.6 and 25 kgCO_2/m^3 and from MED units is between 7 and 17.6 kgCO_2/m^3.
- The RO process is cost-effective owing to the continuous reduction of water production costs. In 2017 for RO units with a larger scale and a generation capacity of more than 4×10^4 m^3/day, production costs varied in the range of 0.8–1.2 \$/m^3, which is expected to decrease by more than 60% over the next 20 years to reach 0.3–0.5 \$/m^3 [167].

The RO process is a pressure-driven process in which salt is separated from water by a semipermeable membrane through a solution-diffusion mechanism [52]. In this process, saline water is directed to the membrane with high pressure to overcome the osmotic pressure. Freshwater is collected from the permeate side and concentrated brine is rejected [74]. The operating pressure in SWRO is between 77 and 55 bar and in brackish water RO is between 15 and 30 bar [168]. The RO process consists of four parts [169]:

1. Pretreatment with useful chemicals that can include percolation and sterilization to decrease scaling and fouling,
2. High-pressure pump (HPP) to provide the pressure difference required to push water along the membrane,
3. Semipermeable membrane to separate freshwater from saline water, and
4. Posttreatment with chemicals to produce high-quality freshwater.

Owing to the low SEC of the RO process compared to other desalination processes, RO-based renewable energies have received a lot of attention. The SEC consists of two parts [166]: (1) The energy needed for the RO process itself that is related to factors, such as water quality, membrane efficiency, pump efficiency, recovery rate, and energy recovery device (ERD) employed and varies between 1.7 and 2.5 kWh/m^3. (2) Energy needed for secondary processes, including feedwater pumping, pretreatment, and unit electrical services ranging from 0.3 to 1.5 kWh/m^3. SEC reduction is achieved through innovation in membrane material valleys, ERDs, and more efficient pumps [170]. Commercialized membranes are currently capable of 99.8% desalination with the flow rate of 0.16–1.2 m^3/m^2/day over a period of more than 30 years [95]. ERDs use the energy remaining in the brine to pressurize the feed [171]. About 2% increase in pump efficiency results in a significant reduction in SEC, especially for a feedwater with high salinity [170]. Feed salinity, permeate quality, recovery rate, and feed temperature are operational parameters, which have effects on the required pressure values and thus energy consumption [172]. The required pressure can be supplied both by electricity and by mechanical pumps [52]. PV and solar thermal techniques are proper to supply electrical and mechanical power in RO units; however, in the case of thermal techniques, a thermal energy unit or thermal energy–driven pumps are required for pressurization of the feed [52].

4.4.4.1 Photovoltaic reverse osmosis (PV-RO)

A schematic of a PV-powered RO desalination system is displayed in Fig. 4.25. The PV-driven RO desalination system was first introduced in the 1980s, and since then has become a mature market leader among solar desalination technologies [173]. The solar PV is proper as a driver for RO units owing to the following reasons [166]:

- The modularity of PV modules makes it possible to run them with the RO system at various scales and the capacity of these modules can be boosted after initial installation.

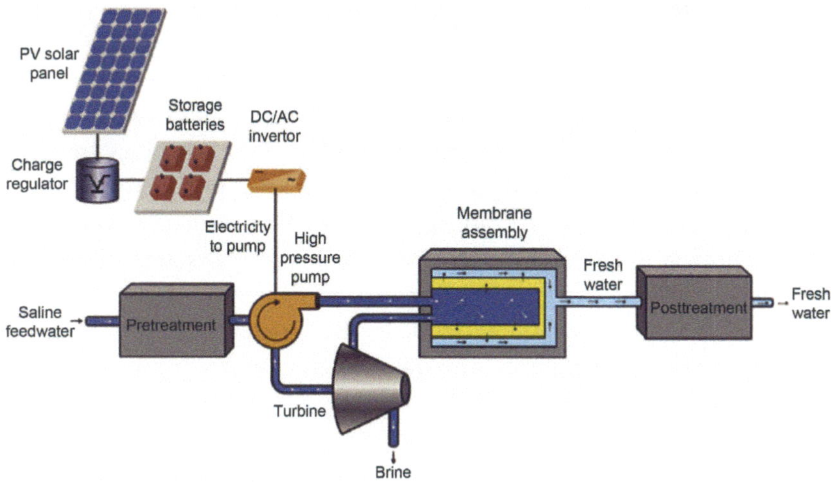

Figure 4.25 A picture of PV-RO desalination unit [48].

- PV modules need little maintenance and have a lifespan of longer than 20 years.
- Areas that require a lot of water consumption generally have a lot of solar radiation that makes PVs adapt to the intended use.
- The predictability of the daily/monthly/yearly solar energy facilitates planning for unit performance during any time period.
- Water storage capacity means mitigate the need for energy storage (e.g., during the night or solar lulls).

The SEC of PV-RO prototypes varies between 1.1 and 16.3 kWh/m^3, which is related to the system size, battery life, feed source, pretreatment, and ERD type [174]. The water cost of a PV-RO system is about 15.6 \$/m^3 when the membrane life is 5 years [74]. Reductions in PV costs, depending on solar insolation, type of source water used, system size, and government policies, make PV-RO systems more feasible [175]. There are many studies on the feasibility of PV-RO systems and different configurations have been proposed to increase the feasibility of such systems. Studies showed that the use of PV with batteries in the RO system is not suitable owing to large costs [174]. In battery-less PV mode, various techniques can be used, including a direct connection between the PV and the DC motor to start the high-pressure RO pump, the use of a supercapacitor as an electrical regulator, or the use of a controlled DC/DC converter. The performance of a PV-RO unit in two modes namely using the battery as an energy storage system and connecting directly to the PV array showed that in the case of no need to charge the battery or charge controller in direct connection mode, the system is less complex [176].

Technoeconomic analyses for various configurations of PV-RO systems, such as RO unit which is alternately driven by diesel engine, RO unit which is directly connected to PV array, and RO unit which works by combining PV and diesel engine, shows that PV-RO systems are more economically feasible than diesel-equipped systems, provided that there would be sufficient solar insolation [175,177]. However, the widespread implementation of PV-RO on both small and large scales generally is limited owing to high energy costs that can be overcome by boosting the energy efficiency of PV-RO systems. In the case of the PV system, the problem of the variability of the solar energy source must be solved through the development of energy storage devices or batteries. In the case of RO systems, energy efficiency should also be improved further through better-performing membranes, ERDs, and more efficient pumps [52]. Upgraded process design has also a considerable role in improving the efficiency of RO systems [52]. These matters are discussed in the following subsections.

4.4.4.1.1 Improving performance of PV modules used in RO desalination

Various techniques have been employed to enhance the performance of PV modules used in RO desalination. For example, solar trackers are used for maximization of the collection of solar energy and accordingly, water production increases by nearly 30% [178]. When single and biaxial trackers are used simultaneously according to the maximum power point tracking (MPPT) algorithm, the annual water generation increases by 33% and 36%, respectively [179,180]. MPPT adjusts the load of the RO unit so that the PV cell voltage is equal to the voltage required to achieve the maximum power at the incident solar radiation and the corresponding cell temperature [181]. In arid regions, power generation decreases owing to the accumulation of dust and sand on the surface of the PV cell. Accumulation of dust on the panels leads to an increase in panel temperature, reduced solar radiation, and also physical damages. To overcome this problem, several PV cleaning methods have been proposed, including mechanical, PV coating, or electrostatic techniques [166].

Ambient temperature affects the PV performance considerably, because with increasing PV temperature, power conversion efficiency linearly decreases [182]. Removing heat from the PV plate and transferring it to the feedwater before entering the RO process simultaneously reduce the PV temperature and increase the RO process efficiency. Open circuit voltage of solar panels and output power increase at lower PV temperatures, which could be coincided with higher membrane permeability at a higher feedwater temperature [183]. This method reduces energy consumption by 28% [184]. On the other hand, for each degree of increase in feedwater temperature, the membrane flux increases by 3% [185]. Removing heat from PV and transferring it to the feedwater in a PV-RO process, in the case of using silicon PV cells wherein the electrical efficiency decreases in the range of 0.2%–0.5% per degree of temperature, and improvement of membrane hydrophilicity will lead to a 40% reduction in energy costs [52]. Of course, in this technique, the negative consequences of warmer feedwater must be considered. Warmer water causes bacteria to grow on the membrane surface, resulting in biofouling aggravation. In addition, at high temperatures, along with increasing water permeability, salt permeability also increases and produces low quality permeate [52].

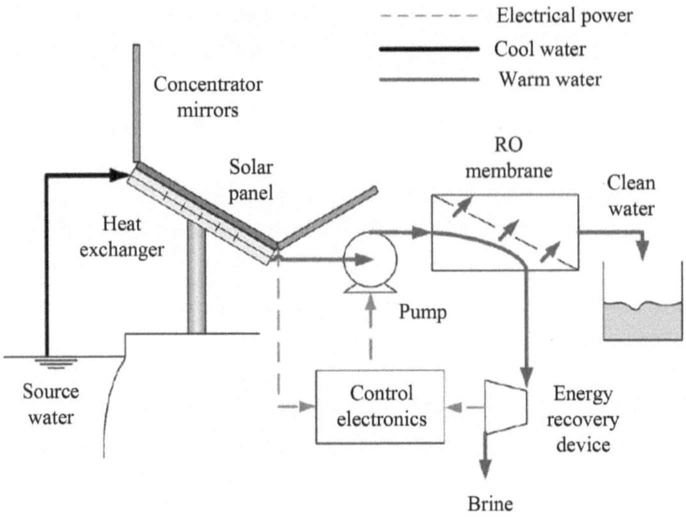

Figure 4.26 PV-RO system equipped by panel cooling and concentrating mirrors [183].

Using solar mirrors to boost the accumulated solar radiation along with PV panel cooling and feedwater heating, improves water generation by about 57% (see Fig. 4.26). Concentrated mirrors cannot be used without a cooling system, as the PV panels overheat and their efficiency decreases [183].

4.4.4.1.2 Membranes and configurations for PV-RO

In the RO process, the membrane acts as the heart of the desalination. Water flows through the semipermeable membranes and salt remains behind the membranes. The performance of the membrane depends on the feedwater pressure, which must be greater than the osmotic pressure required for the separation operation. In simple RO units, the HPP provides the flow at a constant pressure. However, in PV-RO systems, or in general any RE RO system, owing to unsteady input power and hence unsteady operation, feed pressure and discharge rate are variable [166]. If the membrane operates under steady state conditions recommended by the manufacturer, it will have a lifespan of about 5 years [186]. Continuous start-ups, shutdowns, flow changes, and pressure fluctuations create unfavorable conditions for the membrane and mechanical wear has negative effects on the life and performance of the membrane [187]. Accordingly, the cost of water production is affected by the variable performance of the membrane, owing to the cost of membrane replacement [186].

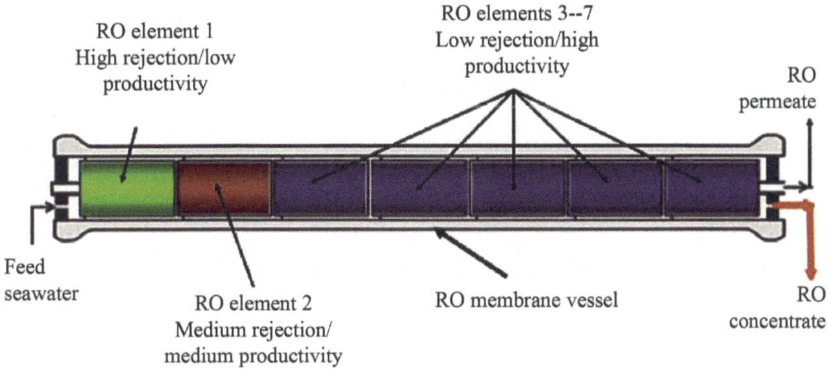

Figure 4.27 Hybrid membrane configuration [167].

To reduce RO costs, a PV-HEMRO system based on the hydro electromagnetic process (HEM) has been introduced to reduce the salt concentration by up to 30% before entering the RO process [188]. HEM, as a pretreatment process, reduces the salt concentration and pressure required for the RO process through the electromagnetic power that causes the separation of anions and cations.

In simple SWRO systems that use identical membranes in the process, the first membrane generates 25% of permeate and consumes 25% of the available energy. This causes the nonuniform distribution of energy in the vessel. This also makes the first element more prone to fouling. However, with a gradual increase in feed salinity, the other elements are more exposed to concentration polarization [52]. With a hybrid membrane configuration called hybrid membrane interstage design, a more uniform distribution of flux can be achieved by combining high permeability/low salt rejection, medium permeability/medium salt rejection, and low permeability/high salt rejection membranes (see Fig. 4.27) [52]. Other feasible methods in plant design, including three-center RO system, low-recovery plant, and split-partial two-pass RO system designs, can reduce RO energy consumption in a PV-RO system [52].

4.4.4.1.3 Energy recovery devices for PV-RO

In RO systems, HPPs represent the largest share of the system's energy consumption and operating costs [52]. A large amount of energy needed for the pressurization of the feed is released from the brine flow. This recovery from the brine flow through ERDs, especially in SWRO systems that require a large amount of pressure, reduces the SEC of the

PV-RO system [52]. Using ERD reduces energy consumption by up to 60%. Thus the SEC of RO systems with ERD reaches below 5 kWh/m^3 [166]. The first ERDs employed had an efficiency of 77%, but with the advent of novel strategies, the efficiency of newer pumps reached 97%, resulting in 40% energy savings [189]. ERDs are divided into two categories: centrifugal and isobaric [52,166]. Francis turbine, Pelton wheel, and turbocharger are the first creation or centrifugal ERDs [52]. Isobaric ERDs also include piston-type work exchangers and rotary pressure exchangers (PX) [52]. Centrifugal ERDs are known with high flow rates, limited capacity, and energy efficiency of about 82% [190]. Isobaric ERDs, on the other hand, have become more popular owing to their higher efficiencies of nearly 97%, lower energy consumption, decoupled performance of HPP, and smaller size than centrifugal ERDs [190].

Four commercial ERD technologies are: Pelton turbine, PX, axial piston motor (APM), and the Clark pump, also known as pressure intensifier. Each of these technologies is proper for a minimum brine flow rate [173]. In a small-scale SWRO system under partial-load conditions that are proper for use with the PV system, both the Clark pump and the APM pump result in a considerable decrease in energy consumption [191]. Among isobaric commercial technologies, PX devices have several advantages such as modularity, scalability, low maintenance, low need for HPPs, and lower return time [52]. In PX devices, the high-pressure brine current contacts the low-pressure feed, thus transferring brine energy at a high level of efficiency (see Fig. 4.28) [192].

In small-scale PV-SWRO units, the use of ERDs is not economically justified, because it increases the cost of water produced [193]. The world's first large-scale PV-SWRO power plant, Al Khafji, is powered by PX technology provided by Energy recovery Inc. (ERI), as part of a 1.4×10^6 \$ contract [52]. ERI has supplied its PX ERDs to more than 600 RO plants online or under construction, with capacities of $10^2 - 1.4 \times 10^5\ m^3/day$ [52]. The Qatar Solar Technologies Polysilicon Project online PV-RO power plant with a capacity of $1.2 \times 10^3\ m^3/day$ uses fluid equipment development turbochargers for energy recovery [52]. Turbochargers are centrifugal ERD, inexpensive, and easy to implement [52]. They have a hydraulic turbine, which transmits brine energy to the feed stream by means of a rotor (see Fig. 4.28). In commercialized RO power plants, sizing and selection of ERDs are done according to the optimal operating point of the power plant to operate the ERD at its optimized efficiency during normal operation [194]. Employing

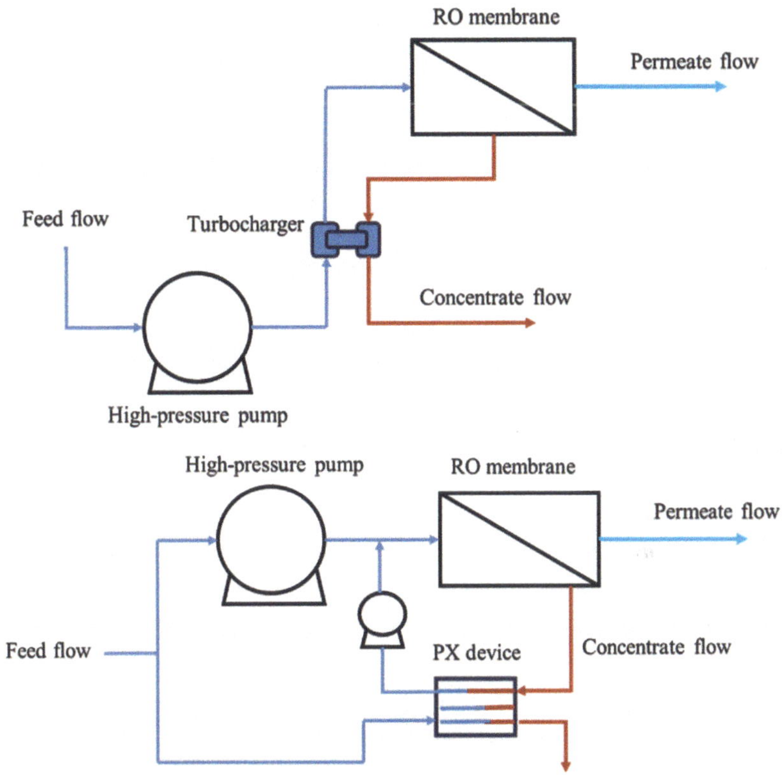

Figure 4.28 Flow diagrams for an RO system [192].

renewable energy in RO desalination poses new issues in relation to the variable performance of ERDs [166]. Despite the two PV-RO power plants listed with ERD and the high potential of them, ERD in PV-RO systems is used rarely, which needs more attentions [52].

4.4.4.2 Solar thermal-driven reverse osmosis system

In a solar thermal RO (ST-RO) system, heat can be received by the collector and transferred to the power conversion unit (PCU) or/and heat storage module [52]. The PCU consists of a power cycle that provides electrical or mechanical energy to the RO system [195]. One of the most considered power cycles is ORCs. Note that when ORC is combined with RO, the ORC can employ cold feedwater as a heat sink that heats the feedwater and improves the membrane flux [196]. Other power cycles, such as steam Rankine cycles, Stirling engines, and Brayton cycles, are employed in solar heating systems. The Rankine cycle is a good choice for ST-RO systems owing to its simplicity and has been the most prominent power cycle for ST-RO at this moment [52].

An ORC is a thermodynamic power cycle and can be employed to convert heat into mechanical energy using organic working fluid [74]. An ORC has pumps, evaporators, turbines, and condensers [197]. An ORC-based ST-RO system has a solar field, the Rankine cycle, and RO unit(s) [174]. In the solar field, flat plate, evacuated tube, and/or PTC are employed to supply thermal energy of the Rankine cycle [174]. The organic operating fluid can be pressurized and injected into the evaporator. It is heated inside the evaporator to evaporate through heat exchange with a high-temperature fluid. After that, the generated steam is expanded inside the turbine and the necessary mechanical power is provided to start the HPP of the RO unit. Meanwhile, to increase the permeability of the membrane, saline water can be preheated in the condenser [74]. This process is shown in Fig. 4.29.

Optimizing solar-driven ORCs in order to increase their efficiency plays an important role in creating a vision for scaling up ST-RO systems [52]. Advances in the design and efficiency of the solar thermal power cycles are also effective in creating the prospect for ST-RO scaling up [198].

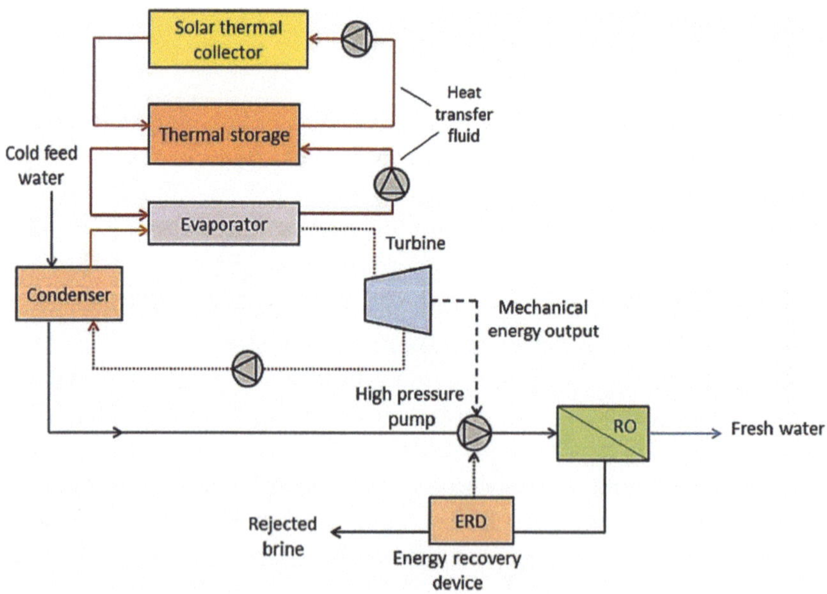

Figure 4.29 Solar ORC-driven RO system [74].

Solar ORC efficiency depends on solar collector efficiency and ORC efficiency, in other words, on some variables such as the type of working fluid and operating temperature [52]. Concentrating solar collectors including parabolic trough, linear Fresnel, and compound parabolic collectors (CPC), can be employed to improve the overall efficiency of solar-driven ORCs [74]. The organic working fluid employed in the ORC must be chosen in proportion to the temperature level of the heat source obtained from the various solar collectors. Siloxane MM is suitable for ORC with PTC. Other operating fluids, including Solakatherm SES36, isobutene, isopentane, R245ca, and R245fa, have been proposed for ORC with stationary solar collectors [74].

4.4.4.3 Comparison between PV-RO and ST-RO

Unlike the PV-RO system, that is rather mature and can be found at many scales, ST-RO systems are in the preliminary stages [74]. In recent years, PV-RO units have been reported from small to medium scale with the capacity in the range of 0.2–200 m^3/day for seawater and brackish water desalination. In contrast, ST-RO units have been tested for desalination systems with larger scales, a capacity of $1.186 \times 10^3 - 5 \times 10^4$ m^3/day. However, no power plants have been reported being serviced [74]. Analyses show that the cost of water production in the PV-RO system is 8.855 $\$/m^3$, while in the solar-driven ORC-RO system is 13.78 $\$/m^3$ [199]. The cost of the solar-driven ORC system is expected to decrease by 30% after maturing and commercialization [199]. The levelized cost of electricity for ORC is 27% lower than that of a PV-powered system with energy storage [200]. Table 4.2 lists the RO-based solar power plants with the capacity and type of solar technology used. Al Khafji is the world's first solar-powered plant with a large scale based on PV-RO. Water production capacity in Al Khafji varies between 16 and 60,000 m^3/day [52].

4.5 Solar disinfection

Currently, approximately 2 billion people in the world consume fecal-contaminated drinking water, and 2.3 billion people lack adequate sanitation. This condition leads to water-borne diseases. With more than 2.2 million deaths annually, it is the leading cause of death, especially in developing countries. Cholera, typhoid fever, dysentery, hepatitis A virus, etc. are among the most common water-borne diseases. In fact, in developing countries, most of the wastewater is discharged into the

Table 4.2 Specification of installed or underinstallation solar-powered RO plants [52].

Location	Capacity (m^3/day)	Feedwater	Solar system	Kind of collector
United States	3.168×10^3	BW (3×10^3–2×10^4 ppm)	Concentrated solar power	Parabolic trough
Tunisia	1.8×10^3	BW	Photovoltaic	-
Saudi Arabia	6×10^4	SW (2×10^4–5×10^4 ppm)	Photovoltaic	—-
Mexico	8.4×10^2	BW	Photovoltaic	—
Mexico	4.8×10	BW	Photovoltaic	—
Madagascar	1.6×10	BW	Not provided	—
Spain	4.8×10^2	BW	Concentrated solar power	Parabolic trough
Spain	7.2×10^2	BW	Concentrated solar power	Parabolic trough
Brazil	3.6×10^3	BW	Photovoltaic	—
Algeria	1.577×10^3	WW	Solar thermal	Parabolic trough
United States	7.5×10	BW	Photovoltaic	—
Qatar	1.2×10^4	SW	Photovoltaic	—
Vanuatu	9.6×10	SW	Photovoltaic	—
Vanuatu	9.6×10	SW	Photovoltaic	—

BW, Brackish water; *SW*, seawater; *WW*, wastewater.

environment without any treatment. As a result, it pollutes surface waters and leads to the transmission of water-borne diseases. In addition, the spread of pathogens and the prevalence of water-borne diseases are related to environmental and climatic conditions. It is predicted that with the increase in the frequency and severity of tropical storms, droughts, and floods caused by climate change in the future, the health problems related to drinking water will intensify [201].

In order to provide safe drinking water, various disinfection technologies have been developed and used. The purpose of disinfection is to remove pathogenic microorganisms that cause water-borne disease. In general, disinfection is done by two general methods: (1) physical methods including flocculation and sedimentation, filtration, ultraviolet (UV) radiation, and pasteurization, and (2) chemical techniques including chlorination, chloramination, chlorine dioxide treatment, and ozonation. These commonly used processes are energy intensive and require a significant amount of capital, expertise, and infrastructure. Currently, water treatment technologies are used on the large and medium scales and have been successful in maintaining public health against water-borne diseases [201]. Energy and water are two essential and inseparable sources, without which the other cannot be produced or supplied. Recently, energy–water nexus has received a lot of attention due to the increasing energy demand in the water sector. The water disinfection process is no exception. Conventional disinfection processes consume 0.25–1 kWh/m^3 of energy, which accounts for about 2%–3% of the world's energy consumption [202]. However, as demand for high-quality water and drinking water increases, energy consumption in the water treatment sector will increase. Therefore on the one hand, high energy consumption in water treatment plants and on the other hand, the energy crisis following the increase in energy consumption in the world and increasing greenhouse gas emissions, highlights the need to use sustainable technologies in water treatment processes. Meanwhile, solar energy is one of the most efficient sources of sustainable energy for disinfection.

Solar disinfection is not a new technology. Downes and Bluent [203] first proposed the bacterial effect of sunlight in the late 1870s and the relationship between parameters, including solar radiation intensity, solar exposure duration, and wavelength with the inactivation of bacteria. In many developing countries, where drinking water supply is not possible due to lack of local electricity network, high electricity costs, and lack of access and high cost of chemicals for treatment, solar disinfection can be used as a low-cost, electricity- and chemical-independent solution [201]. However, its most widespread use is in rural and remote areas with low access to safe drinking water and high access to solar radiation.

4.5.1 Solar water disinfection

Solar water disinfection (SODIS) is an easy, stable, and low-cost method of water treatment that kills pathogenic microorganisms by employing the germicidal effect of UV rays and heat production [201]. Currently, more than 5 million people in more than 50 countries in Asia, Latin America, and Africa use this method daily for drinking water treatment. In this method, water is exposed to sunlight in a transparent glass or plastic container (usually 2 L PET bottles). Exposure time varies from 6 to 48 hours according to the intensity of radiation and resistance of pathogens [204]. Unlike the conventional UV method in wastewater treatment, where UVC rays penetrate directly into the DNA of pathogens and destroy DNA strands, in the SODIS process, UVA rays first form reactive oxygen species in water, then, these species destroy DNA of pathogens and inactivate microbes [205]. Fig. 4.30 shows a simple schematic of the SODIS process.

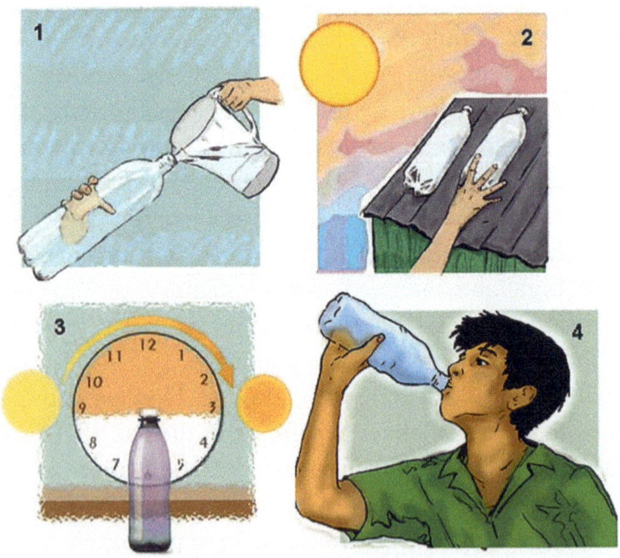

Figure 4.30 Graphical descriptions for SODIS household water treatment method [204].

It is recommended that disinfected water be used within 24 hours to prevent postexposure regrowth. However, disinfection efficiency in this basic system can be increased through the following strategies [204]:

- Putting filled bottles on reflective plates: to increase the absorption of solar energy;
- Blackening the bottom surface of SODIS reactor: to increase solar heat;
- Shaking a two-thirds filled bottle before exposing it to the sun: to increase the level of dissolved oxygen for solar-induced oxidative inactivation processes; and
- Filter water before filling the reactor.

Although conventional plastic bottles are cost-effective, the main drawback is the limited amount of capacity (less than 2 L) of this type of reactor. Glass bottles are one of the best alternatives to plastic ones. Ordinary glass bottles are able to transmit 90% of solar radiation, especially wavelengths in the UVA range [206]. The heaviness of the bottle after filling and the possibility of injury to people after breaking the bottle are problems of this type of reactors. Another option is PET bottles. Undamaged PET bottles pass about 85%–90% of the UVA wavelength and block the UVB wavelength [207]. Under conditions of long-time exposure, the release of chemical compounds in plastic and reaction with water is a major problem for the use of these reactors [204]. The PET bag is also used as a SODIS reactor. In these bags, which are made of low-density polyethylene [208], the exposure area is maximized and the path length for light penetration into water is minimized [204]. To improve the performance, the bag is placed on a black screen [204]. In remote areas with a shortage of PET bottles, the use of SODIS bags with the possibility of easy transportation and storage in large numbers would be a good option [204].

4.5.2 Solutions to promote solar disinfection

In SODIS bottles, sunlight shines only on the upper surface of the bottle so that most of the radiation does not reach the water. In order to increase the radiation received by the bottle, efforts have been made to concentrate sunlight, including the use of reflective surfaces and low-cost concentrating equipment [209]. Other effective solutions to increase the efficiency of solar disinfection include the use of chemical additives such as photocatalysts, sodium percarbonate, lemon juice or pulp, and riboflavin [201]. In this section, two methods of using reflective surfaces and concentrating equipment, and adding photocatalysts to improve the efficiency of solar disinfection are discussed.

4.5.2.1 Thermal enhancement

Since there is a high synergy between optical and thermal inactivation at temperatures above 45°C, many strategies are developed to improve thermal inactivation [207]. In general, the thermal enhancement solutions of SODIS reactor are [204]: blackening the bottle, circulating water on a black surface as a solar energy absorber, and using a collector and solar reflector. In terms of converting solar energy into heat, blackened bottles have the lowest efficiency [210]. Unlike blackened surfaces, reflectors reflect UVA rays even on cloudy days. Thus even on cloudy days, optical inactivation and increase of water temperature for disinfection occur [204]. The use of solar mirrors as reflectors reduces disinfection time to 3–4 hours [211]. Combining the SODIS process with concentrating parabolic collectors (CPC) in less than 6 hours results in complete disinfection of water [212].

4.5.2.2 Photocatalysis

Photocatalysis is one of the most effective technologies employed to mineralize resistant organic compounds and inactivate water pathogens among advanced oxidation processes (AOPs). Photocatalysts are divided into two categories: heterogeneous (semiconductor catalysts for water treatment) and homogeneous (photo-Fenton process) [213]. In the process of heterogeneous photocatalysis, the degradation of stubborn organic matter is performed by the combined action of a semiconductor photocatalyst, an energy source, and highly reactive oxygen species [214]. Among semiconductor photocatalysts, TiO_2 is one of the most widely used photocatalysts in water treatment applications [204]. Solar photocatalytic processes with TiO_2 photocatalyst are used to disinfect water in which UVA-resistant microorganisms are inactivated by the TiO_2 photocatalyst [215].

The most common homogeneous solar photocatalytic process in water treatment is the photo-Fenton process. The Fenton oxidation process is an AOP process, which produces $\dot{OH}$ radicals by the catalytic reaction of H_2O_2 with iron ions (see Fig. 4.31). The photo-Fenton process (Fe^{2+}/H_2O_2/UV vis) has a higher oxidation rate, lower iron consumption, and consequently less sludge production than the Fenton reaction [74]. In this process, owing to light radiation in the range of near UV to visible and

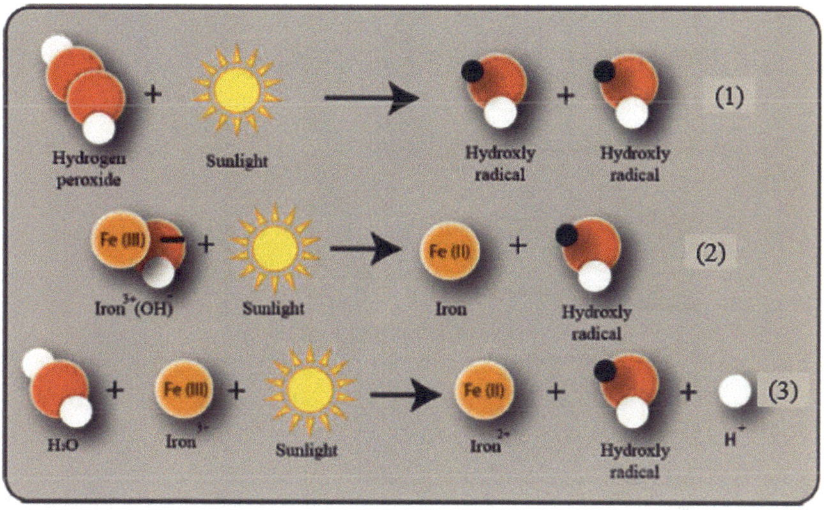

Figure 4.31 Photo-Fenton mechanism [216].

up to 600 nm wavelength, free radicals are formed and cause water disinfection. The nonselectivity of this process has made it possible to remove a wide range of antimicrobial-resistant microorganisms. Parameters affecting the efficiency of this process are: pH, temperature, the concentration of hydrogen peroxide (H_2O_2), and iron and their ratio and the intensity and wavelength of radiant light [216]. The type of light source has been shown to have a significant effect on the disinfection of hydrogen peroxide. As a result, the higher intensity of light radiation leads to a higher disinfection rate [217,218].

In solar photocatalytic disinfection processes, the most common photoreactors are PTC, CPC, and inclined plate collectors (IPC) [74]. PTC and CPC are well-established concentrating collectors. In a PTC reactor, the reactant fluid flows in a transparent tube located on the parabolic focal line. In heterogeneous photocatalytic processes, the PTC reactor is mostly used for suspended photocatalysts. Because PTC reactors only receive direct sunlight, it is best to use single or biaxial tracking systems. In CPC reactors, two curves are used side by side to concentrate sunlight on the absorber tube. CPC is more effective than PTC in concentrating light, as it reflects both direct and scattered radiation on the absorber tube. In this way, sunlight is evenly distributed on the surface of the absorber tube. Hence, these reactors are suitable for suspended and immobilized photocatalysts. IPC collectors are flat or corrugated black

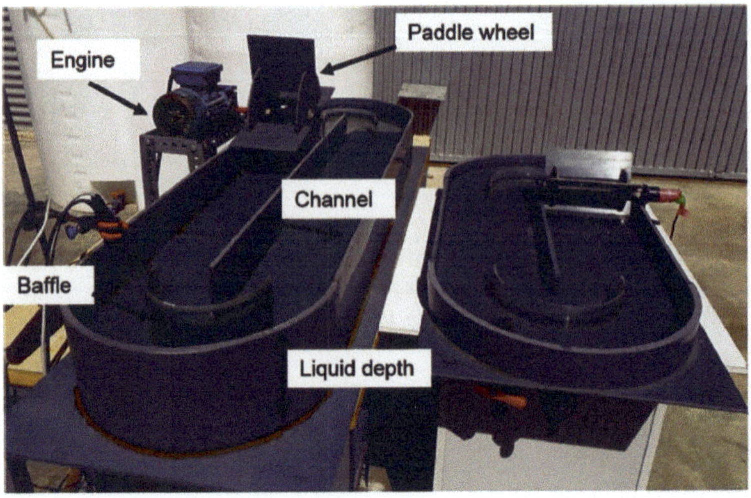

Figure 4.32 Raceway pond reactor [219].

plates with a reactive fluid flowing through a thin film. The flow rate inside the thin film is between 0.1 and 15 L/min, so the IPC reactor requires a larger surface area as compared with PTC and CPC. Therefore this reactor is proper for small-scale applications [74]. Another type of reactor for solar photocatalytic reactions is the open reactors. Soriano-Molina et al. [219] used raceway pond reactors for the solar photo-Fenton reaction. The reactor consisted of a hippodrome-shaped shallow bath in which water flowed through a paddle wheel and was exposed to sunlight (see Fig. 4.32). The benefit of this kind of reactors is the ability to disinfect on a large scale with low treatment costs.

4.5.3 Photovoltaic-solar water disinfection

PV-solar water disinfection (SOLWAT), a hybrid system for disinfection and electricity generation, has been developed according to the use of solar energy. In this system, several conversion mechanisms occur for two final applications (PV to generate electricity and heat, and UV to disinfect). This system has low energy consumption and good performance and is suitable for industrial water treatment systems or remote rural regions that do not have accessibility to drinking water and electricity [220]. The schematic of the SOLWAT system is displayed in Fig. 4.33.

The SOLWAT system performs better than PET bottles in water disinfection owing to the difference in PET bottle configuration (PET, cylindrical and 38 mm of water thickness) compared to the

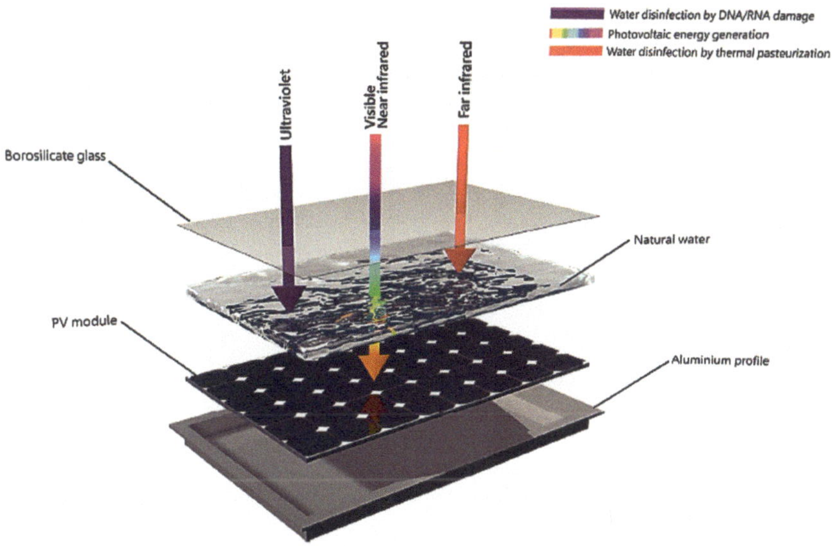

Figure 4.33 The SOLWAT system design and configuration including the solar spectrum usage [220].

SOLWAT system (flat, borosilicate and 18 mm of water thickness) that prevents the penetration of light and reduces the absorption of solar radiation in water. The SOLWAT system can reach a temperature of 45°C and even 50°C in 3–5 hours, in which case the synergy between UV and temperature improves disinfection. On the other hand, in PET bottles, disinfection is performed at temperatures between 20°C and 45°C for 6 hours. In SOLWAT systems, the water layer on the PV module has a cooling effect and therefore eliminates the negative effects of temperature on the module efficiency [220,221].

A V-trough concentrator and a low concentration of hydrogen peroxide (5 mg/L) can be used to enhance the SODIS process in the SOLWAT system [222]. According to the results, in concentrator + water mode, compared to nonconcentrator + water mode, the complete disinfection time of 8 L of water was reduced to 1.25 and 2.5 hours for *Escherichia coli* and *salmonella*, respectively. In terms of power generation efficiency, the output power of the module in concentrator + water mode reached 43 W, which is the highest value compared to the reference mode (26.1 W) and nonconcentrator + water mode (24.1 W).

A photocatalytic treatment system can be combined with a PV system [223]. This system absorbs the solar spectrum more effectively. The UV

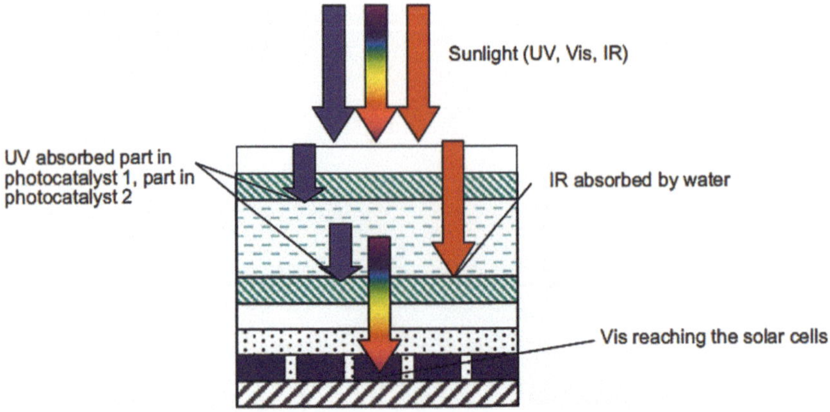

Figure 4.34 Spectral absorbance diagram of an integrated PWAP system, UV is absorbed by each of the photocatalytic layers, far-IR is absorbed by water, and visible and near-IR is absorbed by the photovoltaic solar cells [223].

spectrum is absorbed by the photocatalytic reaction. The visible and near-infrared spectra are absorbed by solar cells, and the far- and near-infrared spectra are absorbed by water (see Fig. 4.34). Since only the two visible and near-infrared spectra reach close to the surface of the PV cell, there is no increase in heat owing to the loss of other solar spectra, and the efficiency of the module increases by up to 35%.

4.6 Environmental impact of generation II technologies

It is commonly thought that the desalination process is a clean process in the supply of drinking water. However, this process, like any other industrial process, has potential environmental effects [56]. Although desalination technologies have many social, economic, and public health profits, they are energy intensive and therefore have a detrimental effect on the environment [224]. Currently, the energy required by the desalination units is supplied from fossil fuels. Consumption of fossil fuels will lead to greenhouse gas emissions and air pollution. On the other hand, in the desalination process along with freshwater stream, a discharge stream called brine is also generated. Brine is a hypersaline solution and contains chemical compounds, therefore it is considered environmentally hazardous. At present, brine disposal takes place in the marine environment, which has exacerbated environmental concerns [56]. In addition, the impingement and entrainment of seawater through submerged pipelines or open intakes along the shoreline remove fish eggs and small marine organisms such as plankton and larvae from seawater. The release of nutrients, organic matter, and organisms along with seawater leads to aggressive pretreatment of water through more chemical additives and more frequent cleaning of filters and membranes; which ultimately affect the brine flow [225].

4.6.1 Brine disposal

Concentrated or reject or brine is a highly concentrated by-product of the desalination process. This liquid stream has a large number of dissolved solids in concentrated form. The quality and quantity of brine depend on the quality of the feedwater, the pretreatment, the type of desalination process, and the water recovery rate. Apart from high salinity, brine contains hazardous pretreatment chemicals (e.g., antiscalants, coagulants, and flocculants), organic matter, and heavy metals. Disposal of brine at sea has devastating effects on the marine environment. These destructive effects include eutrophication, pH fluctuation, and increasing concentrations of heavy metals in the marine environment [226].

Brine salinity is 1.6–2.1 times higher as compared to seawater (~35 g/L). Brine salinity for thermal processes (MSF and MED) is ranged from 55 to 65 g/L TDS and for the RO process is ranged from 60 to 85 g/L TDS. This difference is owing to the higher water recovery rate (40%–45%) of commercial and well-established RO units [56]. According to a study conducted by Gacia et al. [227], even a marginal increase in salinity upsets the osmotic balance of marine species with the environment and has a detrimental effect on marine life. Osmotic imbalance reduces turgor pressure and causes the extinction of marine species in the long-term run [228].

The temperature of the brine is related to the type of desalination process. Thus the brine generated by membrane processes has a temperature equal to the temperature of seawater (22°C) and the brine temperature resulting from thermal processes is 1.37–1.82 times higher than the temperature of seawater [226]. The higher temperature of brine than seawater (30°C–40°C) has destructive influences on marine life, because the toxicity of metals and chemicals boosts with increasing temperature [229].

Heavy metals and chemical residues also have negative effects on marine species. High temperatures in thermal processes cause corrosion of metal equipment. For example, the presence of Cu and Ni elements in brine is owing to the removal of Ni–Cu alloy because of wear in heat exchangers and pumps [230]. In membrane processes, owing to the use of polymeric materials in the membrane, the concentration of heavy metals is less than that to affect marine life [231].

The world currently produces 141.5×10^6 m^3/day (51.7×10^6 m^3/year) of brine, which is 50% more than the total amount of freshwater produced in the world. Seventy percent of the world's brine is produced in the Middle East and North Africa (approximately 10^8 m^3/day) [4]. This is twice the amount of water produced in these areas. This fact shows that the desalination units work in these areas with a very low water recovery ratio of 0.25 [4]. Brine disposal techniques are: surface water and sewer discharges, deep-well injection, evaporation ponds, and land application [226,232]. Choosing the right method depends on several factors. Quantity, quality, and composition of the brine, geographical position of the disposal site, site accessibility, the permissibility of the option, social acceptability, costs, and facility potential are these factors [226,232]. Table 4.3 gives a brief description of each method.

The zero liquid discharge (ZLD) approach, aiming at improving the rate of water recovery by reducing brine production and thus reducing environmental impact, is a method of a brine treatment. The ZLD method recovers 90%–95% of water. Condensed solids are also disposed of in an environmental friendly manner. ZLD consists of three stages: (1) preconcentration, (2) evaporation, and (3) crystallization. In the first step, water recovery and reduction of brine volume are done by membranes. In the next two steps, minimizing the brine volume and solids production is done by thermal methods [226].

4.6.2 Energy consumption and GHGs emission

Energy is a major issue in the environmental assessment of desalination units. The energy can be used for the desalination process, freshwater and brine transportation, unit lighting, office equipment, etc. For industrial-scale desalination units, high energy consumption is considered the main obstacle [56]. The energy demand in desalination processes is related to the desalination technology (thermal or membrane), the type or quality of feedwater (seawater, brackish water, and wastewater), unit design (recovery system design, unit capacity, and energy recovery system efficiency) [233]. In general, membrane technologies (RO) require a large amount of electrical energy, while the total energy required in thermal processes (MSF and MED) owing to the need for both thermal and electrical energies, is higher. In both cases, the required energy is easily supplied from fossil fuels.

Table 4.3 Summary of the brine disposal methods [226].

Technique	Principle	Cost ($/m^3 brine)	Environmental issues
Sewer discharge	Brine rejection occurs in a sewage collection unit.	0.32–0.66	Bacterial growth is inhibited in wastewater treatment plants.
Evaporation pond	Brine evaporation occurs in a pond and remaining salt is gathered.	3.28–10.04	Groundwater is polluted and soil can be salinized.
Surface water discharge	Brine rejects into surface water.	0.05–0.30	The marine environment can be polluted.
Deep-well injection	Brine rejection occurs into perforated subsurface rock formations.	0.54–2.65	Groundwater is polluted and soil can be salinized.
Land application	Brine is employed in the irrigation of salt-tolerant crops and grasses.	0.74–1.95	Soil can be salinized.

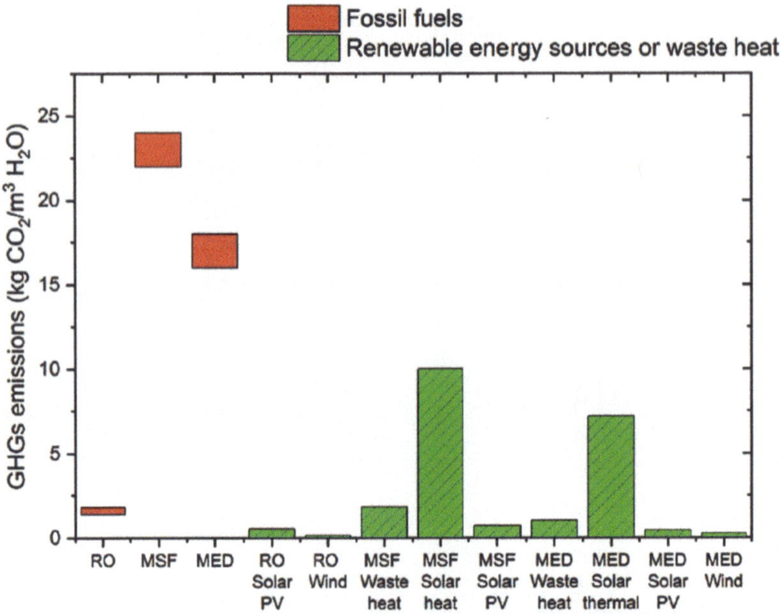

Figure 4.35 GHGs emissions per cubic meter of freshwater generated by the desalination methods [56].

Consumption of fossil fuels is associated with greenhouse gas emitted. Fig. 4.35 illustrates the amount of greenhouse gas emitted per cubic meter of freshwater generated when fossil fuels, renewable energy sources, and waste heat are employed. As can be seen, the emission rate of greenhouse gases from thermal processes is at least 10 times higher than the membrane process. As a result, desalination has a critical role in air pollution [56].

One way to diminish the environmental consequences of high energy consumption in desalination processes is to power desalination units using renewable resources, including solar, geothermal, wind, tidal, or other alternative energy sources, including waste heat from industrial activities. As shown in Fig. 4.35, the emission rate of greenhouse gases per cubic meter of freshwater is significantly reduced when using renewable energy sources [56]. Mannan et al. [234] used life cycle assessment (LCA) to assess the environmental impact of MSF desalination units in Qatar wherein about 75% of the freshwater is supplied by MSF technology. For this purpose, three MSF units with different GORs in terms of climate change, freshwater eutrophication, degradation of the ozone layer, fossil fuel depletion, and human toxicity were examined and the results are shown in Figs. 4.36 and 4.37. Note that in Fig. 4.36, the numbers 1, 2, and 3 indicate the three MSF plants including plants 1, 2, and 3, respectively. GORs of plants 1, 2, and 3 are 8.21, 9.73, and 16.07, respectively. According to LCA results, the greatest environmental impact of desalination is owing to energy consumption. Increasing the GOR reduces the severity of environmental effects. As a result, the amount of carbon dioxide emission owing to energy consumption is 7.32 kg in unit 3 (highest GOR) and 12.6 kg in unit 1 (lowest GOR). Mannan et al. [234] also studied the scenario of reducing carbon dioxide emissions by supplying the energy required by the MSF unit through 5%, 10%, 15%, and 20%

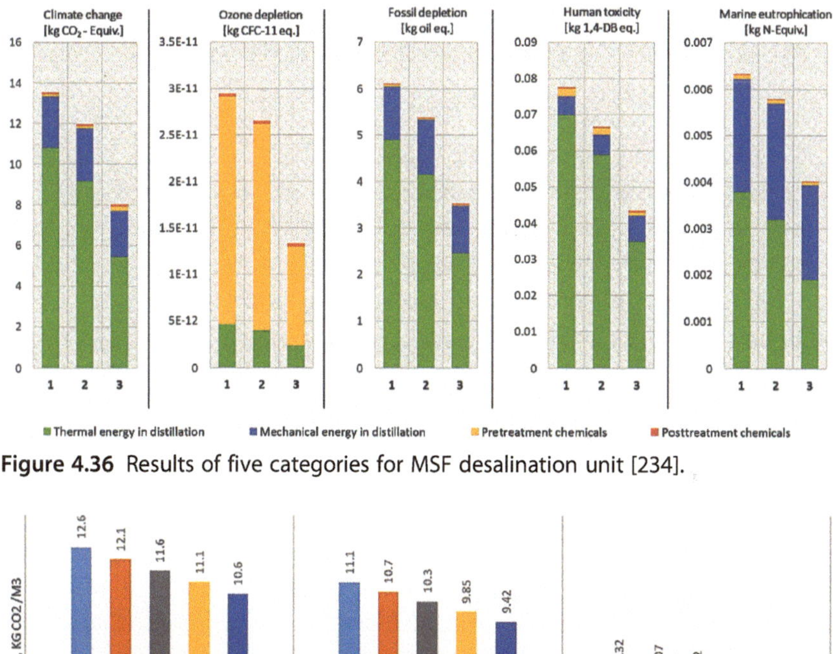

Figure 4.36 Results of five categories for MSF desalination unit [234].

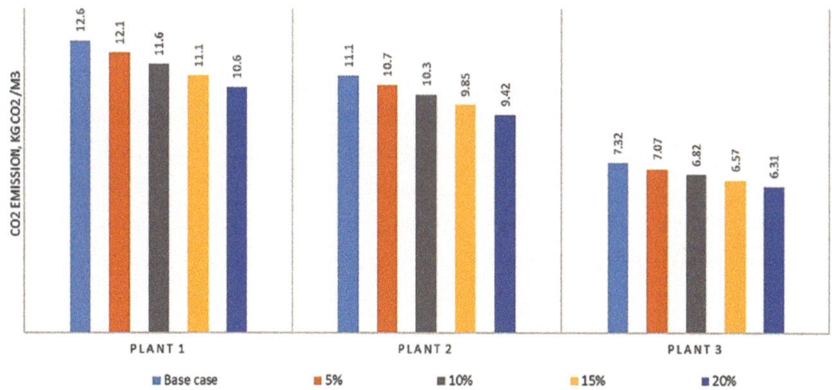

Figure 4.37 Decrease in carbon dioxide emission owing to integrating solar thermal energy for four various scenarios (coupling of 5%, 10%, 15%, and 20% of solar energy) together with the base case data for three MSF units [234].

solar thermal energy. The results show that by increasing the percentage use of solar thermal energy, the amount of carbon dioxide emitted decreases, so that in unit 3, the use of 20% solar energy reduces the carbon dioxide emissions by 13%.

4.7 Conclusions

Solar desalination technology is a growing field of research that has made significant progress in recent years. Increasing drinking water demands, limitations by decarbonization laws, and mitigating the side-effects of global climate change have intensified researchers' efforts to combine desalination processes with renewable energy sources. Meanwhile, solar energy is a reliable and accessible source of energy. Choosing the appropriate technology for solar-powered water treatment is site-specific and dependent on the conditions. In this chapter the second generation of solar water treatment technologies and their environmental effects have been reviewed in detail, and the results could be summarized as below:

- Among direct solar water desalination technologies, solar still technology is one of the low-cost, low-tech, and low-investment methods, which is suitable for remote areas especially in developing countries with low financial supports and access to skilled workers.
- Indirect solar-driven water desalination technologies, including thermal and membrane technologies, are more reliable and technically more mature. In recent years RO technology has received special attention compared with other technologies due to its lower energy demand, lower cost, and available solutions to increase membrane durability.
- In many developing countries, with the problem of drinking water supply for rural and remote areas, solar disinfection can be used as a low-cost, energy-free solution. SODIS method is one of the most common methods of solar disinfection of water in deprived areas due to its simplicity, chemical independence, availability, and cheapness.
- The SEC of conventional solar thermal (MSF and MED) and RO desalination plants exceeds the minimum energy required for desalination. For this reason, desalination is referred to as an energy-consuming process. In particular, SEC demand is higher in thermal processes due to the consumption of both thermal and electrical energy. The SEC is more equivalent to consuming more fossil fuels and thus exacerbating environmental pollution.
- Understanding the sources of energy consumption in desalination technologies and the factors affecting them play a key role in the development of solutions to reduce energy consumption. The development of desalination

technologies in the direction of drinking water supply must be done in a sustainable way, which is fulfilled by combining these technologies with renewable energy sources, such as solar energy.

- Compared with conventional fossil fuel-based desalination plants, the cost of producing water from solar desalination processes is still relatively high due to the high cost of solar equipment. For this reason, the commercialization speed of solar desalination processes is still slow. However, in many cases, the environmental costs of using fossil fuels (including pollutions in the preparation stage, greenhouse gas emissions, and air pollutions) are largely ignored. In addition, the estimated costs of solar desalination plants indicate that they are becoming economically comparable to conventional power plants. However, most solar desalination plants are in developing stages and are limited in the real application. Hence, there are still challenges in the reduction of solar equipment and cost by the development of solar energy technology.
- One of the common techno-economic issues in most desalination technologies is the cost-effective and low environmental impact of wastes (i.e., brines) disposal. Disposal of brines can account for a large portion of the cost of water and has potentially negative effects on the environment. Therefore, in addition to efforts to improve the efficiency and reduce the cost of solar technologies and water treatment processes, the development of new solutions to this issue should be considered in future research studies.

References

[1] B.D. Richter, D. Abell, E. Bacha, K. Brauman, S. Calos, A. Cohn, et al., Tapped out: how can cities secure their water future? Water Policy 15 (3) (2013) 335–363.

[2] J. Schewe, J. Heinke, D. Gerten, I. Haddeland, N.W. Arnell, D.B. Clark, et al., Multimodel assessment of water scarcity under climate change, Proceedings of the National Academy of Sciences 111 (9) (2014) 3245–3250.

[3] M.M. Mekonnen, A.Y. Hoekstra, Four billion people facing severe water scarcity, Science Advances 2 (2) (2016) e1500323.

[4] E. Jones, M. Qadir, M.T. van Vliet, V. Smakhtin, S.-M. Kang, The state of desalination and brine production: a global outlook, Science of the Total Environment 657 (2019) 1343–1356.

[5] M. Qadir, G.C. Jiménez, R.L. Farnum, L.L. Dodson, V. Smakhtin, Fog water collection: challenges beyond technology, Water 10 (4) (2018) 372.

[6] V.G. Gude, Desalination and water reuse to address global water scarcity, Reviews in Environmental Science and Bio/Technology 16 (4) (2017) 591–609.

[7] M.T. van Vliet, M. Flörke, Y. Wada, Quality matters for water scarcity, Nature Geoscience 10 (11) (2017) 800–802.

[8] N.C. Darre, G.S. Toor, Desalination of water: a review, Current Pollution Reports 4 (2) (2018) 104–111.

[9] H. Arafat, Desalination Sustainability: A Technical, Socioeconomic, and Environmental Approach, Elsevier,, 2017.
[10] K.R. Khalilpour, Polygeneration With Polystorage: For Chemical and Energy Hubs, Academic Press, 2018.
[11] D. Harper, Online etymology dictionary (2001). Available from: <http://www.etymonline.com/index.php>, 2015.
[12] H. Cooley, P.H. Gleick, G.H. Wolff, Desalination, with a grain of salt: a California perspective, Pacific Institute for Studies in Development, Environment, and Security, 2006.
[13] E. Delyannis, Historic background of desalination and renewable energies, Solar Energy 75 (5) (2003) 357–366.
[14] S.A. Kalogirou, Seawater desalination using renewable energy sources, Progress in Energy and Combustion Science 31 (3) (2005) 242–281.
[15] H.T. Chua, B. Rahimi, Low Grade Heat Driven Multi-effect Distillation and Desalination, Elsevier, 2017.
[16] P. Simon, Tapped Out: The Coming World Crisis in Water and What We Can Do About It, Welcome Rain Publishers, 1998.
[17] P. Prakash, V. Velmurugan, Parameters influencing the productivity of solar stills – a review, Renewable and Sustainable Energy Reviews 49 (2015) 585–609.
[18] H.T. El-Dessouky, H.M. Ettouney, Fundamentals of Salt Water Desalination, Elsevier, 2002.
[19] M. Telkes, Solar Distiller for Life Rafts, 1945.
[20] O. Buros, The ABCs of Desalting, International Desalination Association, Topsfield, MA, 2000.
[21] S. Loutatidou, O. Musthafa, S. Mavukkandy, S. Chakraborty, A.A. Hassan, Introduction: What Is Sustainable Desalination, Desalination Sustainability, Elsevier, Amsterdam, 2017.
[22] A. Kabeel, T. Arunkumar, D. Denkenberger, R. Sathyamurthy, Performance enhancement of solar still through efficient heat exchange mechanism – a review, Applied Thermal Engineering 114 (2017) 815–836.
[23] K. Srithar, T. Rajaseenivasan, Recent fresh water augmentation techniques in solar still and HDH desalination – a review, Renewable and Sustainable Energy Reviews 82 (2018) 629–644.
[24] G.P. Narayan, M.G.S. John, S.M. Zubair, Thermal design of the humidification dehumidification desalination system: an experimental investigation, International Journal of Heat and Mass Transfer 58 (1–2) (2013) 740–748.
[25] G.P. Narayan, M.H. Sharqawy, J.H. Lienhard, V.S.M. Zubair, Thermodynamic analysis of humidification dehumidification desalination cycles, Desalination and Water Treatment 16 (1–3) (2010) 339–353.
[26] M. Kazemian, A. Behzadmehr, S. Sarvari, Thermodynamic optimization of multi-effect desalination plant using the DoE method, Desalination 257 (1–3) (2010) 195–205.
[27] I.-S. Park, S.-M. Park, J.-S. Ha, Design and application of thermal vapor compressor for multi-effect desalination plant, Desalination 182 (1–3) (2005) 199–208.
[28] H. Baig, M.A. Antar, S.M. Zubair, Performance evaluation of a once-through multi-stage flash distillation system: impact of brine heater fouling, Energy Conversion and Management 52 (2) (2011) 1414–1425.
[29] C. Hanshik, H. Jeong, K.-W. Jeong, S.-H. Choi, Improved productivity of the MSF (multi-stage flashing) desalination plant by increasing the TBT (top brine temperature), Energy 107 (2016) 683–692.
[30] M.A. Jamil, S.M. Zubair, On thermoeconomic analysis of a single-effect mechanical vapor compression desalination system, Desalination 420 (2017) 292–307.
[31] M.A. Jamil, S.M. Zubair, Design and analysis of a forward feed multi-effect mechanical vapor compression desalination system: an exergo-economic approach, Energy 140 (2017) 1107–1120.

[32] S. Zhou, L. Gong, X. Liu, S. Shen, Mathematical modeling and performance analysis for multi-effect evaporation/multi-effect evaporation with thermal vapor compression desalination system, Applied Thermal Engineering 159 (2019) 113759.
[33] H. Jayakody, R. Al-Dadah, S. Mahmoud, Numerical investigation of indirect freeze desalination using an ice maker machine, Energy Conversion and Management 168 (2018) 407–420.
[34] C. Xie, L. Zhang, Y. Liu, Q. Lv, G. Ruan, S.S. Hosseini, A direct contact type ice generator for seawater freezing desalination using LNG cold energy, Desalination 435 (2018) 293–300.
[35] M.N. Khan, C.J. Peters, C.A. Koh, Desalination using gas hydrates: the role of crystal nucleation, growth and separation, Desalination 468 (2019) 114049.
[36] J. Zheng, F. Cheng, Y. Li, X. Lü, M. Yang, Progress and trends in hydrate based desalination (HBD) technology: a review, Chinese Journal of Chemical Engineering 27 (9) (2019) 2037–2043.
[37] J. Kim, K. Park, D.R. Yang, S. Hong, A comprehensive review of energy consumption of seawater reverse osmosis desalination plants, Applied Energy 254 (2019) 113652.
[38] M. Qasim, M. Badrelzaman, N.N. Darwish, N.A. Darwish, N. Hilal, Reverse osmosis desalination: a state-of-the-art review, Desalination 459 (2019) 59–104.
[39] N.A. Qasem, B.A. Qureshi, S.M. Zubair, Improvement in design of electrodialysis desalination plants by considering the Donnan potential, Desalination 441 (2018) 62–76.
[40] G. Gopi, G. Arthanareeswaran, A. Ismail, Perspective of renewable desalination by using membrane distillation, Chemical Engineering Research and Design 144 (2019) 520–537.
[41] A.S. Alsaman, A.A. Askalany, K. Harby, M.S. Ahmed, A state of the art of hybrid adsorption desalination–cooling systems, Renewable and Sustainable Energy Reviews 58 (2016) 692–703.
[42] T. Richter, J. Landsgesell, P. Košovan, C. Holm, On the efficiency of a hydrogel-based desalination cycle, Desalination 414 (2017) 28–34.
[43] C.V. Subban, A.J. Gadgil, Electrically regenerated ion-exchange technology for desalination of low-salinity water sources, Desalination 465 (2019) 38–43.
[44] A.M. Lopez, M. Williams, M. Paiva, D. Demydov, T.D. Do, J.L. Fairey, et al., Potential of electrodialytic techniques in brackish desalination and recovery of industrial process water for reuse, Desalination 409 (2017) 108–114.
[45] P. Tsiakis, L.G. Papageorgiou, Optimal design of an electrodialysis brackish water desalination plant, Desalination 173 (2) (2005) 173–186.
[46] M.A. Abdelkareem, M.E.H. Assad, E.T. Sayed, B. Soudan, Recent progress in the use of renewable energy sources to power water desalination plants, Desalination 435 (2018) 97–113.
[47] A. Giwa, N. Akther, A. Al Housani, S. Haris, S.W. Hasan, Recent advances in humidification dehumidification (HDH) desalination processes: improved designs and productivity, Renewable and Sustainable Energy Reviews 57 (2016) 929–944.
[48] A. Al-Karaghouli, L.L. Kazmerski, Energy consumption and water production cost of conventional and renewable-energy-powered desalination processes, Renewable and Sustainable Energy Reviews 24 (2013) 343–356.
[49] S. Burn, M. Hoang, D. Zarzo, F. Olewniak, E. Campos, B. Bolto, et al., Desalination techniques—a review of the opportunities for desalination in agriculture, Desalination 364 (2015) 2–16.
[50] A.T. Fane, A grand challenge for membrane desalination: more water, less carbon, Desalination 426 (2018) 155–163.
[51] A. Kabeel, M.H. Hamed, Z. Omara, S. Sharshir, Water desalination using a humidification-dehumidification technique—a detailed review, Natural Resources 4 (2013).
[52] F.E. Ahmed, R. Hashaikeh, N. Hilal, Solar powered desalination – technology, energy and future outlook, Desalination 453 (2019) 54–76.

[53] O. Coddington, J. Lean, P. Pilewskie, M. Snow, D. Lindholm, A solar irradiance climate data record, Bulletin of the American Meteorological Society 97 (7) (2016) 1265–1282.

[54] E. Kabir, P. Kumar, S. Kumar, A.A. Adelodun, K.-H. Kim, Solar energy: potential and future prospects, Renewable and Sustainable Energy Reviews 82 (2018) 894–900.

[55] K. Quteishat, M. Abu-Arabi, Available from: http://www.menarec.com/docs/Abu-Arabi.pdf (accessed 28.03.06) Promotion of Solar Desalination in the MENA Region, Middle East Desalination Centre, Muscat, 2006.

[56] A. Panagopoulos, K.-J. Haralambous, Environmental impacts of desalination and brine treatment-challenges and mitigation measures, Marine Pollution Bulletin 161 (2020) 111773.

[57] A. Kasaeian, S. Babaei, M. Jahanpanah, H. Sarrafha, A.S. Alsagri, S. Ghaffarian, et al., Solar humidification-dehumidification desalination systems: a critical review, Energy Conversion and Management 201 (2019) 112129.

[58] S. Gorjian, B. Ghobadian, H. Ebadi, F. Ketabchi, S. Khanmohammadi, Applications of solar PV systems in desalination technologies, Photovoltaic Solar Energy Conversion, Elsevier, 2020, pp. 237–274.

[59] S. Gorjian, B. Ghobadian, T.T. Hashjin, A. Banakar, Experimental performance evaluation of a stand-alone point-focus parabolic solar still, Desalination 352 (2014) 1–17.

[60] T.M. Letcher, Why Solar Energy? A Comprehensive Guide to Solar Energy Systems, Elsevier, 2018, pp. 3–16.

[61] IRENA, Solar PV Costs 2010–2015, 2016.

[62] IRENA, The Power to Change: Solar and Wind Cost Reduction Potential to 2025, 2016.

[63] O.O. Ogbomo, E.H. Amalu, N. Ekere, P. Olagbegi, A review of photovoltaic module technologies for increased performance in tropical climate, Renewable and Sustainable Energy Reviews 75 (2017) 1225–1238.

[64] L. García-Rodríguez, Renewable energy applications in desalination: state of the art, Solar Energy 75 (5) (2003) 381–393.

[65] A. Giwa, A. Yusuf, A. Dindi, H.A. Balogun, Polygeneration in desalination by photovoltaic thermal systems: a comprehensive review, Renewable and Sustainable Energy Reviews 130 (2020) 109946.

[66] V.A. Malviy, R.S. Rana, Comparative Experimental Study of Single Slope Single Basin Solar Still, Advances in Power Generation from Renewable Energy Sources (APGRES), 2019.

[67] V.P. Katekar, S.S. Deshmukh, A review of the use of phase change materials on performance of solar stills, Journal of Energy Storage 30 (2020) 101398.

[68] V.K. Chauhan, S.K. Shukla, J.V. Tirkey, P.K.S. Rathore, A comprehensive review of direct solar desalination techniques and its advancements, Journal of Cleaner Production (2020) 124719.

[69] M.S.S. Abujazar, S. Fatihah, A. Rakmi, M. Shahrom, The effects of design parameters on productivity performance of a solar still for seawater desalination: a review, Desalination 385 (2016) 178–193.

[70] V.P. Katekar, S.S. Deshmukh, A review on research trends in solar still designs for domestic and industrial applications, Journal of Cleaner Production 257 (2020) 120544.

[71] F. Sarhaddi, F.F. Tabrizi, H.A. Zoori, S.A.H.S. Mousavi, Comparative study of two weir type cascade solar stills with and without PCM storage using energy and exergy analysis, Energy Conversion and Management 133 (2017) 97–109.

[72] P.V. Kumar, A. Kumar, O. Prakash, A.K. Kaviti, Solar stills system design: a review, Renewable and Sustainable Energy Reviews 51 (2015) 153–181.

[73] D.D.W. Rufuss, S. Iniyan, L. Suganthi, P. Davies, Solar stills: a comprehensive review of designs, performance and material advances, Renewable and Sustainable Energy Reviews 63 (2016) 464–496.

[74] Y. Zhang, M. Sivakumar, S. Yang, K. Enever, M. Ramezanianpour, Application of solar energy in water treatment processes: a review, Desalination 428 (2018) 116–145.

[75] A. Agrawal, R. Rana, P.K. Srivastava, Heat transfer coefficients and productivity of a single slope single basin solar still in Indian climatic condition: experimental and theoretical comparison, Resource-Efficient Technologies 3 (4) (2017) 466−482.
[76] P. Pal, P. Yadav, R. Dev, D. Singh, Performance analysis of modified basin type double slope multi-wick solar still, Desalination 422 (2017) 68−82.
[77] F.F. Tabrizi, M. Dashtban, H. Moghaddam, Experimental investigation of a weir-type cascade solar still with built-in latent heat thermal energy storage system, Desalination 260 (1−3) (2010) 248−253.
[78] M. Abu-Arabi, Y. Zurigat, H. Al-Hinai, S. Al-Hiddabi, Modeling and performance analysis of a solar desalination unit with double-glass cover cooling, Desalination 143 (2) (2002) 173−182.
[79] H. Tanaka, Experimental study of a basin type solar still with internal and external reflectors in winter, Desalination 249 (1) (2009) 130−134.
[80] A. El-Bahi, D. Inan, A solar still with minimum inclination, coupled to an outside condenser, Desalination 123 (1) (1999) 79−83.
[81] R. Sathyamurthy, H.J. Kennady, P. Nagarajan, A. Ahsan, Factors affecting the performance of triangular pyramid solar still, Desalination 344 (2014) 383−390.
[82] G.M. Cappelletti, An experiment with a plastic solar still, Desalination 142 (3) (2002) 221−227.
[83] A. El-Sebaii, Thermal performance of a triple-basin solar still, Desalination 174 (1) (2005) 23−37.
[84] M. Elashmawy, An experimental investigation of a parabolic concentrator solar tracking system integrated with a tubular solar still, Desalination 411 (2017) 1−8.
[85] T. Arunkumar, R. Jayaprakash, D. Denkenberger, A. Ahsan, M. Okundamiya, H. Tanaka, et al., An experimental study on a hemispherical solar still, Desalination 286 (2012) 342−348.
[86] A. Kabeel, Performance of solar still with a concave wick evaporation surface, Energy 34 (10) (2009) 1504−1509.
[87] N. Srivastava, M. Din, G. Tiwari, Performance evaluation of distillation-cum-greenhouse for a warm and humid climate, Desalination 128 (1) (2000) 67−80.
[88] K. Selvaraj, A. Natarajan, Factors influencing the performance and productivity of solar stills—a review, Desalination 435 (2018) 181−187.
[89] D. Mevada, H. Panchal, K. Kumar Sadasivuni, M. Israr, M. Suresh, S. Dharaskar, H. Thakkar, Effect of fin configuration parameters on performance of solar still: a review, Groundwater for Sustainable Development 10 (2020) 100289.
[90] Z. Omara, M.H. Hamed, A. Kabeel, Performance of finned and corrugated absorbers solar stills under Egyptian conditions, Desalination 277 (1−3) (2011) 281−287.
[91] G. Tiwari, V.B. Rao, Transient performance of a single basin solar still with water flowing over the glass cover, Desalination 49 (3) (1984) 231−241.
[92] Z. Omara, A. Kabeel, A. Abdullah, A review of solar still performance with reflectors, Renewable and Sustainable Energy Reviews 68 (2017) 638−649.
[93] M. Boubekri, A. Chaker, Yield of an improved solar still: numerical approach, Energy Procedia 6 (2011) 610−617.
[94] A. Kuhe, A.O. Edeoja, Distillate yield improvement using a parabolic dish reflector coupled single slope basin solar still with thermal energy storage using beeswax, Global Journal of Research in Engineering (2016).
[95] D. Das, U. Bordoloi, P. Kalita, R.F. Boehm, A.D. Kamble, Solar still distillate enhancement techniques and recent developments, Groundwater for Sustainable Development 10 (2020) 100360.
[96] S.J.P. Gnanaraj, S. Ramachandran, Optimization on performance of single-slope solar still linked solar pond via Taguchi method, Desalination and Water Treatment 80 (2017) 27−40.

[97] K.H. Nayi, K.V. Modi, Pyramid solar still: a comprehensive review, Renewable and Sustainable Energy Reviews 81 (2018) 136–148.
[98] J. Eze, O. Ojike, Comparative evaluation of rectangular and pyramid-shaped solar stills using saline water, International Journal of Physical Sciences 7 (31) (2012) 5202–5208.
[99] R. Sathyamurthy, P. Nagarajan, H. Kennady, T. Ravikumar, V. Paulson, A. Ahsan, Enhancing the heat transfer of triangular pyramid solar still using phase change material as storage material, Frontiers in Heat and Mass Transfer (FHMT) 5 (1) (2014).
[100] T. Rajaseenivasan, K.K. Murugavel, Theoretical and experimental investigation on double basin double slope solar still, Desalination 319 (2013) 25–32.
[101] J. Tang, C. Wang, W. Xie, Y. Xia, T. Yu, Z. Chen, Study on the heat and mass transfer performance of a tubular still enhanced by hydrophilic surface modification, Desalination 469 (2019) 114089.
[102] A. Kabeel, S.W. Sharshir, G.B. Abdelaziz, M. Halim, A. Swidan, Improving performance of tubular solar still by controlling the water depth and cover cooling, Journal of Cleaner Production 233 (2019) 848–856.
[103] R. Sathyamurthy, S. El-Agouz, P. Nagarajan, J. Subramani, T. Arunkumar, D. Mageshbabu, et al., A review of integrating solar collectors to solar still, Renewable and Sustainable Energy Reviews 77 (2017) 1069–1097.
[104] Z.S. Abdel-Rehim, A. Lasheen, Experimental and theoretical study of a solar desalination system located in Cairo, Egypt, Desalination 217 (1–3) (2007) 52–64.
[105] V. Velmurugan, J. Mandlin, B. Stalin, K. Srithar, Augmentation of saline streams in solar stills integrating with a mini solar pond, Desalination 249 (1) (2009) 143–149.
[106] S. Rai, G. Tiwari, Single basin solar still coupled with flat plate collector, Energy Conversion and Management 23 (3) (1983) 145–149.
[107] S. Kumar, A. Dubey, G. Tiwari, A solar still augmented with an evacuated tube collector in forced mode, Desalination 347 (2014) 15–24.
[108] L. Zuo, Y. Zheng, Z. Li, Y. Sha, Solar chimneys integrated with sea water desalination, Desalination 276 (1–3) (2011) 207–213.
[109] M. Gaur, G. Tiwari, Optimization of number of collectors for integrated PV/T hybrid active solar still, Applied Energy 87 (5) (2010) 1763–1772.
[110] A.M. Manokar, D.P. Winston, A. Kabeel, R. Sathyamurthy, Sustainable fresh water and power production by integrating PV panel in inclined solar still, Journal of Cleaner Production 172 (2018) 2711–2719.
[111] S. Abdallah, O. Badran, M.M. Abu-Khader, Performance evaluation of a modified design of a single slope solar still, Desalination 219 (1–3) (2008) 222–230.
[112] H.K.S. Abad, M. Ghiasi, S.J. Mamouri, M. Shafii, A novel integrated solar desalination system with a pulsating heat pipe, Desalination 311 (2013) 206–210.
[113] A. Abdullah, Improving the performance of stepped solar still, Desalination 319 (2013) 60–65.
[114] N. Fathi, P. McDaniel, S.S. Aleyasin, M. Robinson, P. Vorobieff, S. Rodriguez, et al., Efficiency enhancement of solar chimney power plant by use of waste heat from nuclear power plant, Journal of Cleaner Production 180 (2018) 407–416.
[115] S. Hu, D.Y. Leung, J.C. Chan, Impact of the geometry of divergent chimneys on the power output of a solar chimney power plant, Energy 120 (2017) 1–11.
[116] A.M. Manokar, D.P. Winston, A. Kabeel, S. El-Agouz, R. Sathyamurthy, T. Arunkumar, et al., Integrated PV/T solar still—a mini-review, Desalination 435 (2018) 259–267.
[117] H. Al-Hinai, M. Al-Nassri, B. Jubran, Effect of climatic, design and operational parameters on the yield of a simple solar still, Energy Conversion and Management 43 (13) (2002) 1639–1650.
[118] A.S. Nafey, M. Abdelkader, A. Abdelmotalip, A. Mabrouk, Parameters affecting solar still productivity, Energy Conversion and Management 41 (16) (2000) 1797–1809.

[119] H. Singh, G. Tiwari, Monthly performance of passive and active solar stills for different Indian climatic conditions, Desalination 168 (2004) 145–150.
[120] H.E. Fath, M. El-Samanoudy, K. Fahmy, A. Hassabou, Thermal-economic analysis and comparison between pyramid-shaped and single-slope solar still configurations, Desalination 159 (1) (2003) 69–79.
[121] K.K. Murugavel, K.K. Chockalingam, K. Srithar, Progresses in improving the effectiveness of the single basin passive solar still, Desalination 220 (1–3) (2008) 677–686.
[122] A. Ghoneyem, A. Ileri, Software to analyze solar stills and an experimental study on the effects of the cover, Desalination 114 (1) (1997) 37–44.
[123] K. Ranjan, S. Kaushik, N. Panwar, Energy and exergy analysis of passive solar distillation systems, International Journal of Low-Carbon Technologies 11 (2) (2016) 211–221.
[124] I. Sarbu, C. Sebarchievici, A comprehensive review of thermal energy storage, Sustainability 10 (1) (2018) 191.
[125] N. Khan, S. Dilshad, R. Khalid, A.R. Kalair, N. Abas, Review of energy storage and transportation of energy, Energy Storage 1 (3) (2019) e49.
[126] A. Dinker, M. Agarwal, G. Agarwal, Heat storage materials, geometry and applications: a review, Journal of the Energy Institute 90 (1) (2017) 1–11.
[127] Z. Andrássy, Z. Szantho, Thermal behaviour of materials in interrupted phase change, Journal of Thermal Analysis and Calorimetry 138 (6) (2019) 3915–3924.
[128] J.P. Da Cunha, P. Eames, Thermal energy storage for low and medium temperature applications using phase change materials—a review, Applied Energy 177 (2016) 227–238.
[129] K. Manickam, P. Mistry, G. Walker, D. Grant, C.E. Buckley, T.D. Humphries, et al., Future perspectives of thermal energy storage with metal hydrides, International Journal of Hydrogen Energy 44 (15) (2019) 7738–7745.
[130] J. Chiew, C. Chin, J. Jia, W. Toh, Z. Gao, Effects of void spaces in a phase change material based thermal energy storage system, Energy Procedia 143 (2017) 559–565.
[131] R. Sathyamurthy, S. El-Agouz, V. Dharmaraj, Experimental analysis of a portable solar still with evaporation and condensation chambers, Desalination 367 (2015) 180–185.
[132] D. Kantesh, Design of solar still using phase changing material as a storage medium, International Journal of Scientific & Engineering Research 3 (12) (2012) 1–6.
[133] P. Sivasamy, A. Devaraju, S. Harikrishnan, Review on heat transfer enhancement of phase change materials (PCMs), Materials Today: Proceedings 5 (6) (2018) 14423–14431.
[134] R. Singh, S. Sadeghi, B. Shabani, Thermal conductivity enhancement of phase change materials for low-temperature thermal energy storage applications, Energies 12 (1) (2019) 75.
[135] G.P. Narayan, M.H. Sharqawy, E.K. Summers, J.H. Lienhard, S.M. Zubair, M.A. Antar, The potential of solar-driven humidification–dehumidification desalination for small-scale decentralized water production, Renewable and Sustainable Energy Reviews 14 (4) (2010) 1187–1201.
[136] D.U. Lawal, N.A. Qasem, Humidification-dehumidification desalination systems driven by thermal-based renewable and low-grade energy sources: a critical review, Renewable and Sustainable Energy Reviews 125 (2020) 109817.
[137] F. Nematollahi, A. Rahimi, T.T. Gheinani, Experimental and theoretical energy and exergy analysis for a solar desalination system, Desalination 317 (2013) 23–31.
[138] M. Ben-Amara, I. Houcine, A.-A. Guizani, M. Maalej, Efficiency investigation of a new-design air solar plate collector used in a humidification–dehumidification desalination process, Renewable Energy 30 (9) (2005) 1309–1327.
[139] C. Yamalı, İ. Solmuş, Theoretical investigation of a humidification-dehumidification desalination system configured by a double-pass flat plate solar air heater, Desalination 205 (1–3) (2007) 163–177.

[140] E.K. Summers, M.A. Antar, Design and optimization of an air heating solar collector with integrated phase change material energy storage for use in humidification–dehumidification desalination, Solar Energy 86 (11) (2012) 3417–3429.

[141] F.A. Al-Sulaiman, M.I. Zubair, M. Atif, P. Gandhidasan, S.A. Al-Dini, M.A. Antar, Humidification dehumidification desalination system using parabolic trough solar air collector, Applied Thermal Engineering 75 (2015) 809–816.

[142] M.A. Antar, M.H. Sharqawy, Experimental investigations on the performance of an air heated humidification–dehumidification desalination system, Desalination and Water Treatment 51 (4–6) (2013) 837–843.

[143] A. Kabeel, M.H. Hamed, Z. Omara, S. Sharshir, Experimental study of a humidification-dehumidification solar technique by natural and forced air circulation, Energy 68 (2014) 218–228.

[144] M.S. Mahmoud, T.E. Farrag, W.A. Mohamed, Experimental and theoretical model for water desalination by humidification-dehumidification (HDH), Procedia Environmental Sciences 17 (2013) 503–512.

[145] R. Tariq, N.A. Sheikh, J. Xamán, A. Bassam, An innovative air saturator for humidification-dehumidification desalination application, Applied energy 228 (2018) 789–807.

[146] M.H. Hamed, A. Kabeel, Z. Omara, S. Sharshir, Mathematical and experimental investigation of a solar humidification–dehumidification desalination unit, Desalination 358 (2015) 9–17.

[147] A. Rafiei, A.S. Alsagri, S. Mahadzir, R. Loni, G. Najafi, A. Kasaeian, Thermal analysis of a hybrid solar desalination system using various shapes of cavity receiver: cubical, cylindrical, and hemispherical, Energy Conversion and Management 198 (2019) 111861.

[148] G. Wu, C. Kutlu, H. Zheng, Y. Su, D. Cui, A study on the maximum gained output ratio of single-effect solar humidification-dehumidification desalination, Solar Energy 157 (2017) 1–9.

[149] A.J. Toth, Modelling and optimisation of multi-stage flash distillation and reverse osmosis for desalination of saline process wastewater sources, Membranes 10 (10) (2020) 265.

[150] T. Rajaseenivasan, K. Srithar, Potential of a dual purpose solar collector on humidification dehumidification desalination system, Desalination 404 (2017) 35–40.

[151] Z. Rahimi-Ahar, M.S. Hatamipour, Y. Ghalavand, Solar assisted modified variable pressure humidification-dehumidification desalination system, Energy Conversion and Management 162 (2018) 321–330.

[152] Z. Rahimi-Ahar, M.S. Hatamipour, Y. Ghalavand, Experimental investigation of a solar vacuum humidification-dehumidification (VHDH) desalination system, Desalination 437 (2018) 73–80.

[153] J.-h Wang, N.-y Gao, Y. Deng, Y.-l Li, Solar power-driven humidification–dehumidification (HDH) process for desalination of brackish water, Desalination 305 (2012) 17–23.

[154] A. Giwa, H. Fath, S.W. Hasan, Humidification–dehumidification desalination process driven by photovoltaic thermal energy recovery (PV-HDH) for small-scale sustainable water and power production, Desalination 377 (2016) 163–171.

[155] F. Alnaimat, J.F. Klausner, Solar diffusion driven desalination for decentralized water production, Desalination 289 (2012) 35–44.

[156] J.F. Klausner, Y. Li, R. Mei, Evaporative heat and mass transfer for the diffusion driven desalination process, Heat and Mass Transfer 42 (6) (2006) 528–536.

[157] F. Alnaimat, J.F. Klausner, R. Mei, Transient dynamic response of solar diffusion driven desalination, Applied Thermal Engineering 51 (1–2) (2013) 520–528.

[158] M.C. Garg, Renewable energy-powered membrane technology: cost analysis and energy consumption, Current Trends and Future Developments on (Bio-) Membranes, Elsevier, 2019, pp. 85–110.

[159] I. Darawsheh, M. Islam, F. Banat, Experimental characterization of a solar powered MSF desalination process performance, Thermal Science and Engineering Progress 10 (2019) 154–162.
[160] M.T. Ali, H.E. Fath, P.R. Armstrong, A comprehensive techno-economical review of indirect solar desalination, Renewable and Sustainable Energy Reviews 15 (8) (2011) 4187–4199.
[161] S. Moustafa, D. Jarrar, H. El-Mansy, Performance of a self-regulating solar multi-stage flash desalination system, Solar Energy 35 (4) (1985) 333–340.
[162] I.B. Askari, M. Ameri, A techno-economic review of multi effect desalination systems integrated with different solar thermal sources, Applied Thermal Engineering (2020) 116323.
[163] H. Sharon, K. Reddy, A review of solar energy driven desalination technologies, Renewable and Sustainable Energy Reviews 41 (2015) 1080–1118.
[164] D.-C. Alarcón-Padilla, L. García-Rodríguez, Application of absorption heat pumps to multi-effect distillation: a case study of solar desalination, Desalination 212 (1–3) (2007) 294–302.
[165] Z. Zhang, Z. Hu, H. Xu, X. Dai, J. Wang, W. Jiao, et al., Theoretical analysis of a solar-powered multi-effect distillation integrated with concentrating photovoltaic/thermal system, Desalination 468 (2019) 114074.
[166] M.T. Mito, X. Ma, H. Albuflasa, P.A. Davies, Reverse osmosis (RO) membrane desalination driven by wind and solar photovoltaic (PV) energy: state of the art and challenges for large-scale implementation, Renewable and Sustainable Energy Reviews 112 (2019) 669–685.
[167] N. Voutchkov, Energy use for membrane seawater desalination—current status and trends, Desalination 431 (2018) 2–14.
[168] R. Dashtpour, S.N. Al-Zubaidy, Energy efficient reverse osmosis desalination process, International Journal of Environmental Science and Development 3 (4) (2012) 339.
[169] E.J. Okampo, N. Nwulu, Optimisation of renewable energy powered reverse osmosis desalination systems: a state-of-the-art review, Renewable and Sustainable Energy Reviews 140 (2021) 110712.
[170] A. Karabelas, C. Koutsou, M. Kostoglou, D. Sioutopoulos, Analysis of specific energy consumption in reverse osmosis desalination processes, Desalination 431 (2018) 15–21.
[171] R.W. Baker, Membrane Technology and Applications, John Wiley & Sons, 2012.
[172] A. Riedinger, C. Hickman, Considerations of energy consumption in desalination by reverse osmosis, Desalination 40 (3) (1982) 259–270.
[173] J. Rheinländer, D. Geyer, Photovoltaic reverse osmosis and electrodialysis, Seawater Desalination, Springer, 2009, pp. 189–211.
[174] S. Shalaby, Reverse osmosis desalination powered by photovoltaic and solar Rankine cycle power systems: a review, Renewable and Sustainable Energy Reviews 73 (2017) 789–797.
[175] A.M. Bilton, R. Wiesman, A. Arif, S.M. Zubair, S. Dubowsky, On the feasibility of community-scale photovoltaic-powered reverse osmosis desalination systems for remote locations, Renewable Energy 36 (12) (2011) 3246–3256.
[176] E.S. Mohamed, G. Papadakis, E. Mathioulakis, V. Belessiotis, A direct coupled photovoltaic seawater reverse osmosis desalination system toward battery based systems—a technical and economical experimental comparative study, Desalination 221 (1–3) (2008) 17–22.
[177] A. Helal, S. Al-Malek, E. Al-Katheeri, Economic feasibility of alternative designs of a PV-RO desalination unit for remote areas in the United Arab Emirates, Desalination 221 (1–3) (2008) 1–16.
[178] B.S. Richards, A.I. Schäfer, Design considerations for a solar-powered desalination system for remote communities in Australia, Desalination 144 (1–3) (2002) 193–199.

[179] D.B. Riffel, P.C. Carvalho, Small-scale photovoltaic-powered reverse osmosis plant without batteries: design and simulation, Desalination 247 (1–3) (2009) 378–389.
[180] M. Thomson, D. Infield, A photovoltaic-powered seawater reverse-osmosis system without batteries, Desalination 153 (1–3) (2003) 1–8.
[181] S. Kumarasamy, S. Narasimhan, S. Narasimhan, Optimal operation of battery-less solar powered reverse osmosis plant for desalination, Desalination 375 (2015) 89–99.
[182] F.M. Zaihidee, S. Mekhilef, M. Seyedmahmoudian, B. Horan, Dust as an unalterable deteriorative factor affecting PV panel's efficiency: why and how, Renewable and Sustainable Energy Reviews 65 (2016) 1267–1278.
[183] L.C. Kelley, S. Dubowsky, Thermal control to maximize photovoltaic powered reverse osmosis desalination systems productivity, Desalination 314 (2013) 10–19.
[184] H.D. Raval, S. Maiti, Ultra-low energy reverse osmosis with thermal energy recovery from photovoltaic panel cooling and TFC RO membrane modification, Desalination and Water Treatment 57 (10) (2016) 4303–4312.
[185] R. Singh, Hybrid Membrane Systems for Water Purification: Technology, Systems Design and Operations, Elsevier, 2006.
[186] J. Carta, J. Gonzalez, V. Subiela, Operational analysis of an innovative wind powered reverse osmosis system installed in the Canary Islands, Solar Energy 75 (2) (2003) 153–168.
[187] P. Cabrera, J.A. Carta, J. Gonzalez, G. Melian, Wind-driven SWRO desalination prototype with and without batteries: a performance simulation using machine learning models, Desalination 435 (2018) 77–96.
[188] A.B. Chaabene, K. Ouelhazi, A. Sellami, Novel design and modeling of a photovoltaic hydro electromagnetic reverse osmosis (PV-HEMRO) desalination system, Journal of Desalination and Water Treatment 66 (2017) 36–41.
[189] S.A. Urrea, F.D. Reyes, B.P. Suárez, A. Juan, Technical review, evaluation and efficiency of energy recovery devices installed in the Canary Islands desalination plants, Desalination 450 (2019) 54–63.
[190] R.L. Stover, Seawater reverse osmosis with isobaric energy recovery devices, Desalination 203 (1–3) (2007) 168–175.
[191] E. Dimitriou, E.S. Mohamed, C. Karavas, G. Papadakis, Experimental comparison of the performance of two reverse osmosis desalination units equipped with different energy recovery devices, Desalination and Water Treatment 55 (11) (2015) 3019–3026.
[192] M.A. Jones, Systems Modeling and Economic Analysis of Photovoltaic (PV) Powered Water Pumping Brackish Water Desalination for Agriculture, 2015.
[193] R.G. Romero, A PV Powered BWRO Desalination Installation with Energy Recovery and Hydro-Pneumatic Energy Storage, 2013.
[194] K. Paulsen, F. Hensel, Introduction of a new energy recovery system—optimized for the combination with renewable energy, Desalination 184 (1–3) (2005) 211–215.
[195] A.M. Delgado-Torres, L. García-Rodríguez, B. Peñate, A. Juan, G. Melián, Water Desalination by Solar-Powered RO Systems, Current Trends and Future Developments on (Bio-)Membranes, Elsevier, 2019, pp. 45–84.
[196] C. Li, Y. Goswami, E. Stefanakos, Solar assisted sea water desalination: a review, Renewable and Sustainable Energy Reviews 19 (2013) 136–163.
[197] F. Banat, H. Qiblawey, Q. Al-Nasser, Economic evaluation of a small RO unit powered by PV installed in the village of Hartha, Jordan, Desalination and Water Treatment 3 (1–3) (2009) 169–174.
[198] E. Gençer, R. Agrawal, Synthesis of efficient solar thermal power cycles for baseload power supply, Energy Conversion and Management 133 (2017) 486–497.
[199] D. Manolakos, E.S. Mohamed, I. Karagiannis, G. Papadakis, Technical and economic comparison between PV-RO system and RO-Solar Rankine system. Case study: Thirasia Island, Desalination 221 (1–3) (2008) 37–46.

[200] V.R. Patil, V.I. Biradar, R. Shreyas, P. Garg, M.S. Orosz, N. Thirumalai, Techno-economic comparison of solar organic Rankine cycle (ORC) and photovoltaic (PV) systems with energy storage, Renewable Energy 113 (2017) 1250–1260.
[201] N. Pichel, M. Vivar, M. Fuentes, The problem of drinking water access: a review of disinfection technologies with an emphasis on solar treatment methods, Chemosphere 218 (2019) 1014–1030.
[202] V.G. Gude, Energy and water autarky of wastewater treatment and power generation systems, Renewable and Sustainable Energy Reviews 45 (2015) 52–68.
[203] A.M.W. Downes, T.P. Blunt, Researches on the effect of light upon bacteria and other organisms, Proceedings of the Royal Society of London 26 (1877) 488–500. Available from: https://doi.org/10.1098/rspl.1877.0068.
[204] K.G. McGuigan, R.M. Conroy, H.-J. Mosler, M. du Preez, E. Ubomba-Jaswa, P. Fernandez-Ibanez, Solar water disinfection (SODIS): a review from bench-top to roof-top, Journal of Hazardous Materials 235 (2012) 29–46.
[205] P. Kalt, C. Birzer, H. Evans, A. Liew, M. Padovan, M. Watchman, A solar disinfection water treatment system for remote communities, Procedia Engineering 78 (2014) 250–258.
[206] A. Acra, Y. Karahagopian, Z. Raffoul, R. Dajani, Disinfection of oral rehydration solutions by sunlight, Lancet (London, England) 2 (8206) (1980) 1257–1258.
[207] K. McGuigan, T.M. Joyce, R.M. Conroy, J. Gillespie, M. Elmore-Meegan, Solar disinfection of drinking water contained in transparent plastic bottles: characterizing the bacterial inactivation process, Journal of Applied Microbiology 84 (6) (1998) 1138–1148.
[208] P. Dunlop, M. Ciavola, L. Rizzo, J. Byrne, Inactivation and injury assessment of Escherichia coli during solar and photocatalytic disinfection in LDPE bags, Chemosphere 85 (7) (2011) 1160–1166.
[209] C. Navntoft, E. Ubomba-Jaswa, K. McGuigan, P. Fernández-Ibáñez, Effectiveness of solar disinfection using batch reactors with non-imaging aluminium reflectors under real conditions: natural well-water and solar light, Journal of Photochemistry and Photobiology B: Biology 93 (3) (2008) 155–161.
[210] J.A. Duffie, W.A. Beckman, Solar Engineering of Thermal Processes, 2006.
[211] A. Martín-Domínguez, M.T. Alarcón-Herrera, I.R. Martín-Domínguez, A. González-Herrera, Efficiency in the disinfection of water for human consumption in rural communities using solar radiation, Solar Energy 78 (1) (2005) 31–40.
[212] E. Ubomba-Jaswa, P. Fernández-Ibáñez, C. Navntoft, M.I. Polo-López, K.G. McGuigan, Investigating the microbial inactivation efficiency of a 25 L batch solar disinfection (SODIS) reactor enhanced with a compound parabolic collector (CPC) for household use, Journal of Chemical Technology & Biotechnology 85 (8) (2010) 1028–1037.
[213] M.N. Chong, B. Jin, C.W. Chow, C. Saint, Recent developments in photocatalytic water treatment technology: a review, Water Research 44 (10) (2010) 2997–3027.
[214] S. Ahmed, M. Rasul, W.N. Martens, R. Brown, M. Hashib, Heterogeneous photocatalytic degradation of phenols in wastewater: a review on current status and developments, Desalination 261 (1–2) (2010) 3–18.
[215] S. Malato, P. Fernández-Ibáñez, M.I. Maldonado, J. Blanco, W. Gernjak, Decontamination and disinfection of water by solar photocatalysis: recent overview and trends, Catalysis Today 147 (1) (2009) 1–59.
[216] K. O'Dowd, S.C. Pillai, Photo-Fenton disinfection at near neutral pH: process, parameter optimization and recent advances, Journal of Environmental Chemical Engineering (2020) 104063.
[217] T.W. Ng, T. An, G. Li, W.K. Ho, H.Y. Yip, H. Zhao, et al., The role and synergistic effect of the light irradiation and H_2O_2 in photocatalytic inactivation of *Escherichia coli*, Journal of Photochemistry and Photobiology B: Biology 149 (2015) 164–171.

[218] E. Ortega-Gómez, M.B. Martín, B.E. García, J.S. Pérez, P.F. Ibáñez, Wastewater disinfection by neutral pH photo-Fenton: the role of solar radiation intensity, Applied Catalysis B: Environmental 181 (2016) 1–6.
[219] P. Soriano-Molina, J.G. Sánchez, S. Malato, P. Plaza-Bolaños, A. Agüera, J.S. Pérez, On the design and operation of solar photo-Fenton open reactors for the removal of contaminants of emerging concern from WWTP effluents at neutral pH, Applied Catalysis B: Environmental 256 (2019) 117801.
[220] N. Pichel, M. Vivar, M. Fuentes, Results from a first optimization study of a photovoltaic and solar disinfection system (SOLWAT) for simultaneous energy generation and water purification, Energy Conversion and Management 176 (2018) 30–38.
[221] N. Pichel, M. Vivar, M. Fuentes, Performance study of a hybrid photovoltaic and solar water disinfection system considering climatic variations over a year, Energy Conversion and Management 144 (2017) 312–321.
[222] Y. Wang, Y. Jin, Q. Huang, L. Zhu, M. Vivar, L. Qin, et al., Photovoltaic and disinfection performance study of a hybrid photovoltaic-solar water disinfection system, Energy 106 (2016) 757–764.
[223] M. Vivar, I. Skryabin, V. Everett, A. Blakers, A concept for a hybrid solar water purification and photovoltaic system, Solar Energy Materialsand Solar Cells 94 (10) (2010) 1772–1782.
[224] S. Lattemann, T. Höpner, Environmental impact and impact assessment of seawater desalination, Desalination 220 (1–3) (2008) 1–15.
[225] I. Sola, Y. Fernández-Torquemada, A. Forcada, C. Valle, Y. del Pilar-Ruso, J.M. González-Correa, et al., Sustainable desalination: long-term monitoring of brine discharge in the marine environment, Marine Pollution Bulletin 161 (2020) 111813.
[226] A. Panagopoulos, K.-J. Haralambous, M. Loizidou, Desalination brine disposal methods and treatment technologies—a review, Science of the Total Environment 693 (2019) 133545.
[227] E. Gacia, O. Invers, M. Manzanera, E. Ballesteros, J. Romero, Impact of the brine from a desalination plant on a shallow seagrass (*Posidonia oceanica*) meadow, Estuarine, Coastal and Shelf Science 72 (4) (2007) 579–590.
[228] R. Einav, K. Harussi, D. Perry, The footprint of the desalination processes on the environment, Desalination 152 (1–3) (2003) 141–154.
[229] H. Li, A. Shi, M. Li, X. Zhang, Effect of pH, temperature, dissolved oxygen, and flow rate of overlying water on heavy metals release from storm sewer sediments, Journal of Chemistry 2013 (2013) 434012.
[230] A. Panagopoulos, M. Loizidou, K.-J. Haralambous, Stainless steel in thermal desalination and brine treatment: current status and prospects, Metals and Materials International 26 (10) (2020) 1463–1482.
[231] J. Kucera, Reverse Osmosis: Industrial Processes and Applications, John Wiley & Sons, 2015.
[232] T. Mezher, H. Fath, Z. Abbas, A. Khaled, Techno-economic assessment and environmental impacts of desalination technologies, Desalination 266 (1–3) (2011) 263–273.
[233] K. Elsaid, E.T. Sayed, M.A. Abdelkareem, A. Baroutaji, A. Olabi, Environmental impact of desalination processes: mitigation and control strategies, Science of the Total Environment 740 (2020) 140125.
[234] M. Mannan, M. Alhaj, A.N. Mabrouk, S.G. Al-Ghamdi, Examining the life-cycle environmental impacts of desalination: a case study in the State of Qatar, Desalination 452 (2019) 238–246.

CHAPTER 5

Solar-driven water treatment: Generation III - Low technology readiness

Amr Omar[1], Qiyuan Li[2], David Saldivia[3], Amir Nashed[4] and Bac Van Dang[1]
[1]School of Mechanical and Manufacturing Engineering, University of New South Wales, Sydney, NSW, Australia
[2]School of Chemical Engineering, University of New South Wales, Sydney, NSW, Australia
[3]School of Photovoltaic and Renewable Energy, University of New South Wales, Sydney, NSW, Australia
[4]KarmSolar Group, Cairo, Egypt

5.1 Introduction

5.1.1 Generation III technology overview and background

The enduring field of solar thermal-driven desalination has recently been refreshed with the emergence of new membrane-based water treatment technologies, such as membrane distillation (MD), electrodialysis (ED), and capacitive deionization (CDI). These new desalination technologies have been shown to achieve high production rates in compact modules while using low-grade energy sources.

Since low-grade heat can be easily obtained from the sun and converted to low-grade thermal energy or high-grade electrical energy, the above-mentioned technologies can become solarized. Besides, harvesting solar energy to drive these new compact water treatment technologies can help small communities to access affordable clean water and achieve the United Nations sustainable development goals.

Currently, the dominating desalination technology by capacity is reverse osmosis (Fig. 5.1), with a share of 64% followed by the multi-stage flash and multi-effect distillation technologies, each has a share of 23% and 8%, respectively. The technology with the lowest global desalination capacity is ED (4%), whereas MD and CDI do not even come into the picture (<1% share in the market) since they are still in the laboratory development stage [i.e., low technology readiness–level (TRL) or Generation III technologies].

Solar-Driven Water Treatment
DOI: https://doi.org/10.1016/B978-0-323-90991-4.00001-3

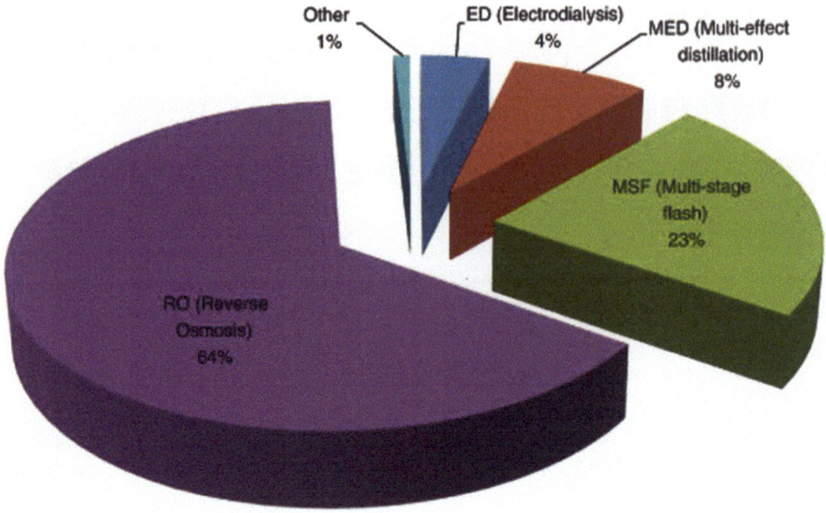

Figure 5.1 Global desalination technologies share by capacity [1]. *Reproduced with permission from F.A. AlMarzooqi, A.A. Al Ghaferi, I. Saadat, N. Hilal, Application of capacitive deionisation in water desalination: a review, Desalination 342 (2014) 3–15.*

TRL is an important aspect to understand if these technologies are ready to be market deployed. There is no doubt that photovoltaic (PV)-driven reverse osmosis (RO) can be considered on top of the TRL; however, these new membrane-based technologies can benefit from the low-grade heat obtained from the sun. Thus, solar-driven MD, ED, and CDI have shown a steady volume of research efforts since the late 19th century with the invention of new membrane manufacturing methods. Nevertheless, these aforementioned technologies coupled with solar energy are still under optimization at theoretical and laboratory levels; thus, they are yet low TRL technologies—Generation III Technologies.

5.2 Solar-driven membrane distillation

5.2.1 Historical background

MD is one of the (emerging) *thermal-driven membrane-based* separation processes. MD process mimics biological membranes by separating salts and other impurities from a saline solution using the partial vapor pressure difference across the membrane.

The first MD was patented by Bodell in 1963 before the first published MD paper in the international journal *Industrial & Engineering*

Chemistry Process Design Development in 1967. In the same year, a second patent was filed by Weyl, claiming an improved feed recovery method. A year later, in 1968, a third patent was filed by Bodell (again) as a continuation of his first patent. This patent provided novel methods to desalt seawater economically. After this, interest in the MD process has faded quickly due to the MD's limited production compared to other membrane-based technologies, such as reverse osmosis.

It was only in the 1980s, the academic communities went back to show interest in MD due to the advancements in membrane manufacturing. A fourth patent was filed by Cheng and Wiersma in 1982, proposing to use composite materials for the membranes to promote a hydrophobic layer. At the same time, the Swedish National Development Corp. Developed the plate-and-frame module to ease the stack of flat sheet membranes, which would increase the MD capacity. Enka AG presented the polypropylene (PP) hollow fiber membranes in tubular modules to increase the membrane packing density in 1984. Since then, numerous research studies have been published, increasing the academic interest rather than the industrial interest.

In 1986 at the "Workshop on Membrane Distillation," the terminology for MD was finally defined, where any MD process should have the following characteristics [2]:

- The membranes used should be porous and not wetted by the process solution.
- Only vapor is transported across the membrane pores, and no capillary condensation should take place inside these pores.
- The membranes must be inert and not alter the vapor equilibrium of the process solution's components.
- At least one side of the membrane should be in direct contact with the process solution.
- The main driving force of the distillation operation is the partial pressure gradient in the vapor phase from each solution's component.

MD has been the subject of worldwide academic studies by many researchers. These studies developed the MD technology to be used in different applications, such as pretreatment and posttreatment in desalination, water treatment process, water purification, removal of salt concentrations in biomedical applications, etc. Initially, MD was developed for small-scale applications due to its simple operation and the ability to use low-temperature sources, making it a perfect fit to be coupled with solar technologies.

MD is very suitable for compact solar-driven water treatment units for small to medium production range (i.e., $<10\,m^3$/day). It was in 1985 where the spiral wound membranes were designed, allowing the integration with solar energy. However, it was not until 2003, the MEMDIS project was kicked off by the Fraunhofer Institute for Solar Energy Systems that began developing MD modules power by solar energy. Their first design was able to desalinate seawater and produce drinking water with a production of $\sim 0.1\,m^3$/day. However, with further development of their design, they were able to have a capacity of $2.0\,m^3$/day. These designs will be discussed in more detail later in this chapter.

5.2.2 Membrane distillation design and operation

The MD process consists of three layers: hot feed solution side, hydrophobic, permeable membrane layer, and permeate side. The distillation happens along with the feed-membrane boundary layer, where the permeate vapor diffuses across the membrane pores from the hot feed side to the permeate side, as shown in Fig. 5.2. These membranes are hydrophobic membranes that do not allow the liquid solution to go through the membrane pores due to the hydrophobic surface tension forces.

Thermal boundary layers are created on both sides of the membrane as the feed solution evaporation occurs at the membrane surface. This results in a heat transfer resistance due to the temperature difference between the liquid–vapor interface (i.e., at the membrane surface) and the feed solution bulk. This can be measured using the temperature polarization coefficient (Θ) across the heat transfer boundary layer, as shown:

$$\Theta = \frac{T_{f-m} - T_{p-m}}{T_f - T_p} \tag{5.1}$$

T_{f-m} and T_{p-m} are the feed-membrane and permeate-membrane temperatures, respectively, while T_f and T_p are the bulk feed and bulk permeate temperatures, respectively.

A highly efficient module design has a high-temperature polarization coefficient (i.e., approaches one), indicating low mass transfer resistance. This can be achieved by promoting turbulence/mixing throughout the feed flow using spacers and baffles. For instance, the temperature polarization coefficient is in the range of 0.90–0.97 for spacer-filled channels compared to 0.5–0.76 for channels without spacers [3]. This is in line

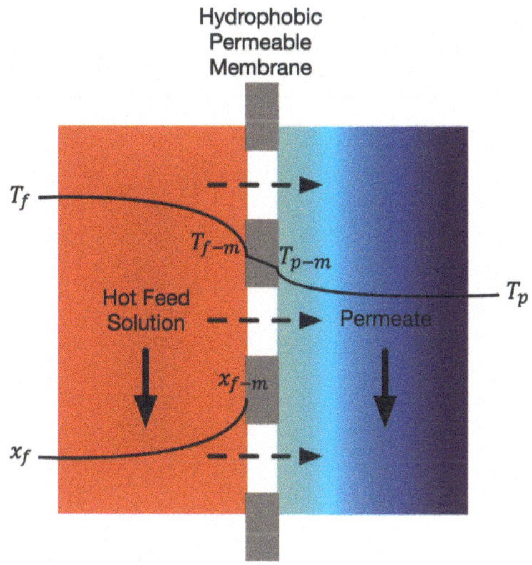

Figure 5.2 Membrane distillation schematic.

Table 5.1 Membrane distillation heat transfer correlations for laminar flow [5].

Correlation	
$Nu = 1.86\left(\frac{Re\,Pr}{L/D}\right)^{1/3}$	Eq. (5.2)
$Nu = 4.36 + \frac{0.036RePr(D/L)}{1 + 0.0011(RePr(D/L))^{0.8}}$	Eq. (5.3)
$Nu = 0.13Re^{0.64}Pr^{0.38}$	Eq. (5.4)
$Nu = 1.95\left(\frac{Re\,Pr}{L/D}\right)^{1/3}$	Eq. (5.5)
$Nu = 0.097Re^{0.73}Pr^{0.13}$	Eq. (5.6)
$Nu = 3.66 + \frac{0.104RePr(D/L)}{1 + 0.0106(RePr(D/L))^{0.8}}$	Eq. (5.7)

with introducing baffles to induce flow mixing, enhancing the heat transfer coefficient by 20%–28% [4].

It is important to accurately measure the heat transfer and mass transfer models of the MD process. There are different heat transfer models derived based on experimental assumptions, which provides a good estimation of the MD's heat transfer coefficients. Some of these correlations are summarized in Tables 5.1 and 5.2.

Table 5.2 Membrane distillation heat transfer correlations for turbulent flow [5].

Correlation	
$Nu = \left(1 + \frac{6D}{L}\right)\left(\frac{(f/8)Re\,Pr}{1.07 + 12.7(f/8)^{1/2}\left(Pr^{2/3} - 1\right)}\right)$	Eq. (5.8)
$Nu = \left(1 + \frac{6D}{L}\right)\left(\frac{(f/8)(Re - 1000)Pr}{1 + 12.7(f/8)^{1/2}\left(Pr^{2/3} - 1\right)}\right)$	Eq. (5.9)
$Nu = 0.023\left(1 + \frac{6D}{L}\right)Re^{0.8}Pr^{1/3}$	Eq. (5.10)
$Nu = 0.036Re^{0.8}Pr^{1/3}\left(\frac{D}{L}\right)^{0.055}$	Eq. (5.11)
$Nu = \left(1 + \frac{6D}{L}\right)\left(\frac{(f/8)Re\,Pr}{1.2 + 13.2(f/8)^{1/2}\left(Pr^{2/3} - 1\right)}\right)$	Eq. (5.12)
$Nu = \left(1 + \frac{6D}{L}\right)\left(\sqrt{\frac{f}{8}}\right)\left[12.48Pr^{2/3} - 7.853Pr^{1/3} + 3.613\ln(Pr) + 5.8 + 2.78\ln\left(\frac{Re\sqrt{f/32}}{45}\right)\right]^{-1}$	Eq. (5.13)
$Nu = 0.027\left(1 + \frac{6D}{L}\right)Re^{0.8}Pr^{1/3}\left(\frac{\mu_{bulk}}{\mu_{sur}}\right)^{0.14}$	Eq. (5.14)

Note: $f = (0.79\ln(Re) - 1.64)^{-2}$ for all correlations.

The heat transfer coefficient is directly proportional to the feed temperature and feed velocity but inversely proportional to the feed salinity. Thus, a higher heat transfer coefficient can be achieved (i.e., higher permeate flux) at high Reynolds number (*Re*) and Prandtl number (*Pr*). Generally, the feed temperature has the most impact on the heat transfer coefficient. Hence, it is important to operate the MD process at a high feed temperature (e.g., 60°C–90°C) without compromising the operating cost. As such, MD is a perfect fit to be coupled with *solar energy* to provide "free" sustainable heat to the feed solution.

The mass transport across the membrane is described by mass transfer models based on the Dusty Gas models, such as the Knudsen model, the Poiseuille model, the Knudsen–Poiseuille transition models, and the molecular diffusion model. Generally, these models suggest that the mass flux (J) can be written as a linear function of the partial vapor pressure difference across the membrane, as shown:

$$J = C(P_2 - P_1) \tag{5.15}$$

where C is the MD coefficient, which is a function of the membrane properties, permeate vapor properties, and feed-membrane temperature.

5.2.3 Membrane materials

The MD performance is directly associated with the overall membrane morphology as it provides the means of the separation process. There are specific membrane characteristics that are needed to aid a good MD performance. For example:

- Excellent chemical resistance to feed solutions;
- Good thermal stability at high temperatures (e.g., 100°C);
- Low thermal conductivity to reduce the heat loss from the feed side to the permeate side;
- High liquid entry pressure[1] and narrow pore size distribution to avoid wetting.

The main requirement of the MD membranes is to be hydrophobic and not to be wetted by the aqueous solutions, while only vapor and noncondensable gases are allowed to cross the membrane boundaries through its pores. Thus, these membranes must be manufactured from polymers with low surface energies and high permeability. In general, permeability is proportional to the membrane pore size (10 nm−1 μm) and porosity (70%−80%) but inversely proportional to the membrane thickness and pore tortuosity. These membranes need to be manufactured as thin as possible to provide the shortest transport distance. In contrast, narrow pore size is important to repel the large liquid particles but allow the small vapor particles (i.e., avoid membrane wetting).

The selection of the membrane materials depends on the aqueous solution and the current application. The choice of the membrane is a compromise between mechanical and thermal stability using thicker membranes but lower permeability or using thinner membranes with higher permeability. In addition, some membrane materials need to be selectively treated to increase hydrophobicity and not react with the aqueous solution while increasing the membrane lifetime. For instance, polytetrafluoroethylene (PTFE) membranes are naturally hydrophobic with good chemical resistance and high thermal stability. However, PTFE membranes are difficult and expensive to manufacture. Thus, polypropylene (PP) and polyvinylidene fluoride (PVDF) membranes are preferred. Table 5.3 presents a list of the commercially available membranes commonly used in MD.

[1] Liquid entry pressure is defined as the minimum transmembrane pressure required for the aqueous feed solution to penetrate the membrane pores.

Table 5.3 Commercially available membranes commonly used in MD [6].

Manufacturer	Membrane trade name	Material	Thickness (μm)	Average pore size (μm)
3M	3M#	PP	<100	[a]
Gore	Gore	PTFE[b]	<50	[a]
		PTFE[b]	400	2.00
Millipore	GVHP	PVDF	110	0.45
	HVHP	PVDF	140	0.22
Enka (Akzo)	Enka	PP	100	0.20
		PP	140	0.10
	Accurel #	PP (Tube)	150	0.43
Gelman Inst. Co.	TF#	PTFE[b]	60	0.20
		PTFE[b]	60	0.45
		PTFE[b]	60	1.00
Hoechst-Celanese	Celgard #	PP	25	0.02
	Celgard X-20	PP (Tube)	25	0.03

[a]Wide range of pore sizes have been given.
[b]PTFE membranes need fabric support that is not included in the given thickness values.

Table 5.4 Membrane arrangements comparison.

Membrane configuration	Packing density (m^2/m^3)	Cleaning difficulty	Fouling resistance	Manufacturing cost
Flat sheet	45–150	Easy	Moderate	High
Tubular	6–120	Great	High	Very high
Spiral wound	150–380	Poor	Low	Moderate
Hollow fiber	150–1500	Poor	Very Low	Low

Membranes can be manufactured in four different arrangements: (1) flat sheet membranes, (2) tubular membranes, (3) spiral wound membranes, and (4) hollow fiber (capillary) membranes. Flat sheet and tubular membranes are commonly used due to their simple operation, but they have the least packing density than spiral wound and hollow fiber membranes. Generally, MD modules must have a high packing density for maximum production while ensuring compactness. Table 5.4 compares the different membrane arrangements in terms of packing density, cleaning difficulty, fouling resistance, and manufacturing cost.

Flat sheet membranes are commonly used in MD as it is attached to the two sides of a rigid plate stacked within a pressurized support framework. These plates contain grooved channels to allow the vapor to flow within, while the aqueous feed solutions are in contact with the outside surface of the membranes. These plates can be restacked next to each other to recycle the vapor energy and transfer it to the feed solution in a multi-effect configuration (i.e., similar to the multi-effect distillation process).

Although tubular membranes are expensive and have a very low packing density, they are the easiest to clean, making them common for pretreatment applications. In this membrane arrangement, the feed enters the tube from one end, where the vapor permeates radially through the membrane, leaving a high saline solution inside the tube as brine. Multiple tubes can be used in series (i.e., a longer membrane-active length) or in parallel (i.e., larger membrane area but lower feed velocity) to increase the permeate flux.

As spiral wound membranes are commonly used in RO modules due to their high packing density, the MD process has also benefited from this type of membrane. The hot feed solution and cold fluid solution (i.e., usually cold water) flow tangentially in a counter-flow arrangement to the membrane surface in an alternate evaporator—cooling channel. The feed solution is introduced from the middle of the module and travels axially along the porous membranes, where the permeate is diffused into the permeate channel. From the other side of the membrane, the cold solution travels axially, condensing the permeate vapor through an impermeable condenser foil glued onto the cooling side of the membrane. The evaporation—condensation process occurs along the membrane length before the concentrated brine leaves the module at the opposite end of the feed. With such a novel membrane configuration, spiral wound membranes possess a high packing density, making them cost-effective. However, due to their complex geometry, these membranes are difficult to clean and susceptible to fouling. In addition, the spiral wound membranes have a high pressure, which increases the pumping costs.

The hollow fiber membranes are similar to tubular membranes but composed of many hair-like fibers, where the feed is in the shell side, and the permeate vapor is diffused into the fibers. This arrangement is very similar to the tube and shell heat exchanger, but with having membranes instead of solid pipes. Compared to other membrane configurations, hollow fiber membranes have the highest packing density, resulting in a significantly higher permeate flux.

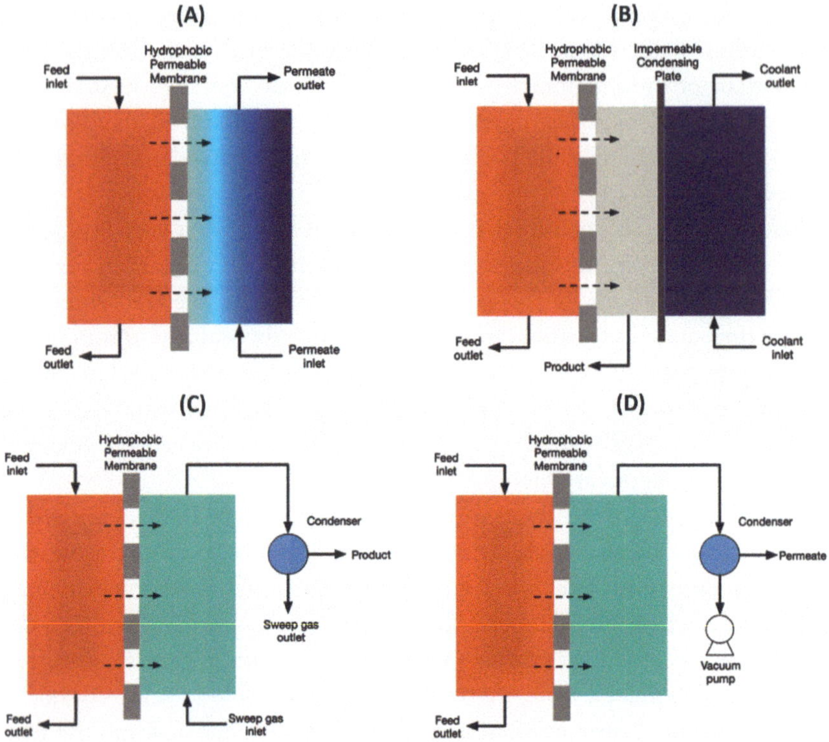

Figure 5.3 Membrane distillation configurations: (A) direct contact membrane distillation; (B) air gap membrane distillation; (C) sweeping gas membrane distillation; (D) vacuum membrane distillation.

5.2.4 Module configuration

There are four common MD configurations: (1) direct contact membrane distillation (DCMD), (2) air gap membrane distillation (AGMD), (3) sweeping gas MD, and (4) vacuum MD. These four MD configurations are shown in Fig. 5.3.

DCMD is the most commonly used, where the hot feed solution and the cold permeate are in direct contact with the membrane. The hot feedwater evaporates and diffuses across the membrane, where the permeate vapor is collected and condensed inside the permeate channel as it gets in contact with a cold fluid (e.g., cold water) flowing on the other side of the membrane. Since the hot feed solution and the permeate vapor are in direct contact with the membrane, conductive heat losses are significant.

The AGMD is similar to DCMD but uses an air gap between the permeate and the cold side. This increases the MD's thermal efficiency by limiting conductive heat loss. However, this extra gap increases the mass transfer resistance due to the larger vapor diffusion length. Thus, AGMD has a relatively lower flux compared to other MD technologies.

Instead of using an air gap to reduce the conductive heat losses, an inert gas can sweep the permeate vapor out of the MD module (i.e., sweeping gas MD) before condensing the vapor outside of the module. This increases the MD's thermal efficiency but requires an external condenser to condense the permeate vapor, and another process to remove the inert gas from the permeate vapor.

Finally, the VMD configuration applies vacuum conditions in the permeate side to pull out the permeate vapor and increase the membrane's partial vapor pressure difference. As such, the VMD has a significantly higher flux compared to other MD technologies. As a trade-off, however, VMD designs have extremely high specific energy consumption.

5.2.5 Solar-driven membrane distillation configurations

In recent years, solar-driven MD systems have become the interest of many researchers as they can represent a sustainable solution in regions with a high correlation between high solar resources and water shortage. As mentioned previously, MD technology requires low-grade thermal energy and can tolerate fluctuating and intermittent operating conditions, making it the best fit for solar integration. Generally, solar thermal collectors [e.g., flat plate solar collectors (FPC) or evacuated tube collectors (ETC)] or direct solar-to-electric collectors (e.g., PV panels) can be used to supply thermal energy to drive the MD process or electric energy to operate the pumps, respectively. As such, there are various coupling techniques to drive a MD process using solar energy.

The most straightforward solar-driven MD design is to collect heat using a solar thermal collector (e.g., FPC or ETC) to heat the feedwater before entering the MD module, as shown in Fig. 5.4. In 1991 the first experimental rig with this configuration was built by Hogan et al. from the University of New South Wales [7]. They used a 3 m^2 FPC coupled with a hollow fiber MD module to produce a 0.05 m^3/day with specific total energy consumption (i.e., including thermal and electrical energy) of 55.6 kWh/m^3. The authors have also reported that their rig has a calculated flux of 17 L/m^2/day (i.e., liters per square meter of collector area per day), comparable to solar multi-stage flash and multi-effect distillation plants.

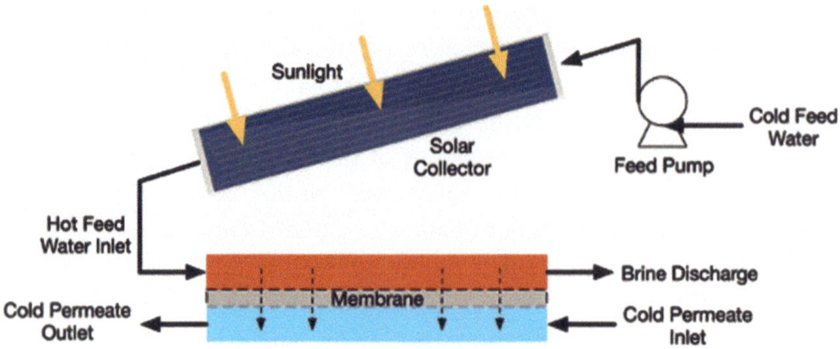

Figure 5.4 Solar thermal-driven membrane distillation configuration.

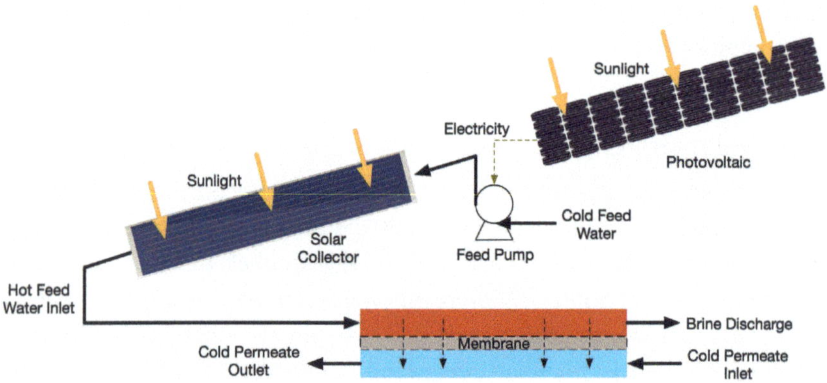

Figure 5.5 Solar thermal + photovoltaic-driven membrane distillation configuration.

In order to make the MD system fully solarized, solar thermal collectors can be used to heat the feedwater, and PV panels can be used to provide the electrical energy needed by the pumps, as shown in Fig. 5.5. This configuration was successful through the SMADES EU-funded project with four compact autonomous solar-powered MD plants with a large production capacity of 100 L/day [8]. Three of these compact systems were installed in Egypt, Jordan, and Morocco, consisting of two FPC (i.e., 5.73 m^2 total solar collector area), one PV module with a DC/AC converter (i.e., 106 W), one spiral wound-based DCMD module (i.e., 10 m^2 membrane area), feed pumps, and feed and distillate storage tanks. While the fourth compact system (i.e., the largest system with a production capacity of 1.0 m^3/day) was installed in the city of Aqaba, Jordan, on the Red Sea coast consisting of 72 m^2 FPC collectors, 14 m^2 PV module (i.e., 1.44 kW), four spiral wound-based DCMD modules (i.e., 40 m^2

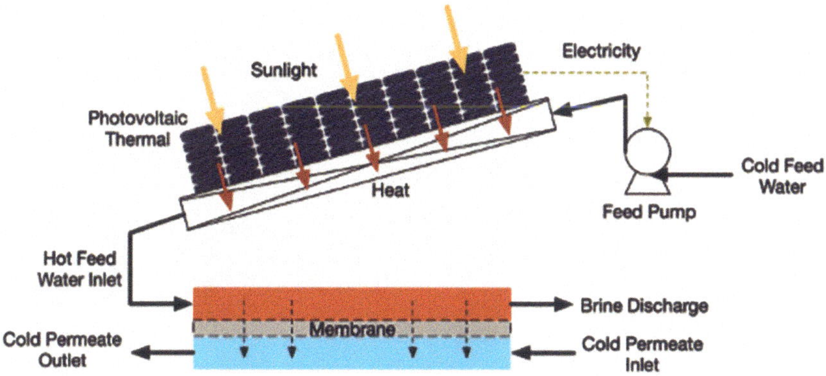

Figure 5.6 Photovoltaic thermal-driven membrane distillation configuration.

total membrane area), a 3 m^3 heat storage capacity, 300 Ah battery storage, feed pumps, and feed and distillate storage tanks. This system produced 600–800 L/day with a specific total energy consumption of 200–250 kWh/m^3 due to the natural fluctuations of solar radiation.

Although the aforementioned solar thermal-driven MD configuration is commonly studied in the literature, it requires two different solar collectors to provide thermal heat and electrical energy separately. Thus, coupling a MD process with a photovoltaic thermal (PVT) would be a better solution to provide the necessary thermal energy (to heat the feedwater) and electrical energy (to run the feed pumps). A schematic of this configuration is shown in Fig. 5.6.

PVT collectors can provide electricity and waste heat as a by-product, providing the electrical power needed for the feed pump and the heat required to start the distillation process, respectively. This can significantly reduce the initial investment cost and the area required by the solar collectors compared to the conventional PV panel and solar thermal collector shown in Fig. 5.5. In addition, since solar radiation can fluctuate throughout the day, which will subsequently affect the electricity and heat delivered by the PVT collector, it will not significantly affect the performance of the MD module. For instance, Murase et al. have shown that an AGMD driven by a PVT collector can provide a constant electrical efficiency greater than 9%, even during low solar radiation intensity [9]. This indicates that the PVT-MD system is suitable for transient operation. Furthermore, it was proven that this system could be highly feasible with an average permeate flux of 7.1 LMH (i.e., liters per square meter per hour of freshwater production) at a low-temperature difference of less than 20°C provided by the

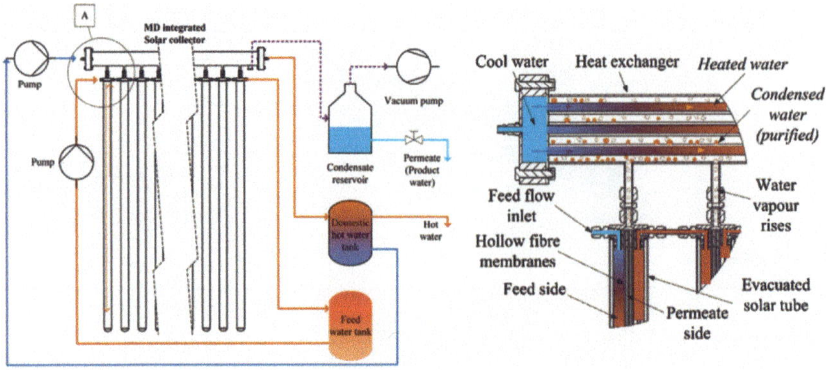

Figure 5.7 Integrated solar-driven membrane distillation [12]. *Reproduced with permission from Q. Li et al., An integrated, solar-driven membrane distillation system for water purification and energy generation, Applied Energy 237 (2019) 534–548.*

PVT collector [10]. This result provides further support that the MD technology can operate with a low-grade heat source that is collected from a PVT collector. Nevertheless, the MD technology can still be coupled with a concentrated PVT collector that can provide high temperature (e.g., 75°C–80°C), which can double the desalination yield and utilize up to 85% of the total available solar radiation [11].

Another novel method to operate an MD process with solar energy is to submerge the membranes inside the solar collector for continuous heating along the membrane length, as shown in Fig. 5.7. Generally, tubular or hollow fiber membranes are packed inside the evacuated tube collector by replacing its heat pipe. Similar to the configuration shown in Fig. 5.4, solar energy is converted into thermal energy, which locally heats the feedwater as the evaporation process starts through the submerged membranes inside the tubes. This configuration opens up a new research area in solar-driven MD.

In this design, the membranes are submerged in the stagnate feedwater, limiting the permeate flux due to the high temperature and concentration polarization coefficients. Nevertheless, to avoid reaching the solar collector's stagnation temperature (i.e., the maximum possible temperature the collector can achieve), internal circulation is applied, which reduces the effect of temperature and concentration polarizations, achieving a higher permeate flux than the conventional submerged MD systems. Another limitation of this design is that it requires a large amount of solar radiation to heat the feedwater before the distillation process takes place. Thus, feed preheating would improve the production rate. This can be achieved by using some of the evacuated tube arrays as a preheater.

In recent years, the literature has been saturated with solar-driven MD technologies. Numerous studies have demonstrated the above-mentioned concepts by experimentally testing different configurations, while others have developed detailed numerical studies to investigate the feasibility of solar-driven MD. A summary of these studies is listed in Table 5.5.

One key opportunity (and challenge) of solar-driven MD is that it can be separately coupled to a solar collector, which eases the operation and maintenance but has low thermal efficiency or integrating the MD technology inside the solar collector, which has a higher thermal efficiency but hard to maintain. Although most of the studies used the former integrating method, there may be some opportunity to have a compact system integrated into buildings. Nevertheless, all of these studies have concluded that the solar-driven MD can alleviate the water stress in rural arid regions due to the focal advantage of the MD process; *it requires a low-temperature source to operate.*

5.3 Solar-driven electrodialysis

5.3.1 Historical background

ED is an electrically driven process, where an electrical field is applied between two electrodes—a cathode and an anode—forcing ions to pass through ion exchanger membranes (IEMs). The IEMs are used inside the ED cell to separate charged species (i.e., ions) from uncharged matters and an aqueous solution. ED has been utilized for over 50 years for the production of potable water from brackish water sources. It also has been widely used in seawater desalination, chemical wastewater treatment, food, medicine, and other fields. For instance, in combination with bipolar membranes or ion-exchange resins, ED has been implemented in industrial chemical processes for the production of acids and alkali recovery from corresponding salts, or in the food and drug industry, and wastewater treatment with the production of high-quality industrial water.

The principle of ED desalination was first presented in 1890 by Maigrot and Sabates as a combination of electrolysis and dialysis for harmful substances removal that interferes in sugar manufacturing [35]. After 1930, ion-exchange materials based on polymer materials were manufactured for the ED process. In 1939, Manegold and Kalauch employed an ion-selective membrane in an ED stack consisting of three compartments. Then, multi-section ED was introduced in 1940 by Meyer and Strauss to reduce energy losses. Later, Juda and McRae of Ionics Inc. in 1950, Wyllie and Patnode of Gulf Research in 1950, and Rohm in 1953 developed the ED process to achieve low electric resistance,

Table 5.5 Summary of the solar-driven membrane distillation available in the literature.

Author	Configuration method	Collector type	Collector area (m^2)	Membrane distillation technology	Membrane area (m^2)	Flux (LMH)	Production (L/day)	Energy consumption (kWh/m^3)	GOR	Unit cost of water (USD/m^3)
Hogan et al. [7]	Solar thermal-driven MD	FPC	1.5–5.0	Hollow fiber membranes, DCMD	0.17	0.71	2.0	55.6	*NA*	3355–4425 USD
		ETC	1.5–5.0							2920–3606 USD
Koschikowski et al. [13]	Solar thermal-driven MD	FPC	6.0	PTFE, Spiral wound membranes, AGMD	8.0	0.42	120–160	140–200	4–6	*NA*
Zwijnenberg et al. [14]	Integrated solar-driven MD	Solar Dew	0.1	Tubular membranes, DCMD	0.32	*NA*	1.5	*NA*	*NA*	*NA*
Banat et al. [8]	Solar thermal + PV-driven MD	FPC, PV	FPC 5.73, PV 0.86	PTFE, Spiral wound membranes, AGMD	10.0	0.5	120	200–300	0.3–0.9	15.0
Rommel et al. [15]	Solar thermal-driven MD	FPC	6.0	PTFE, AGMD	7–10	*NA*	10–30	100–200	*NA*	*NA*
Banat and Jwaied [16]	Solar thermal + PV-driven MD	FPC, PV	FPC 5.73–72, PV 106 W–1440 W	DCMD	10–40	*NA*	100–500	*NA*	*NA*	15–18
Fath et al. [17]	Solar thermal + PV-driven MD	FPC, PV	FPC 2.06, PV *NA*	Spiral wound membranes, AGMD	*NA*	*NA*	64	*NA*	*NA*	*NA*
Wang et al. [18]	Solar thermal-driven MD	FPC	8.0	Hollow fiber membranes, VMD	0.09	32.19	*NA*	*NA*	*NA*	*NA*
Mericq et al. [19]	Solar thermal-driven MD	*NA*	90.0	Flat sheet membranes, VMD	0.354	141.2	617	*NA*	*NA*	*NA*
	Integrated solar-driven MD					157.7				
Saffarini et al. [20]	Solar thermal-driven MD	FPC	*NA*	PTFE, Flat sheet membranes, DCMD	7.0	*NA*	700	*NA*	*NA*	12.7
				PTFE, Flat sheet membranes, AGMD						18.26
				PTFE, Flat sheet membranes, VMD						16.02
Chang et al. [21]	Solar thermal-driven MD	FPC	47.6 89.6	Flat sheet membranes, AGMD	11.5 23.0	*NA*	500 1000	758.87 713.76	*NA*	5.92 5.16

Hughes et al. [22]	Solar thermal + CPVT-driven MD	ETC, CPVT	CPVT aperture area 1 m^2, receiver area 0.002 m^2	AGMD	*NA*	3.4	*NA*	*NA*	*NA*	*NA*
Shim et al. [23]	Solar thermal-driven MD	FPC	4.7	PTFE, Flat sheet membranes, DCMD	0.06	26.5–28.48	*NA*	896–1433	0.44–0.7	*NA*
Duong et al. [24]	Solar thermal-driven MD	FPC	22.6	Flat sheet membranes, DCMD	7.2	0.82	142	*NA*	*NA*	*NA*
Karanikola et al. [25]	Solar thermal-driven MD	FPC, PV	FPC 1050–1500, PV 12–29	DCMD	39–42	*NA*	10,000	1017–1536	*NA*	21–29.7
			FPC 700, PV 50–51	SGMD	32–33			719–729		17.7–18.6
			FPC 650, PV 4	VMD	28			650–660		14.3–14.4
Moore et al. [26]	Solar thermal + PV-driven MD	FPC, PV	FPC 27, PC 4	Hollow fiber membranes, SGMD	5.15	*NA*	240	*NA*	*NA*	84.7
Li et al. [12]	Integrated solar-driven MD	ETC	1.6	PP, Hollow fiber membranes, VMD	0.2	4–10	4.0	*NA*	*NA*	10.0
Zarzoum et al. [27]	Solar thermal + PV-driven MD	FPC, PV	FPC 2.0, PV 1.63	Spiral wound membranes, DCMD	10.0	*NA*	86	*NA*	*NA*	8.55
Deng et al. [28]	Solar thermal + PV-driven MD	ETC, PV	ETC 1.82, PV 1.62	PVDF, Hollow fiber membranes, VMD	0.1	6.26	*NA*	79.87	*NA*	*NA*
Li and Lu [29]	Solar thermal + PV-driven MD	ETC, PV	ETC 4.14, PV 4.92	PVDF, Hollow fiber membranes, SGMD	0.6	*NA*	9.98–23.26	*NA*	0.82	18.34
Rabie et al. [30]	Solar thermal-driven MD	FPC	2.5–5.0	Flat sheet membranes, AGMD	1.0	4.8	1.35–2.7 L/h	*NA*	*NA*	*NA*
Zhao et al. [31]	PVT MD	PVT	1.64	PVDF, DCMD	0.004	4–12	*NA*	*NA*	*NA*	*NA*
Bamasag et al. [32]	Integrated solar-driven MD	ETC	0.1	PP, Hollow fiber membranes, VMD	0.0092	5.9–11.1	*NA*	*NA*	0.44–0.521	*NA*
Kim et al. [33]	Solar thermal-driven MD	FPC	200.0	PVDF, Hollow fiber membranes, VMD	480	*NA*	*NA*	16.6	47.0	0.97
Miladi et al. [34]	Solar thermal-driven MD	FPC	70.0	PVDF, Hollow fiber membranes, VMD	4.0	*NA*	217–598	671–699	0.93–1.01	*NA*

NA, Not Available

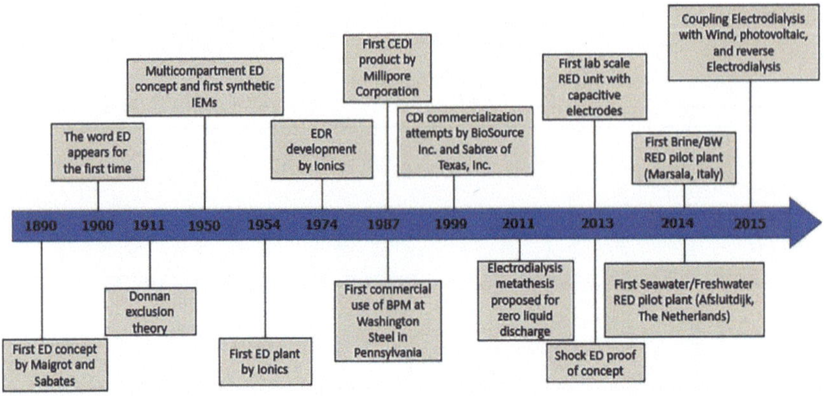

Figure 5.8 Timeline of the most critical developments for ED and related processes [35]. *Reproduced with permission from S. Al-Amshawee, M.Y.B.M. Yunus, A.A.M. Azoddein, D.G. Hassell, I.H. Dakhil, H.A. Hasan, Electrodialysis desalination for water and wastewater: a review, Chemical Engineering Journal 380 (2020) 122231.*

high selectivity, and stability. Therefore, the rapid developments have qualified ED to be scaled up and used in the wastewater industry. It is proposed to use several hundred pairs of the cation-exchange membrane (CEM) and the anion-exchange membrane (AEM) in one ED stack arranged between electrodes for 100–20,000 m^3/day with 1000–5000 mg/L salinity (i.e., total dissolved solids).

The first commercial ED was manufactured in the 1950s for brackish water desalination. While in 1960, Asahi Co. demonstrated for the first time salt production via a monovalent selective membrane. Fig. 5.8 shows ED's developments through time, from the first significant milestones to the recent applications existing so far (e.g., laboratory scale, pilot plant, commercialization attempts). To date, ED is a worldwide recognized technology and leading electro-driven membrane process for salts production and ions separation using IEMs.

5.3.2 Electrodialysis design and operation

An ED module is typically constituted by a train of selective membranes, ending at both sides with an electrode compartment (see Fig. 5.9). The feed flow is split into different streams by two different types of membranes. The basic unit of an ED module is a *cell pair*, consisting of two different selective membranes, anion-exchange membranes (AEMs) and CEMs, and two different channels, a concentrate and a dilute channel. While the AEMs allow exchanging of only negatively charged ions, CEMs only allow cation exchange. When an applied electrical potential is applied in the electrodes, a

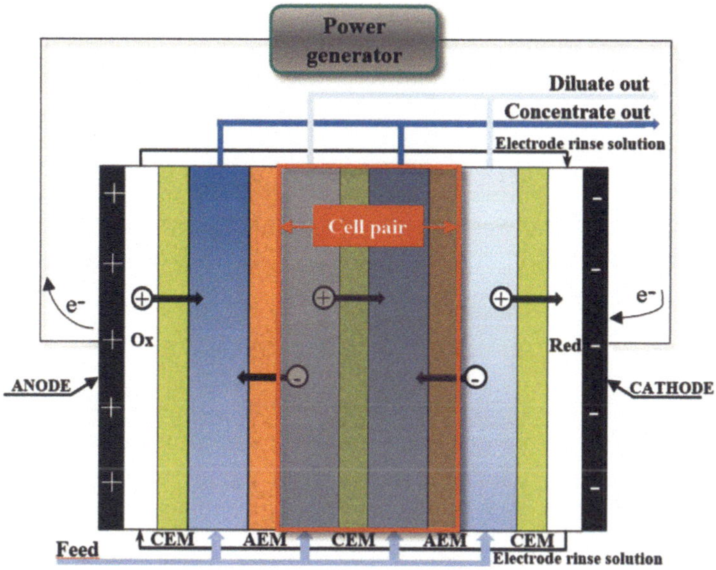

Figure 5.9 A detailed schematic of the ED system [36]. *Reproduced with permission from A. Campione, L. Gurreri, M. Ciofalo, G. Micale, A. Tamburini, A. Cipollina, Electrodialysis for water desalination: a critical assessment of recent developments on process fundamentals, models and applications, Desalination 434 (2018) 121–160.*

reduction reaction happens at the cathode, generating hydroxide (OH −) ions in the cation compartment. To conserve electro-neutrality, cations from the feed compartment will cross the CEM. Similarly, oxidation at the anode generates protons (H +) in the anion compartment (increased acidity in the anode compartment, represented by a decrease in the pH), and anions will be transferred through the AEMs. This initial charge in the electrodes generates a transport of ions from some of the feed streams toward the neighbor streams through the selective membranes. As the cation and anion membranes are intercalated, some of the streams will receive both types of ions, while others will transfer them. As a result of these, some channels increase their salt concentration, while others become diluted. Hence, the feed solution results in two streams, a dilute and a concentrated effluent. For instance, cationic species like NH^{4+}, K^{+}, and Na^{+} pass through CEM towards the cathode compartment. While anionic components like PO^{4-}, SO^{4-}, and CL^{-} are transferred to the anode section through AEM, which provides electrical balance within the matrix.

The *cell pair* unit can be repeated several times, creating a stack of membranes and channels. It could contain from a few cell pairs for laboratory scale-up to over 500 for industrial applications. The active area of a single membrane goes from 10^{-2} m^2 for laboratory units to around 1 m^2 for industrial scale [36].

Inside the ED stack, diffusion (concentration gradient), migration (electric field), and convection (hydrodynamic movement) are the three major forces that transports the negatively and positively charged ions. The flux of an ion *i* can be expressed as:

$$\vec{J_i} = -D_i\vec{\nabla}C_i - z_iFD_iC_i\vec{\nabla}\phi + C_i\vec{u} \tag{5.16}$$

where D_i is diffusion coefficient of *i* in solution (m^2/s), C_i is the concentration of *i* (M), z_i is the valence, *F* is the Faraday's constant, ϕ is the electrical potential, and $\vec{u}$ is the velocity. It is useful to consider only the effects of diffusion and migration forces, which can be expressed mathematically:

$$J_i = -D_i\frac{dC_i}{dx} + \frac{t_iI}{z_iF} \tag{5.17}$$

where D_i is diffusion coefficient of *i* in solution (m^2/s), C_i is the concentration of *i* (M), *I* is the current (A), t_i is the transport number of *i* in membrane, and F is Faraday constant. The ion diffusion coefficient (D_i) can be measured with the help of the Stokes–Einstein equation:

$$D_i = -\frac{RT\lambda_{i,m}}{|z_i|F^2} \tag{5.18}$$

Herein, *R* is the universal gas constant, *T* is the test temperature (K), and $\lambda_{i,m}$ is the limiting ionic conductivity (S cm/mol). The diffusive process works in the direction of the concentration gradient; therefore, the migration process should exceed the diffusion one to produce a diluted product. In consequence, the electrical consumption is proportional to the initial salt concentration and the required final concentration.

ED technology has been employed to process brackish water having salinity higher than 15,000 mg/L, while the industrial scale of ED can process 2500–5000 mg/L with an energy consumption ranging from 0.49 to 0.91 kWh/m^3 [37]. ED is currently not used for industrial-scale seawater desalination as it is considered too energy-intensive. To reduce the energy consumption, a multi-stage ED investigated by Doornbusch et al. showed that 510 mM NaCl (representing the salinity of seawater) could be desalinated at an energy consumption of 3.6 kWh/m^3 using a four-stage ED system [38]. Optimizing the energy consumption showed even lower values down to 2.2 kWh/m^3 for desalination using multi-stage ED, mainly by using an increased flow velocity in later stages and implementing less water-permeable membranes in the two latter stages [39].

ED allows for the separation of anions and metallic cations, possessing the advantages of removing contaminant metals and simultaneously recovering water. Buzzi et al. investigated the possibility of applying ED to treat acid mine drainage and recover water [40]. It was found that ED can remove metallic cations (e.g., Fe, Al, Mn, Pb, Zn, Cu) with greater than 97% of efficiency. Luiz et al. (2017) obtained 96% salt removal from biorefineries wastewater by bench-scale batch operated ED. The wastewater had 380 g COD/L, 72.4 mS/cm conductivity, and 5600–572,000 PtCo color level [35]. Another group of researchers fed ED with cyanide-free copper plating wastewater, and it delivered five to six times concentrate stream with 80% extraction and 90% demineralization [35]. Moreover, Merkela et al. used ED application to recover sodium hydroxide from the high alkaline solution [41]. It can also accomplish a selective separation of monovalent ions (e.g., NO^{3-}, CL^{-}, NH^{4+}, K^{+}, Na^{+}) against multivalent ions (e.g., PO_4^{3-}, SO_4^{2-}, Mg^{2+}, Ca^{2+}) for the production of irrigation water, with the use of monovalent perm-selective IEM.

5.3.3 Membrane and electrode materials

The electrodes can be made of steel 314, titanium, titanium coated with ruthenium oxide, titanium plated with iridium, platinum-plated iridium, titanium coated with titanium and ruthenium oxides ($70RuO_2/30TiO_2$), or graphite [35]. Studies have investigated the quality of various electrodes [e.g., activated carbons, alumina and silica nanocomposites, graphene, mesoporous carbons, carbon aerogel, carbide-derived carbons (CDCs), CNTs, and CNFs].

Vermaas et al. (2013) investigated the use of capacitive electrodes instead of conventional ones. It has an active carbon layer (e.g., coal and wood) to adsorb/desorb ions, and a current collector to release/capture free electrons [35]. Thus it converts an ionic current in the ED stack into an electric current passing through the external electrical circuit without a redox reaction. Moreover, it does not require toxic materials (e.g., Cl_2), but carbon layer saturation is their main drawback. Coal and wood are sent to a pyrolysis process to produce activated carbon with high porosity.

Then, a poly (vinylidene fluoride) binds the activated carbon to maintain its structure. If we compare activated carbon with other electrode materials (e.g., carbon aerogel, carbon cloth, carbon felt, carbon paper), we find activated carbon is more suitable because of its high surface area and pore size distribution.

The oxygen-depolarized cathode is another electrode introduced to minimize the energy consumption of hydrochloric acid electrolysis. It delivers oxygen reduction (i.e., hydroxyl ions generation) instead of hydrogen generation at the cathode and saves 30% of the energy demand. Other studies found the distance between the membrane layer and the cathode, and current density has a significant impact on the electrode performance.

The core of ED technology is the structure and performance of the ion-exchange membrane. IEMs are arranged in series and parallel between two electrodes inside the ED stack to separate two streams and perform a selective mass transport between the feeds. Fig. 5.10 shows the matrix of AEM and CEM with fixed charges on its body and mobile ions in the solution. AEM with positively charged groups allows anions to pass through but preclude cations going through the membrane. CEM with negatively charged groups permits the passage of cations but rejects the passage of anions.

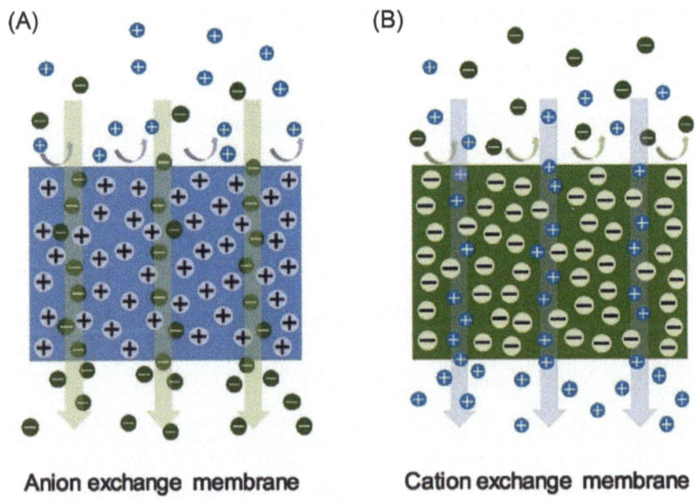

Figure 5.10 Schematic of (A) AEM in which fixed positive charge is uniformly distributed in the polymer matrix and (B) CEM in which fixed negative charge is uniformly distributed in the polymer matrix [42]. *Reproduced with permission from J. Jang, Y. Kang, J.-H. Han, K. Jang, C.-M. Kim, I.S. Kim, Developments and future prospects of reverse electrodialysis for salinity gradient power generation: influence of ion exchange membranes and electrodes, Desalination 491 (2020) 114540.*

IEMs can be defined as thin polymeric films (e.g., polysulfone, polyethylene, polystyrene) with charged groups (i.e., negative or positive charged ions). They are principally classified based on their charged groups into cation exchange membrane (CEM) and anion-exchange membrane (AEM). CEM possesses negative charges [e.g., phosphonic acid (single-bond PO_3H^-), phosphoryl (single-bond PO_3^{2-}), the carboxylic acid (single-bond COO^-), sulfonic acid (single-bond SO^{3-}), $C_6H_4O^-$] covalently bonded to the polymer backbone of the membrane, which blocks negative and coions, and allows positive ions to pass [42]. On the other hand, AEM has positive charges [e.g., quaternary amine (single-bond NR^{3+}), tertiary amine (single-bond NR^2H^+), secondary amine (single-bond NRH^{2+}), ammonium (single-bond NH_3^+), SR_2^+, PR_3^+] and does the opposite work of CEM.

IEMs can be classified into homogeneous and heterogeneous structures depending on their fabrication method and physical structure. In homogeneous membrane, the ion-exchange part is directly provided into the polymer matrix during the fabrication process. Heterogeneous IEMs, on the other hand, have nonuniform charge distribution resulted from their preparation method of blending or melting the charged ion-exchange resin into the uncharged binding polymer, followed by pressing or casting into the membrane. Therefore its mechanical strength is higher than the homogeneous membrane, whereas electrochemical properties such as ion exchange capacity (IEC), perm-selectivity, and areal resistance are comparatively lower than that of the homogeneous membrane. The most desired properties of ion-exchange membranes are high perm-selectivity, low electrical resistance, good mechanical and form stability, high chemical and thermal stability, and low production costs.

The results of a comprehensive characterization of the transport properties of IEMs from three manufactures (Astom, Japan; Shchekinoazot, Russia; and Fujifilm, The Netherlands) are reported in Table 5.6. Homogeneous Neosepta AMX and CMX (Astom), heterogeneous MA-41 and MK-40 (Shchekinoazot), and AEM Type-I, AEM Type-II, AEM Type-X, as well as CEM Type-I, CEM Type-II, and CEM Type-X produced by the electrospinning method (Fujifilm) were studied.

Furthermore, cation- and anion-exchange membranes are classified as strong acid and strong base or weak acid and weak base membranes depending on the dissociation of the charged groups. Strong acid membranes contain sulfonic acid as charged groups. In weak acid membranes,

Table 5.6 Summary of the properties of the ion-exchange membranes in ED system (data provided by the manufacturers) [43].

Materials	Thickness (μm)	Perm-selectivity (%)	Permeability (mL/$m^2 \cdot h \cdot bar$)	IEC (meq/g)	Resistance ($\Omega \cdot cm^2$)
Fujifilm AEM type 10	146	96.0	8	2.85	1.29
Fujifilm CEM type 10	155	97.6	8	2.90	2.02
Fumatech AEM FAB-PK-130	130	>95	*NA*	0.8	(<4)

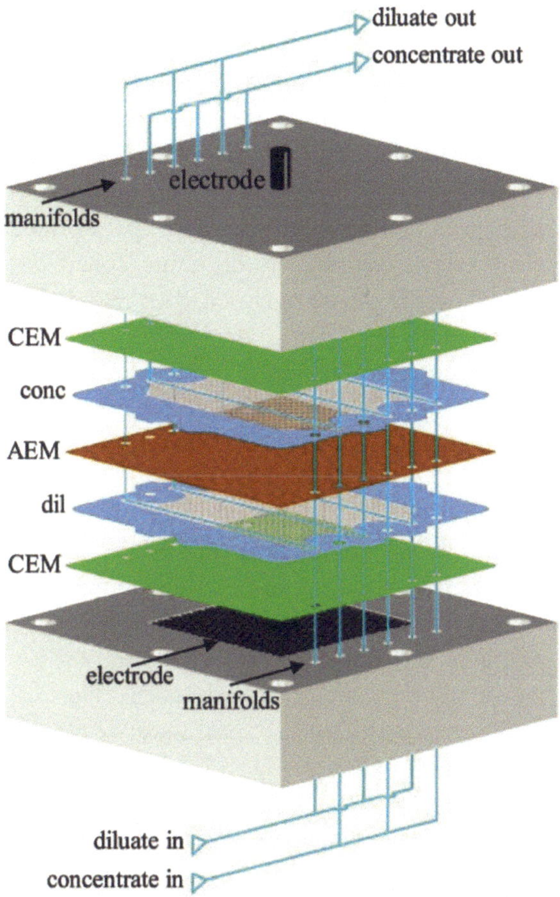

Figure 5.11 ED stack in the sheet flow arrangement of a laboratory-scale unit with net spacers. *Reproduced with permission from A. Campione, L. Gurreri, M. Ciofalo, G. Micale, A. Tamburini, A. Cipollina, Electrodialysis for water desalination: a critical assessment of recent developments on process fundamentals, models and applications, Desalination 434 (2018) 121–160.*

the carboxylic acid is the charge-carrying group. Quaternary and tertiary amines are the fixed charges in strong and weak base ion-exchange membranes, respectively.

5.3.4 Module configurations

An ED unit composed of several cell pairs is usually called an ED stack or stage. The basic components of an ED stack are depicted in Fig. 5.11. In conventional electrodialyzers with the classical plate-and-frame equipment (see Fig. 5.11), net spacers are located between membranes to create the feed channels, maintaining a fixed distance between membranes, and providing mixing promotion in the fluid. The intermembrane distance in ED stacks typically ranges from ~0.3 to ~2 mm. Gaskets along the channels' perimeter are located to seal the channels and guide the fluid through them. Holes in spacers and membranes create special ducts for the two hydraulic circuits acting as manifolds to distribute/collect the solutions to/from the channels. The ED device is closed with two endplates and compressed by bolts and nuts [36].

The electrode compartments contain the two electrodes and the electrodic solution to convert the ionic current into electric current. NaCl has been highly used in the past; however, to avoid the production of Cl_2, the use of salts with SO_4^{2-} ion formation is preferred.

Several stacks of ED units can be connected in series or parallel to increase the production or the salt removal process. In addition, several pairs of electrodes can be arranged to meet the required electricity consumption. Therefore the plant configuration is defined by the number of hydraulic stages (liquid flows in series or parallel) and the number of electrical stages (the number of anode/cathode pairs). Fig. 5.12 shows four hydraulic stages in series with two electrical stages in parallel, that is, one cathode is paired with two anodes. Fig. 5.13 shows four hydraulic stages, each with its electrical stage.

The number of cell pairs can also be distributed between stages to vary the flow velocity, the membrane surface and/or the type of membrane.

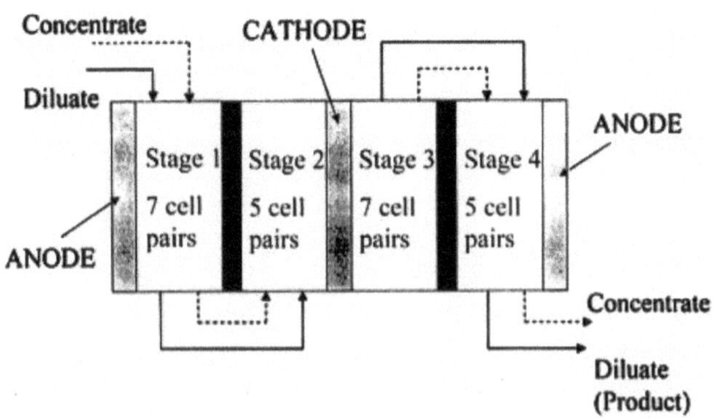

Figure 5.12 Example of multi-stage ED with four hydraulic stage and two electrical stages [44]. *Reproduced with permission from H.M.N. AlMadani, Water desalination by solar powered electrodialysis process, Renewable Energy 28 (12) (2003) 1915–1924.*

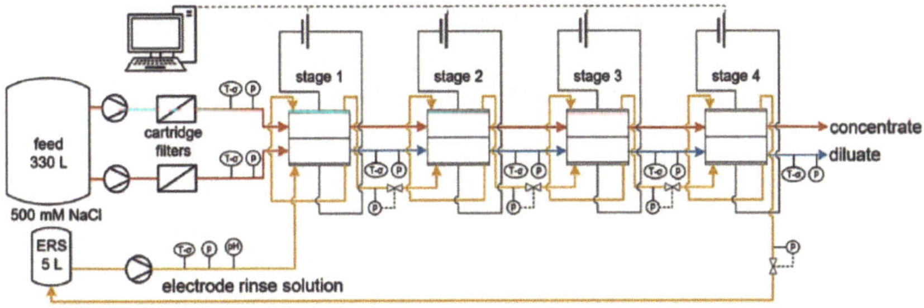

Figure 5.13 Scheme of multi-stage ED system with four stacks hydraulically connected in series. *Reproduced with permission from G.J. Doornbusch, M. Bel, M. Tedesco, J.W. Post, Z. Borneman, K. Nijmeijer, Effect of membrane area and membrane properties in multi-stage electrodialysis on seawater desalination performance, Journal of Membrane Science 611 (2020) 118303.*

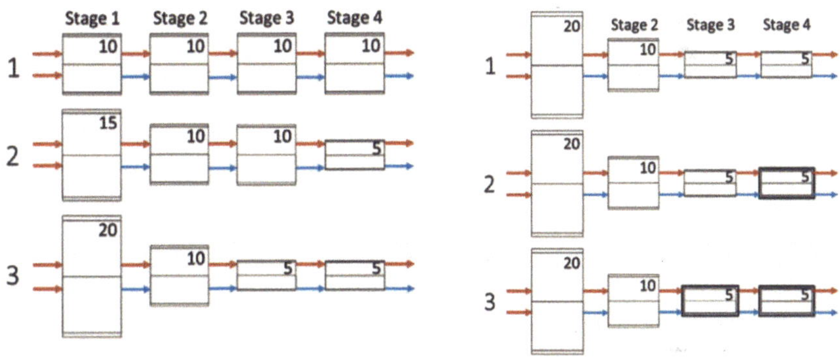

Figure 5.14 Examples of different configurations for a four-stage ED. Left: different ways to arrange 40 cell pairs. Right: the optimized arrange from left, with different type of membranes in each part. *Reproduced with permission from G.J. Doornbusch, M. Bel, M. Tedesco, J.W. Post, Z. Borneman, K. Nijmeijer, Effect of membrane area and membrane properties in multi-stage electrodialysis on seawater desalination performance, Journal of Membrane Science 611 (2020) 118303.*

The reduction in the number of cells increases the velocity flows and reduces the membrane surface, while the type of membrane allows higher or lower permeability. Fig. 5.14 shows the results from an optimization process developed by Doornbusch et al. [39] for the setup shown in Fig. 5.13. They varied the number of cell pairs, the membrane area, and the type of membrane (between two options) for each stage. The optimized design (showed in Fig. 5.14 right) allows to reduce the energy consumption in high salinity water sources, such as seawater.

5.3.5 Solar photovoltaic-driven electrodialysis configurations

Since the ED process requires a direct current (DC) for operation, it makes an excellent fit to be integrated with PV arrays without an inverter. Generally, a PV system consists of several stacks of PV modules, batteries, a regulator or controller, an inverter (DC−AC), and loads. The PV modules can be connected in series and/or parallel, while the solar radiation incident on the surface of the PV array is transformed into electricity (DC). The generated electric charge is then sent to the batteries, protected by the regulator from overcharging or excessive discharge. The batteries store energy that can be used as electrical backup when the energy demand is high, or the solar hours are minimal. In this case, the inverter transforms the DC into alternating current for those devices that work with the latter. These PV systems can be used in remote sites for the self-sufficiency of electrical power reliably and autonomously.

ED powered by PV solar energy represents one of the most promising configurations for solar desalination of brackish water [37]. An ED unit consists of the following essential components: pretreatment system, membrane stack, low-pressure circulation pump, power supply for DC (rectifier or PV system), and posttreatment system. In a typical photovoltaic electrodialysis (PV-ED) system, electrodes are connected to an outside PV source in a saltwater container, removing water salinity as water passes through the ion-selective membranes positioned between the two electrodes, as shown in Fig. 5.15.

The integration of ED with solar PV is one of the very promising configurations that have been recently developed for brackish water desalination. The plant installed in Fukue City, Japan, in 1990 is the largest solar-powered ED plant with average distillate production of 200 m^3/day. The plant was operative for two years and consisted of a 65 kW_p PV array with 1200 Ah storage [46]. The distillate production ranged from 130 to 370 m^3/day with energy consumption of 0.6−1 kWh/m^3. The PV plant fed two subsystems, an electrodialyzer (200 V DC, with batteries) and a deep well pump (200 V AC), making this system completely autonomous.

Kuroda et al. [47] presented a novel PV-ED system for seawater, utilizing lead batteries. The system works in two distinct stages. During sun hours, the system provides enough voltage to work with seawater, and the output is stored as partially desalinated water (PDW). Later, during off-sun hours or with lower solar radiation, the system is powered by

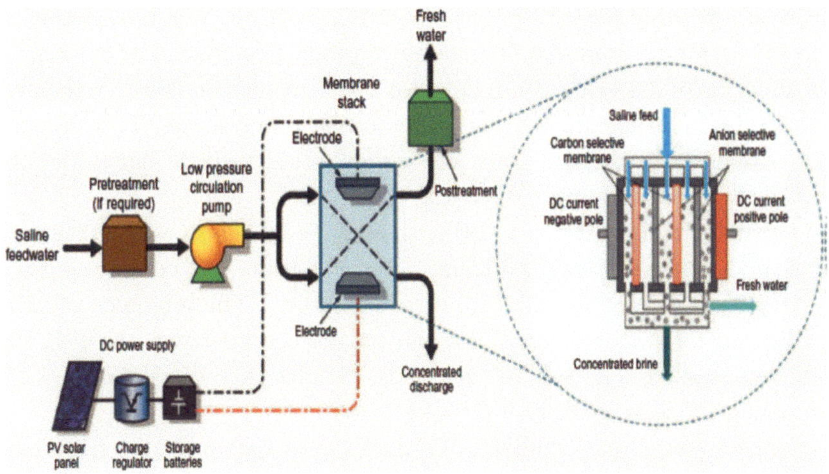

Figure 5.15 Schematic diagram of a PV-ED system [45]. *Reproduced with permission from A. Albloushi et al., Chapter 7—Renewable energy-powered membrane systems for water desalination, in: A. Basile, A. Cassano, A. Figoli (Eds.), Current Trends and Future Developments on (Bio-) Membranes, Elsevier, 2019, pp. 153–177.*

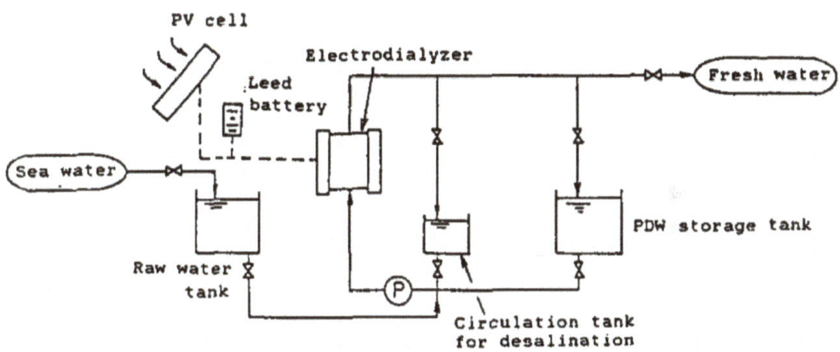

Figure 5.16 Schematic diagram of a PV-ED system. *Reproduced with permission from O. Kuroda et al., An electrodialysis sea water desalination system powered by photovoltaic cells, Desalination 65 (1987) 161–169.*

batteries, and the PDW is fed in the system to obtain freshwater. The additional battery is included in the system, also as a system stabilizer. A schematic diagram of this system is presented in Fig. 5.16. The resulted freshwater had TDS lower than 400 mg/L.

Another plant installed by AlMadani [44] tested ED directly with PV, without batteries. In this work, they reached a maximum of 95% of salt removal for groundwater and 99% for NaCl solution. When operating at a lower flow rate, around 0.56 m^3/day of production instead of the nominal capacity of 1.14 m^3/day. They also checked that for high flow rates, an increase in temperature means a higher salt removal rate. This suggests that a hybrid PV-thermal solar ED system could provide better performance than the conventional PV-ED system.

Peñate et al. [48] tested a 100% isolated PV-ED plant with a capacity of 4 m^3/h in the Canarias Islands. The PV array was connected to a Battery bank of 800 Ah and 7 hours of autonomy. The influence of solar irradiation, feedwater salinity, and the polarity change over the DC PV field was studied. The system provided a desired freshwater salinity for a wide range of solar irradiation (250–1100 W/m^2).

Gonzalez et al. [47] studied a PV purification system for the treatment of brackish water (TDS = 5482 mg/L) with a high concentration of arsenic (2.04 mg/L) in the north of Chile. The purification system combines ion-exchange and adsorption technologies (column filtration system) with an ED system. The ED process consumption was between 2.16–2.86 kWh/m^3 with salt removal above 95% and arsenic removal above 99%.

In a case study in the Canary Islands, the sustainability of PV-ED integrated system for freshwater production through energy and environmental considerations was investigated. PV-ED was compared with a grid-connected system. Energy consumption of 0.49–0.91 kWh/m^3 was taken as the reference for brackish water desalination (2500–5000 mg/L) by PV-ED.

Table 5.7 summarizes some of the installed PV-ED plants worldwide. From the data presented in Table 5.7, it can be concluded that despite the wide range of plant capacities (i.e., 1–200 m^3/day), most of the installations were of small scale and are used mostly for brackish water desalination worldwide. The SEC reported is between 0.4 and 4 kWh/m^3, while most of the plants used PV alone, without any storage.

One of the main advantages of ED coupled with intermittent RE generator (as PV and wind) compared with RO is the first is more flexible in its operation, allowing a better coupling with these sources of energy in poly-generation systems. Campione et al. [57] investigated the coupling of ED with wind and solar energy using dynamic simulations. The results show high flexibility in both long-term and short-term operational fluctuation, for flow rate ranging from 230 to 920 m^3/day with oscillations in output concentration lower than $\pm$ 10%.

Table 5.7 PV-powered ED desalination plants installed worldwide.

Authors	Location	Capacity (m^3/day)[2]	Feed TDS (mg/L)	SEC (kWh/m^3)[a]	PV system	ED configuration
Adiga [49]	Thar Desert, India	1	5000 (BW)	1 kWh/kg_{salt}	0.45 kW_p (80 V)	42 cell pairs (300 × 600 mm per cell)
Cited by Ali et al. [50]	Spencer Valley, Mexico	2.8	1000 (BW)	0.82	1 kW_p w/tracking, 2.3 kW_p stationary	250 cell pairs
Kuroda et al. [47]	Ohsima Island, Japan	10	(SW)	8.5	25 kW_p + 1200 Ah battery.	0.238 m^2 per cell
Ishimaru [46]	Fukue City, Japan	200	700 (BW)	0.6–1	65 kW_p PV + 1,200 Ah battery	*NA*
Lichtwadt and Remmers [51]	New Mexico, Mexico	18	900 (BW)	0.8	2.3 kW_p + 600 Ah battery	*NA*
AlMadani [44]	Isa, Bahrain	0.19–1.14	3300 (BW)	*NA*	0.132 kW_p (12 V)	24 cell pairs (230 × 250 mm per cell). 4 hyd. /2 elec. stages
Ortiz et al. [52]	Alicante, Spain	1.32	2000 (BW)	1.33–1.47	0.152–0.304 kW_p (80 V)	80 cell pairs (550 cm^2 per cell area). Batch system.
Ortiz et al. [53]	Alicante, Spain	0.75 m^3/h	2300–5100 (BW)	0.7–1.7	38.45 W_p	70 cells (0.05 m^2 per cell).
Peñate et al. [48]	Canary Islands, Spain	4 m^3/h	2240–3392 (BW)	0.618	3.69 kW_p + 800 Ah battery	340 cell pairs. 4 hyd. /2 elec. Stages (only 1 stack used with PV)
Círez et al. [54]	Zaragoza, Spain	4.3	580–10,000 (BW)	1.25–4.79	0.103–0.308 kW_p(14 V)	10 cell pairs (0.02 m^2 per cell)
Gonzalez et al. [55]	Loa River, Chile	*NA*	5482 (BW)	2.16–2.86	*NA*	20 cell pairs (0.02 m^2 per cell)
He et al. [56]	Chelluru, India	6 m^3/d (0.85 m^3/h)	1350 (BW)	1.51–1.81	40 m^2 + 16 kWh batteries	56 cell pairs

BW, Brackish water; *SW*, seawater, *NA*, *Not Available*
[a]Unless other units indicated.

Solar PV-ED system has also been presented as an energy and cost-effective means of desalinating groundwater in rural India [58]. Within the salinity range of groundwater in India, ED requires less specific energy than RO (75% less at 1000 mg/L and 30% less at 3000 mg/L). At 2000 mg/L, this energetic scaling translates to a 50% lower PV power system cost for ED versus RO. PV-ED has the potential to expand the reach of desalination units for rural India. To prevent scaling in ED, the modern processes utilize inverters that reverse periodically (about every 20 minutes) the polarity of the electric field, causing a reversal in the flow of ions and cleaning the membrane surface. This process, called ED reversal, shows recoveries higher than other membrane technologies applied for brackish water desalination, such as RO, with average results higher than 90% [45].

He et al. [56] developed a methodology to optimize a PV-ED plant for rural India. The optimization process was carried out, that is, numerically, and later tested experimentally with a 6 m^3/day plant in Chelluru, India. The estimated capital cost and lifetime cost of this system are 34% and 45% lower, respectively, compared to a PV-ED conventional system. This means an estimated LCOW of 1.87 USD/m^3, which is lower than the current water price of RO-produced water in Indian villages (2.4–3.6 USD/m^3). A 2011 study [59] compared techno-economically PV- and diesel-powered ED desalination systems. It is estimated that capital cost of the PV-ED option is 30% higher than the conventional fuel ones, with a cost of water to be 3 USD/m^3 for a 50 m^3/day capacity plant. Different reports estimate the cost of water for PV-ED systems in the range of 3–16 USD/m^3 [50]. These values are one of the main barriers to the widespread implementation of this technology. Therefore, most studies have focused on PV-ED feasibility in remote areas, where grid electricity or water supply is not possible.

Most of the researchers used a PV-driven ED process only for brackish water, as they are not economically competitive for seawater applications because of the large quantities of salt, the high cost of electrodes and ion-exchange membranes, and the relatively short lifetime of membranes when working in a high-density electric field. In fact, a small-scale (i.e., 10 m^3/day) PV-driven ED plant in seawater application requires the recirculation of seawater several times before the desired water quality is obtained [60]. While the main identified barriers for PV-ED preventing a more extensive market penetration match the intermittent output of renewable energies with water demand, the short membrane lifespan, and the high production cost of freshwater compared to its low market price as a commodity [61].

In conclusion, PV-ED is a promising technology like a solar brackish water treatment technology in remote areas. It offers better technical and economical results than the conventional solar desalination technology, such as PV-RO. It has proven to be a reliable alternative with high salt removal rates ($>85\%$) and relatively lower specific energy consumption ($<2\,kWh/m^3$). It is also a promising alternative to match intermittent generators (such as PV and wind) in poly-generation plants; however, the high capital costs are the main limiting factor to its implementation for large-scale plants.

5.4 Solar-driven capacitive deionization

5.4.1 Historical background

CDI is an electrochemical-based desalination technology, where salt ions are captured by applying porous electrodes [62]. CDI is an electrosorption method using a combination of a sorption media and an electrical field to separate ions and charged particles. While a capacitor is a device made of one or more pairs of electrodes, oppositely charged, an electric potential is applied to the electrodes producing positively and negatively charged poles. The electrostatic forces attract the positive ions (cations) to the negative electrode, while the negative ions (anions) are attracted to the positive electrode. Anions and cations are removed from the water and stored in the positively and negatively polarized electrodes, respectively.

Research in CDI desalination can date back to 1833–34, when the first discovery of the dissociation of ions in solution (electrochemical effects) under an electric field was reported by M. Faraday (1833) [1]. In the year 1900 W. T. Gibbs patented his electrolytic device to produce chlorates from water solutions by exposing it to an electric current. This patent served as a starting point for the development of CDI technology for water treatment. The major advances in CDI during the 20th century came through the work of D. D. Caudle's and S. Johnson's groups at the University of Oklahoma [63]. They were the first group to propose the flow-through capacitor system and electrochemical demineralization of water using carbon electrodes in 1966. Y. Oren and A. Soffer later in 1983 demonstrated a cyclic process for water desalination based on CDI through the electrochemical parametric pumping [64]. This was followed, in 1996, by the pioneering work of J. Farmer et al. at Lawrence

Livermore National Laboratory, where they used carbon aerogel and carbon nanotube (CNT) materials as electrodes. Ever since, significant progress and developments have been made in electrode materials, CDI devices, and process optimization [1]. Lee et al. were the first group to use the term Membrane-CDI system for such devices, demonstrating its use in processing power plant wastewater in 2006 [65]. From 2016, the research focus shifted to the design of flow cell configuration and operation, followed by the most recent research on the practical applications of Flow-electrode capacitive desalination (FCDI) [62].

5.4.2 Capacitive deionization design and operation

In the CDI operation, a cycle of ion removal (Fig. 5.17A) and ion discharge (Fig. 5.17B) is alternated to produce two streams of desalinated water and brine. It is important to note that the potential difference should not exceed the point at which the bonds in the water molecule between the hydrogen and the oxygen atoms break (known as the Nernst voltage or the standard electrode potential of water, $E^{\phi}_{\text{cell}} = -1.23\text{V}$) [1]. More specifically, partially desalted water can be obtained by applying a potential difference between the two electrodes, cathode absorbs cations, while anode absorbs anions. Throughout the separation process, the voltage is reduced or reversed to remove the electrodes' ions. This leads to a concentrated salt product stream, which is discharged out of the CDI module [66].

One of the biggest limitations of the CDI technology is that the electrodes get saturated with the oppositely charged ions. At this point, the purification step stops, and the regeneration process is initiated by opposing the sign of the electrode to start the desorption process by putting the

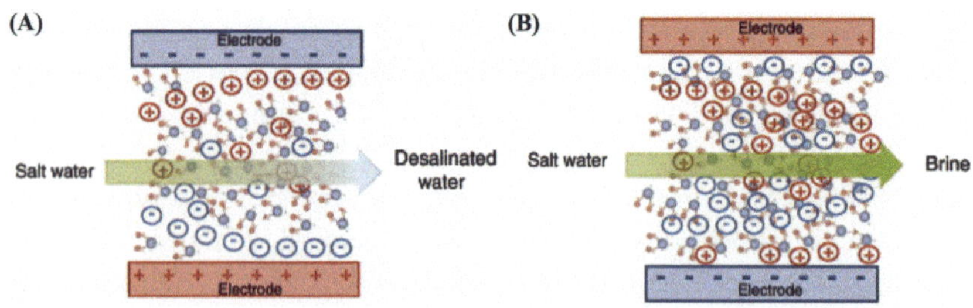

Figure 5.17 CDI (A) purification and (B) regeneration processes [1]. *Reproduced with permission from F.A. AlMarzooqi, A.A. Al Ghaferi, I. Saadat, N. Hilal, Application of capacitive deionisation in water desalination: a review, Desalination 342 (2014) 3–15.*

absorbed ions from the electrode back to the bulk fluid as the brine solution (see Fig. 5.17B). Unfortunately, once the electrode's electric potential changes its polarity, it will start adsorbing the oppositely charged ions from the bulk fluid. Thus ion desorption and adsorption occur simultaneously during this step, which causes incomplete regeneration of the electrode. Hence, this increases the regeneration time and causes residual ion accumulation, which results in a resistance of ion removal in the next purification stage.

The concept of membrane capacitive deionization (MCDI) is a significant improvement from the original CDI technology with additional ion-exchange membranes to assist in preventing adsorption during the regeneration process (see Fig. 5.18). The cations freely permeate through the CEM toward the double layer within the vicinity of the negative electrode, and the anions similarly permeate through the anion-exchange membrane (AEM) towards the positive electrode. Thus the MCDI increases the charge efficiency and the desalination efficiency through the ion-exchange membrane process.

5.4.3 Electrode materials

The majority of effort and focus on CDI and MCDI research have been developing and synthesizing better electrode materials. One of the key parameters for the commercial realization of CDI is the electrodes' salt adsorption capacity. Electrode material development started in the early 1990s with the introduction of activated carbon material electrodes. M. D. Andelman was amongst the first to use such material in this application, initially through his patent on flow-through capacitors in 1990. Recent developments have introduced novel electrical conductive

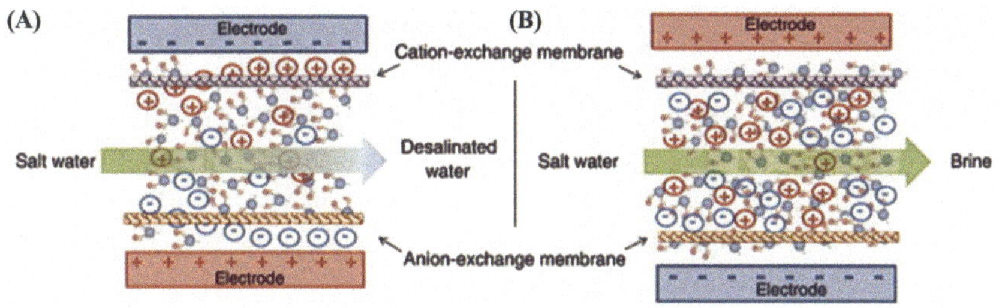

Figure 5.18 MCDI (A) purification and (B) regeneration processes [1]. *Reproduced with permission from F.A. AlMarzooqi, A.A. Al Ghaferi, I. Saadat, N. Hilal, Application of capacitive deionisation in water desalination: a review, Desalination 342 (2014) 3–15.*

materials with surface areas as high as 2000 m^2/g, enhancing the CDI process (CDI requires porous electrodes with extremely high surface areas) [67].

The materials that have been investigated in the literature are activated carbons, carbon aerogel, ordered mesoporous carbon, activated carbon cloth, and nanotubes, graphene family (e.g., graphene sponge and graphene hydrogel), and carbon composite [68]. Table 5.8 summarizes the characteristics of different electrode materials proposed in the literature. Such as physico-chemical properties of electrode materials, including specific surface area (SSA), mean pore diameter (D_p), carbon graphitization degree (I_D/I_G), and charge transfer resistance (R_{ct}). They play important roles in the performance of CDI. For instance, previous studies have reported that a higher I_D/I_G value should be beneficial to the charge transfer in the adsorption process [68]. Nevertheless, carbon materials are the most suitable as capacitor electrodes, since it has a high SSA, good electrical conductivity, and high electrochemical stability.

5.4.4 Module configuration

A conventional CDI cell consists of two porous electrodes positioned on opposite sides of a sandwiched-spacer layer that serves as the water channel. Various configurations for the design of CDI cells were proposed in the literature over the past decades. The CDI research field has progressed enormously with a spring-up of various cell architectures, including Flow-by CDI and MCDI, flow-through CDI, inverted CDI, flow-electrode CDI, hybrid CDI, desalination battery, and cation intercalation desalination (CID), as shown in Fig. 5.19 [66].

Flow-by CDI and MCDI (or CDI with flow-between electrodes) are the most widely studied and utilized cell architectures in the CDI field [89]. Flow-by CDI, MCDI, flow-through CDI, i-CDI, and FCDI remove ions via the non-Faradaic process (i.e., electrostatic and capacitive effects). While flow-by CDI stands out owing to its simple structure without the need of expensive IEMs or pumping energy requirements for flow-electrodes, MCDI presents higher charge efficiency than flow-by CDI at the cost of significantly more expensive cell components [66]. As shown in Fig. 5.20C, flow-through CDI allows for more compact cells and improved desalination rates but suffers from greater feed pressure drops than flow-by CDI [66]. The i-CDI cell (Fig. 5.20D) consists of an anode and a cathode with a net negative surface charge and a net positive

Table 5.8 Characteristics, properties, and performance of electrode materials used in CDI.

Modulus	Electrode property					Equilibrium sorption and capacitance					Ref.
	Type	SSA (m^2/g)	S_m (m^2/g)	D_p (nm)	I_D/I_G	R_{ct} (Ω)	C_{salt} (g/L)	SAC (mg/g)	R_s (mV/s)	C_s (F/g)	
CDI	Activated carbon (AC), Lignin-derived AC and Fluorinated AC	920–2088	—	1.08–2.42	0.9	—	0.47–0.58	7.6–18.5	5	60–192	[69–74]
	Nitrogen-doped carbon spheres, or carbon nanocrystals	842.3–1640	—	8.45	—	0.86	0.2–1.0	10.87–20.63	1–10	220–290.74	[75,76]
	N-C/Fe_3O_4	11.2	—	5.82	—	—	—	—	5	67.4	[77]
	N-CNFA	874.4	—	3.03	0.91	0.5	1.0	17.29	2	356.4	[78]
	e-CNF-PCP	1450.6	—	3.84	1.04	0.70	0.5	16.98	1	284.8	[79]
	graphene oxides (GO)	288–445	—	—	1.10–1.19	—	0.5	17.1–21.58	1	190–200	[80,81]
	ZrO_2/GO	~240	—	—	—	—	0.05	4.8–6.3	10	452	[82]
	Graphene sponge	356	—	-	1.04	0.27	0.5	14.9	5	205.2	[83]
	Graphene hydrogel	517	151	11.8	—	5.64	0.5	49.34	5	~35	[84]
	Graphite felt/AC	621	—	—	—	0.46	2.1	8.7	5	20	[85]
	TiO_2-NT/HCF	—	—	—	0.825	0.15	0.5	13.1	20	238.2	[86]
	NACTZ	3675.5	—	72.2	—	0.05	—	3.98	10	691.8	[87]
	NiCoAl-MMO	83.77	—	—	—	0.80	0.175	108.8	1	1167	[88]
	HZC	642.7	—	—	—	1.45	0.1–0.5	8.57–15.31	1 (5)	243.0 (214.7)	[89]
	HPC	1877	611	6.04	0.92	0.51	0.292	12.63	2	150	[90]
	MnO_2/ACC/SWCNT	—	—	—	—	0.8	0.06	17.8	1	331	[91]
	Ag-AC/MWCNT	488.97	—	—	—	0.42	0.6	5.3	1	9.97	[92]
MCDI	SG-CNT	—	—	—	1.18	—	0.4	9.54	2	117	[93]
	PVA-bound AC	2040	—	—	—	—	0.58	18.0	—	—	[94]
FCDI	AC/CBP	1100	425	1.69	—	—	—	—	—	—	[95]
RCDI	Nafion-coated AC	—	—	—	—	—	2.92	44.5	—	-	[96]

AC, Activated carbons; *ACC*, activated carbon cloth; *CBP*, carbon black powder; C_s, specific capacitance per gram of electrode mass (F/g); C_{salt}, initial salt concentration (g/L); D_p, mean pore diameter (average pore size, nm); FCDI, flow-electrode CDI; *GO*, graphene oxides; *HCAM*, hierarchical carbon aerogel monoliths; *HCF*, hollow carbon fiber; *HPC*, hierarchical porous carbon; *HZCs*, hollow ZIFs-derived nanoporous carbons; I_D/I_G, carbon graphitization degree (-); *MCDI*, membrane CDI; *N-CS*, nitrogen-doped carbon spheres; *NACTZ*, TiO_2/ZrO_2 nanofibers incorporated activated carbon; *N-CNFA*, nitrogen-doped carbon nanofiber aerogels; *NiCoAl-MMO*, NiCoAl-mixed metal oxide; *NMCs*, nitrogen-doped mesostructured carbon nanocrystals; *OMC*, ordered mesoporous carbons; *OMSi-CDC*, ordered mesoporous silicon carbide-derived carbons; *RCDI*, rocking-chair CDI; R_{ct}, charge transfer resistance of electrodes (Ω); R_s, scan rate (mV/s); *SAC*, salt adsorption capacity (mg/g); *SG*, sulfonated graphene; S_m, specific surface area of micropores (m^2/g); *SWCNT*, single-walled carbon nanotubes

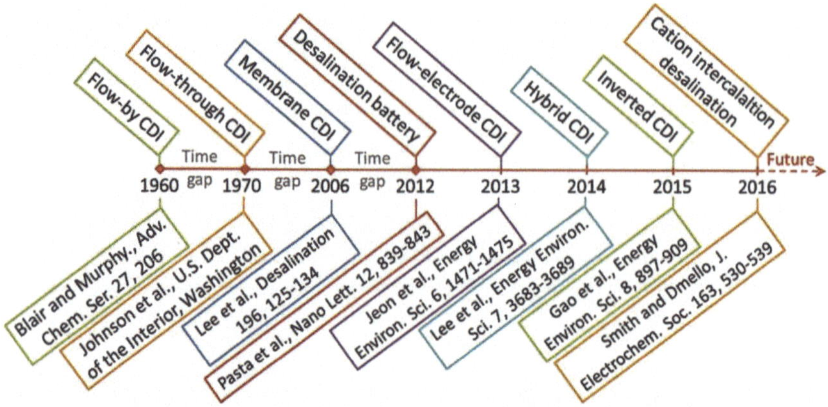

Figure 5.19 Timeline displaying the years when various CDI cell architectures emerged and the corresponding seminal work [66] . *Reproduced with permission from W. Tang, et al., Various cell architectures of capacitive deionization: recent advances and future trends, Water Research 150 (2019) 225–251.*

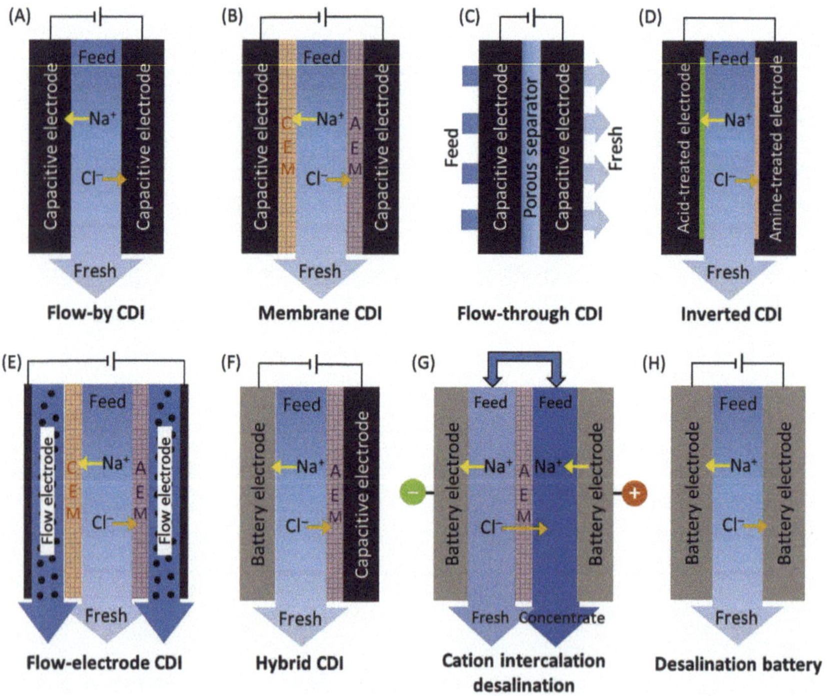

Figure 5.20 Typical cell architectures of various CDI: (A) Flow-by CDI; (B) Membrane CDI; (C) Flow-through CDI; (D) Inverted CDI; (E) Flow-electrode CDI; (F) Hybrid CDI; (G) Cation intercalation desalination; (H) Desalination battery [66]. *Reproduced with permission from W. Tang, et al., Various cell architectures of capacitive deionization: recent advances and future trends, Water Research 150 (2019) 225–251.*

surface charge, respectively. An inverted CDI cell was constructed using amine-treated cathodes and acid-treated anodes, where the cathode is the negatively polarized electrode (COO−) and the anode is the positively polarized electrode (NH^{3+}). Thus, salt is desorbed when a voltage is applied to charge the cell, whereas salt is adsorbed when the applied voltage is reduced. This configuration exhibits a highly efficient, stable, and environmentally friendly salt separation process [97].

FCDI, a relatively recent advance in CDI architecture, provides the possibility of avoiding some of the limitations of fixed-electrode CDI. Instead of utilizing stationary electrodes, flowing electrodes made of carbon suspension are applied as electrodes in the FCDI cell. By continuously replenishing the electrode chamber with regenerated carbon slurry through short-circuited closed-cycle operation, FCDI exhibits high desalting efficiency with the possibility of continuous operation.

For instance, the slurry-type electrode contains typically small particles of activated carbon sized at $\sim 10\ \mu m$ with a high SSA of $\sim 3200\ m^2/g$, leading to higher electrosorption capacity (> 20 mg of NaCl per gram of carbon particles) compared to MCDI (average 1−11 mg/g) [98,99]. Fig. 5.21 illustrates a schematic diagram of the FCDI module, consisting of a spacer chamber (thickness of $\sim 500\ \mu m$) made of a silicone gasket and a nylon sheet (100-mesh, 160×70 mm) between the cation and anion-exchange membranes (CEM Type-I/AEM Type-I, FUJIFILM Europe). Two composite current collectors placed against the ion-exchange membranes were made of graphite paper and acrylic sheets.

To increase the desalination surface area for a conventional two-dimensional cell structure (CDI), stacking in one direction or increasing the cell size in two dimensions is required. Cho et al. developed a novel three-dimensional system from a honeycomb-shaped porous substrate by coating an ion-exchange membrane and a graphene layer within the channels [100]. More recently, Zhang et al. [101] demonstrated that continuous operation could be achieved by coupling microfiltration (MF) module into an FCDI system (see Fig. 5.22) with the following processes occurring simultaneously: (1) ion transport across the IEMs and temporary electro-adsorption in the respectively charged flow-electrodes; (2) flow-electrode regeneration and brine production through charge neutralization in a shared reservoir; (3) brine separation via an MF module immersed in the shared reservoir; and (4) electrolyte replenishment to maintain the volume and carbon content of the flow-electrode stable. An experimental study revealed that the integrated FCDI/MF system was capable of desalting brackish water to

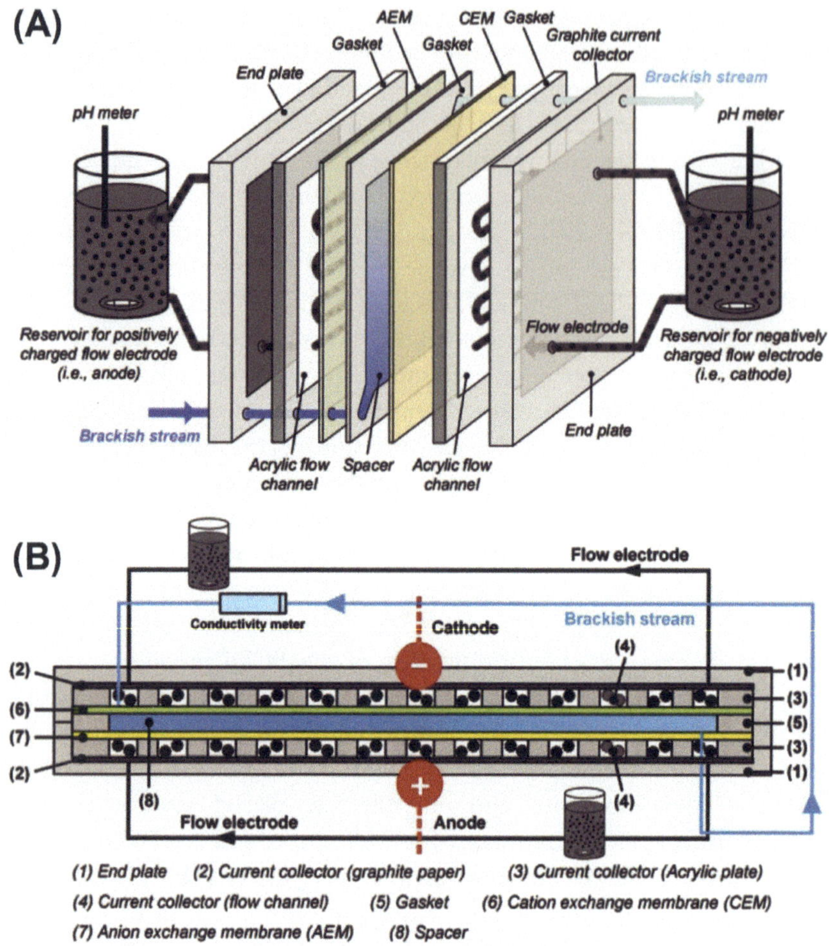

Figure 5.21 Schematic representations of (A) the structure and (B) the cross-section of the flow-electrode capacitive deionization (FCDI) cell [99]. *Reproduced with permission from J. Ma, C. He, D. He, C. Zhang, T.D. Waite, Analysis of capacitive and electrodialytic contributions to water desalination by flow-electrode CDI, Water Research 144 (2018) 296–303.*

a potable level (<0.5 g/L) with extremely high-water recovery rates ($>90\%$) and thermodynamic energy efficiency ($\sim 30\%$). The FCDI/MF system was capable of continuously desalinating brackish water (~ 2 g/L) to portable levels (<0.5 g/L) while sustaining an extraordinary water recovery rate ($\sim 92\%$) and relatively low energy consumption (~ 0.5 kWh/m^3).

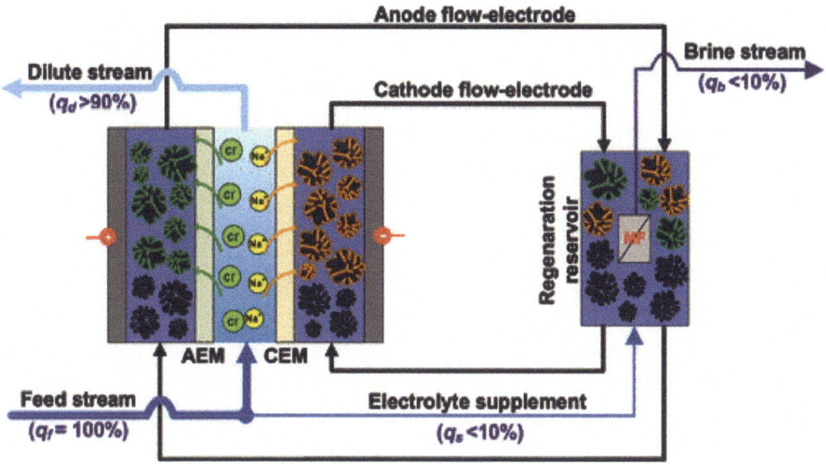

Figure 5.22 Schematic representation of a continuous water desalination system consisting of a FCDI cell and a MF module [101]. *Reproduced with permission from C. Zhang, L. Wu, J. Ma, A.N. Pham, M. Wang, T.D. Waite, Integrated flow-electrode capacitive deionization and microfiltration system for continuous and energy-efficient brackish water desalination, Environmental Science & Technology 53 (22) (2019) 13364–13373.*

In terms of HCDI, desalination battery, and CID, these cells remove ions via the Faradaic process (i.e., reversible redox reactions). Owing to the superior ion sorption capacity of battery electrodes compared to traditional capacitive electrodes, HCDI, desalination battery, and CID receive increasing attention in recent years. Generally, the HCDI cell consists of a Faradaic (i.e., battery) electrode for cation adsorption/desorption and a capacitive electrode for anion adsorption/desorption. An AEM could be placed adjacent to the capacitive electrode to enhance HCDI's performance (see Fig. 5.20F). The desalination battery cell is composed of two different battery electrodes (one for cation adsorption/desorption and the other for anion adsorption/desorption). The CDI cell is constructed by employing Faradaic cation intercalation materials for both electrodes with the electrodes separated by an AEM. CDI cells with battery electrodes (HCDI, desalination battery, and CDI) regularly achieve salt adsorption capacity far exceeding that of CDI cells with capacitive electrodes. However, battery electrodes exhibit a slower salt removal and incur a higher capital cost relative to capacitive electrodes.

5.4.5 Solar photovoltaic-driven capacitive deionization configurations

As an electrically driven separation process capable of removing ions and small charged molecules from the feedwater, CDI is low pressure, low temperature, and low voltage (typically <1.3 V) process that is energy-efficient for brackish water desalination. In view of the low voltages required to induce CDI, MCDI is well suited to be powered by the output of PV modules. The solar-powered CDI is an energy-efficient alternative with low operating voltage and the possibility of energy recovery (most of the energy can be subsequently recovered from the electric double-layer capacitance), making it ideal for brackish water desalination, especially in remote locations where mains power may not be available.

The potential of MCDI to be used in conjunction with renewable energy sources has been conducted by Tan et al. [102]. A PV battery-powered pilot-scale MCDI system has been designed and tested, as shown in Fig. 5.23. This was the first integration of PV with MCDI technologies, capable of electrode charging currents above 100 A and product volumes on the order of 5 m^3/day, which represented a critical step in designing and operating CDI systems in remote areas away from the power grid. This study (specific energy consumption: 0.7−1.1 kWh/m^3; water recovery: 49%−73.2%) shows that PV-powered MCDI water treatment, particularly when combined with energy recovery, is competing against more mature water treatment technologies (e.g., RO).

A pilot-scale MCDI system capable of a production rate of about 4 m^3/day has been developed with energy recovery using a high-efficiency, bi-directional DC−DC converter. It demonstrated that energy recovery from the electrode further reduces the energy consumption of MCDI (<0.4 kWh/m^3 by the MCDI electrode and about 1.1 kWh/m^3 by the complete pilot system) and allows for the possibility of stand-alone operation with the unit powered entirely by renewable energy such as that provided by PV [103]. Furthermore, this PV-driven MCDI pilot system has been investigated for irrigation with brackish groundwater [104]. Results show that the cost of water supply, including bore drilling and pumping, MCDI desalination, PV, brine disposal, and water storage is less than AU$1/$m^3$ for typical field applications with salinity removals

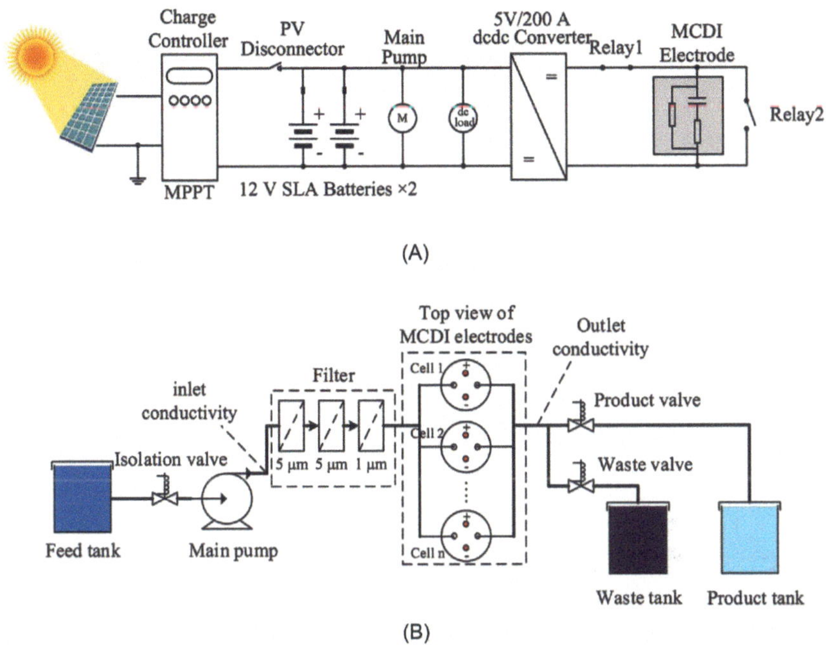

Figure 5.23 (A) Electrical diagram schematic and (B) block diagram of the experimental layout of the prototype MCDI unit [102]. *Reproduced with permission from C. Tan, C. He, W. Tang, P. Kovalsky, J. Fletcher, T.D. Waite, Integration of photovoltaic energy supply with membrane capacitive deionization (MCDI) for salt removal from brackish waters, Water Research 147 (2018) 276–286.*

<2.3 dS/m. It consumes only 0.4 kWh per ton of brackish water treated, and the water production cost about AU$0.90/m^3. With PV installed, the whole system can be operated automatically and remotely in places where solar energy is abundant, but power supply and fresh water are limited [104] (Table 5.9).

In conclusion, with advantages of low energy consumption, low environmental impact (i.e., using renewable solar energy, without the use of any added chemicals or the generation of hazardous substances), simple equipment structure, and convenient operation, solar-driven CDI have emerged as a promising alternative to established water desalination technologies such as reverse osmosis, ED, and thermal distillation, especially for desalination of waters with low to medium salinity.

Table 5.9 Summary of the solar-driven CDI available in the literature.

Author	PV area/power	CDI technology	Desalination ΔTDS (feed TDS -permeate TDS) (mg/L)	Production (L/day)	Energy consumption (kWh/m^3)	Unit cost of water (USD/m^3)
Tan et al. [102]	1 kW	MCDI	800 mg/L	5000	0.7–1.1	N/A
Tan et al. [103]	1 kW	MCDI	600–750 mg/L	4000	< 0.4 by CDI, 1.1–1.2 by total system	N/A
Bales et al. [104]	12 pcs of 375 W	MCDI	bore 3.2–1.2 dS/m on-crop	20,000	0.4 by CDI	~0.70

NA, Not Available

5.5 Economic development potential of low technology readiness–level technologies

5.5.1 Technology comparison

The conducted review on the low TRL technologies (e.g., MD, ED, and CDI) has shown that there are many studies in the literature that have investigated the potential to hybridize these water separation technologies with solar energy. Some studies have investigated new solar designs and configurations, while others focused on the techno-economic performance of the overall system. In general, the advantages and disadvantages of these aforementioned technologies are summarized in Table 5.10.

One key opportunity of the reviewed water treatment technologies is that they can be coupled using different methods, as was shown in this chapter and summarized here and presented in Fig. 5.24:

- *Solar thermal-driven MD*: Heating the feedwater using a solar collector before entering the MD module.
- *Solar thermal + PV-driven MD*: Heating the feedwater using a solar collector before entering the MD module while using the electricity generated from the PV module to run the feed pumps.
- *PV thermal-driven MD*: Supplying the heat and the electrical energy required via a PVT collector to the MD module.
- *Integrated solar-driven MD*: Integrating the MD module inside the solar collector (i.e., a submerged MD module), where solar incidence is continuously heating the submerged feedwater.
- *PV-driven ED*: Supplying the electrical energy required by the ED system via PV.
- *PV-driven CDI*: Supplying the electrical energy required by the CDI system via PV collectors.

There are multiple solar-driven configurations for the MD technology, while there is only one solar-driven configuration for ED and CDI technologies (i.e., PV-driven system). MD technology requires both heat and electrical energy, allowing versatility with which solar technology to use. Nevertheless, as the MD technology has a higher specific thermal energy consumption than specific electrical energy consumption, it is favorable to use solar thermal collectors than stand-alone PV. Nevertheless, solar thermal with PV-driven MD system is the most studied technology in the literature, that is, PV thermal-driven MD.

Solar-driven MD has a relatively high cost of water. For example, for a scaled-up solar-powered AGMD (i.e., the yearly production of 24,000 m^3),

Table 5.10 One Key Opportunity Comparative summary between solar-driven LTR technologies.

Water treatment process	Advantages	Disadvantages
Membrane Distillation	• Modular module design • Insensitive to feed concentration • High-quality permeate (e.g., permeate's TDS is less than 10 ppm) • Low working pressure • Module is mostly made of plastic • Ease of operation • Less pretreatment is required • Ability to cope with solar intermittency	• Membranes have low endurance • Requires external heat to start the distillation process • A cooling process is needed, that is, can be internal or external, to condense the permeate vapor produced • High possibility of corrosion • Low productivity (e.g., 1–456 m^3/day) • High specific energy consumption (e.g., 30–10,000 $kW_{th}h/m^3$) • High capital cost and unit water cost (e.g., 0.28–20 USD/m^3) • Thermal and electrical energy is required
Electrodialysis	• High salt removal (e.g., 85%–95%) and water production (e.g., 2–145,000 m^3/day) • Less prone to membrane fouling • Less pretreatment is required • Only electrical energy is required • Membranes have a longer lifetime (e.g., 7–9 years) • Can have a low capital cost and unit water cost (e.g., 0.6–1.05 USD/m^3) • Flexible operation. Ability to cope with solar intermittency	• Nonionic substances and microorganisms cannot be desalinated. Require pretreatment • Can have a high specific energy consumption (e.g., 0.5–10 kW_eh/m^3) • Only suitable for small-scale applications • Can only desalinate low salt concentration solutions (e.g., up to 12,000 ppm) • Low product water quality (e.g., 150–500 ppm)
Capacitive deionization	• Low specific energy consumption (e.g., 0.1–2 kW_eh/m^3) • Low maintenance cost • Long-term cyclability of electrode materials • Less prone to fouling • Less pretreatment is required • Only electrical energy is required • Ability to cope with solar intermittency	• Nonionic substances and microorganisms cannot be desalinated • Only suitable for small-scale applications • Low product water quality (e.g., 150–500 ppm)

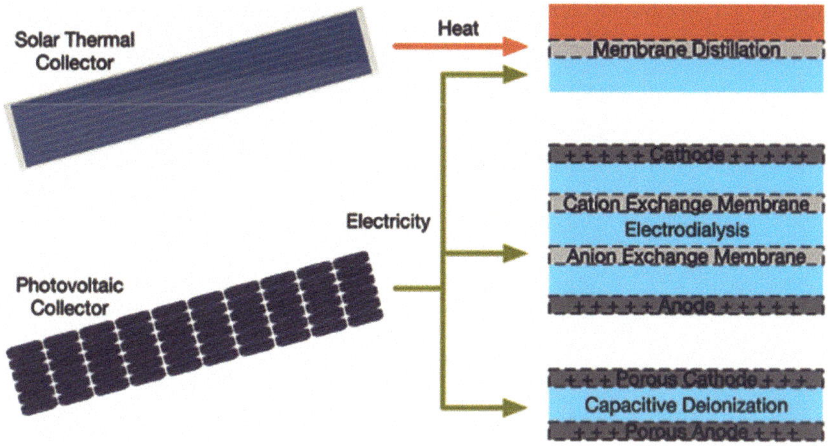

Figure 5.24 Solar-driven combination with LTR water treatment technologies.

the cost of water is 8.9 USD/m^3. However, 70% of this cost is associated to the solar collector of the system. In addition, the membrane lifetime is a key factor in the economics of solar-driven MD. This is the same case for ED and CDI technologies, where the membrane's lifetime and cost limit these membrane water treatment technologies market potentials.

Table 5.11 summarizes some techno-economic parameters for the solar-driven LTR technologies reviewed in this Chapter. As mentioned previously, solar-driven MD technologies are not yet economically ready (knowing that solar thermal collectors are more expensive than PV collectors) compared to PV-driven ED or PV-driven CDI. For example, the average cost of water of solar thermal-driven MD reported in the literature is between ~0.5 and ~20 USD/m^3 compared to PV-driven ED and PV-driven CDI (i.e., ~0.2 to ~2.0 USD/m^3). Furthermore, thermal processes (i.e., MD) generally have a higher specific thermal energy consumption (~15 to ~750 kWh/m^3) due to the thermal energy required compared to the low specific electrical energy of ED and CDI (i.e., ~0.5 to ~6 kWh/m^3).

Fig. 5.25 provides a visual summary of the reviewed solar-driven technologies and how they are compared within each other with respect to their levelized cost of water and their specific energy consumption. Generally, the MD technology is confounded with a very high specific energy consumption, ensuing a high levelized cost of water. Furthermore, although the cost of PV is relatively lower than solar

Table 5.11 Comparative summary between solar-driven LTR technologies.

Parameter	Solar-driven membrane distillation	Solar-driven electrodialysis	Solar-driven capacitive deionization
Solar collector type	Flat plat collectors; evacuated tube collectors; photovoltaic; photovoltaic thermal	Photovoltaic; photovoltaic thermal	Photovoltaic
Operating temperature (°C)	40–90	15–40	15–40
Specific energy consumption (kWh/m^3)	15–750	0.5–2.0	0.5–1.0
Typical production rate (m^3/day)	1–456	1–200	1–5
Unit cost of water (USD/m^3)	0.5–20	1.7–16	0.2–1
Feedwater salinity range	Brackish to seawater	Brackish water	Brackish water
Sensitivity to feedwater salinity	Low	Medium–high	Medium–high
Product water quality (ppm)	1–40	250–500	250–500
Pretreatment requirement	Low	Medium	Medium
Scaling and fouling issues	Medium–high	Low–medium	Low–medium
Maintenance and operation requirements	Low	Medium	Medium
Market potential	Medium–high	High	Low–medium

thermal collectors, the LCOW of PV-driven MD is considered high. The reason for this is that the MD process is a thermal-driven technology, wherein the PV module fails to supply the necessary thermal energy. To mitigate this, and another solar thermal collector or a PVT

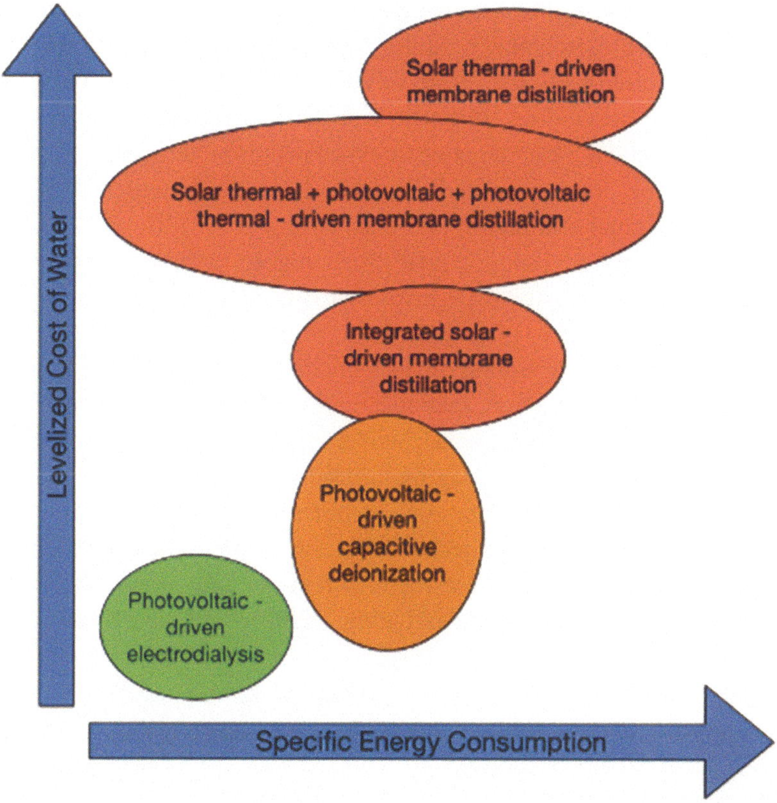

Figure 5.25 Reviewed solar-driven water treatment systems using LTR technologies LCOW and SEC map.

collector is needed. Whereas when the PV collector is used to run an ED or CDI processes, their LCOW is low due to the low cost of the PV collectors. Furthermore, the PV-driven ED process has the least LCOW and SEC compared to PV-CDI since the ED process is a more well-established technology.

5.5.2 Research status

A number of studies on these low TRL technologies have been performed on a laboratory scale using small membrane modules and a few square meters of solar collectors. Most of the publications are concerned with experimental studies and theoretical models, including heat and mass transfer models, to estimate the annual performance of these technologies. Similarly, novel membrane modules and solar integration designs have been studied intensively (i.e., solar-driven MD technology has been studied most in the literature). Nevertheless, there are some repeated studies that have reached almost the same conclusions: these technologies are not ready yet to be commercialized, but they have the potential to provide clean water in rural arid regions.

There are still several potentially exciting research areas within the solar-driven MD, ED, and CDI fields that can be further investigated and outlined as shown.

- A comparative study among these technologies with higher TRL, such as PV-driven reverse osmosis, solar stills, and concentrated solar power plants coupled with well-developed thermal-driven desalination technologies (e.g., multi-effect distillation or multi-stage flash desalination).
- Design and fabrication of new membrane modules that incorporate internal energy recovery methods (i.e., adapt the multi-effect concept). The energy-optimization of these systems would reduce the cost of water of these technologies and allow them to go "*up the ladder*" to a higher TRL.
- The continuous development of selective coatings applied to membranes to reduce its fouling potential in long-hour operation, especially if thermal storage or batteries are installed with the solar collector.
- More attention should be paid to deploy these solar-driven technologies to applications other than water treatment. For instance, mineral extraction from the brine discharge coming from industries or mines. This also includes multidisciplinary innovation of combining these technologies with other separation processes to compensate for their shortcomings.
- Other high-temperature solar collectors (e.g., parabolic solar collectors or solar concentrators) with thermal storage can be tested with MD technologies for longer operation throughout the day. This also includes the use of battery storage with PV modules to drive ED and CDI technologies.
- Incorporate the findings of the more established technologies in the membrane separation process (e.g., reverse osmosis), such as molecular imprints materials, nanotechnology, and novel spacer designs, into the MD technologies. Similarly, the ED and CDI technologies would benefit from the well-developed battery technologies.

- Hybridizing MD, ED, and CDI with other well-established water treatment technologies, such as RO, multi-effect distillation, and multi-stage flash, to overcome the shortcomings of those LTR technologies.

5.5.3 Market considerations and upscaling

One of the best approaches to reducing any desalination cost is minimizing or eliminating the driving energy cost by using a sustainable power source, such as solar energy. However, the establishment cost of solar-based desalination technologies, especially LTR water treatment technologies (i.e., discussed in this chapter), is still considered relatively high due to the limited water production because of the low-capacity factor. Although this reason can be applied to any solar-driven water treatment technology, including solar-driven reverse osmosis, a high TRL water treatment technology, the water treatment technologies discussed in this chapter (i.e., MD, ED, and CDI) are not well-established compared to RO, making them not yet fit to be coupled with solar technologies. Nevertheless, these technologies play an important role in solar-driven water treatment technologies because of their potential flexibility and modular design. For example, operating conditions can be adjusted to compensate for possible changes in the solar energy delivered. Nevertheless, they are still under evaluation although it has been studied for more than 40 years.

Although these technologies seem very promising at the development stage in the research work, several commercialization attempts have failed. As a matter of fact, these water treatment technologies have not yet received adequate focus on scale-up and commercialization. This includes the relatively developed ED technology as there have been multiple pilot operational plants in remote areas but fails in a large-scale installation.

It was shown that MD has a high energy consumption and has difficulties with long-term operation due to membrane wetting and fouling. Hence, solar-driven MD technology has occupied a more academic role rather than being commercialized. However, it is expected that the MD technology will catch up commercially due to its advantages of low sensitivity to feed concentration, very high salt rejection factor, and operating at low temperatures compared to the well-established thermal desalination technologies. Thus, the MD technology commercialization will be achieved by optimizing its thermal energy performance and including them in hybrid systems in conjunction with other desalination technologies powered by solar energy. In fact, the MD's energy consumption has been decreasing throughout the past couple of years due to novel module designs and energy recovery methods.

The integration of ED with PV is limited by economic barriers, especially for seawater applications. As discussed earlier, ED can only remove salts and small amounts of charged organic compounds, while most uncharged compounds remain in the treated water. This is also similar to the CDI technology as they use expensive electrodes and ion-exchange membranes. Furthermore, to make these technologies more cost-effective, suitable energy storage needs to be established to smooth the solar energy output via battery storage. This would help with the intermittent operation and result in good behavior with fluctuating fluxes compared to PV-RO, which has poor performance in intermittent energy input. Thus, simple power management can push PV-ED, in particular, to be readily deployed in the market. This is more for PV-ED than PV-CDI, because of its simple setup, compact design, robustness, and long lifetime of the energy supply system components. As such, there have been a few PV-driven ED water treatment plants installed worldwide, but all in small-scale applications.

Saying all of this, these LTR technologies can only become feasible to be coupled with solar energy when the cost of solar energy is reduced. Currently, there are countless renewable energy investment incentives to expand the renewable energy market. However, up to this date, solar renewable energy is considered an expensive energy form compared to conventional energy sources (e.g., fossil fuels and natural gas). There is no doubt that the choice of energy storage helps with the renewable energy cost, but only in the short term. However, renewable energy systems need to maintain the continuous rise in the technology-efficiency level to be better economically suited with water treatment technologies.

5.6 Conclusion

MD, ED, and CDI are considered to have a low TRL as they are still in the laboratory development stage. However, this has not stopped researchers from investigating different hybrid designs with solar energy. In general, these aforementioned technologies use low-grade energy (i.e., either thermal or electrical energy), which can benefit from becoming solarized.

Since MD technology is a thermal-based desalination process, solar thermal collectors (e.g., FPC and ETC) can be used to provide the necessary thermal energy to the MD process. This is the most straightforward configuration as the solar collectors heat the feedwater before entering the MD module. Furthermore, to make this design completely solar, PV can be integrated with the solar thermal collectors to provide the necessary electrical energy to run the feed pumps. Otherwise, PV thermal collectors can be used to provide thermal and electrical energy simultaneously. Another novel design is the integrated solar-driven MD, where the membranes are submerged inside an evacuated tube. This allows the feedwater to be heated while the distillation process is taking place, achieving a more compact design than the conventional solar thermal-driven MD configuration.

Unlike MD, ED, and CDI technologies only require electrical energy to operate. As such, PV collectors are used to solarizing ED and CDI technologies. Both ED and CDI use charged electrodes and ion-exchange membranes to separate charged ions from the feed. Generally, PV-ED is more established than PV-CDI as the former technology has proven its potential as there are PV-ED plants installed worldwide. Nevertheless, these two technologies would still require more research in their electrode and membrane materials as they are expensive and require regular maintenance.

In general, there are plenty of opportunities for MD, ED, and CDI (LTR) technologies to become solarized by taking advantage of their low-grade energy source. The recent developments in the pilot plants of these solar-driven technologies affirm the new outlook of sustainable growth. However, there is still much room for intensive and continuous research efforts to propel these technologies into the market. After all, it is only recently that PV-driven reverse osmosis has become cost-effective, while the technologies reviewed in this chapter have the potential to be technically reliable as well as financially profitable.

References

[1] F.A. AlMarzooqi, A.A. Al Ghaferi, I. Saadat, N. Hilal, Application of capacitive deionisation in water desalination: a review, Desalination 342 (2014) 3–15. Available from: https://doi.org/10.1016/j.desal.2014.02.031.

[2] K. Smolders, A.C.M. Franken, Terminology for membrane distillation, Desalination 72 (3) (1989) 249–262. Available from: https://doi.org/10.1016/0011-9164(89)80010-4.

[3] A. Tamburini, P. Pitò, A. Cipollina, G. Micale, M. Ciofalo, A thermochromic liquid crystals image analysis technique to investigate temperature polarization in spacer-filled channels for membrane distillation, Journal of Membrane Science 447 (2013) 260–273. Available from: https://doi.org/10.1016/j.memsci.2013.06.043.
[4] Q. Li, et al., Improving the performance of vacuum membrane distillation using a 3D-printed helical baffle and a superhydrophobic nanocomposite membrane, Separation and Purification Technology (2020) 117072. Available from: https://doi.org/10.1016/j.seppur.2020.117072.
[5] J. Phattaranawik, R. Jiraratananon, A.G. Fane, Heat transport and membrane distillation coefficients in direct contact membrane distillation, Journal of Membrane Science 212 (1) (2003) 177–193. Available from: https://doi.org/10.1016/S0376-7388(02)00498-2.
[6] M. Khayet, T. Matsuura, Chapter 2—Membranes used in MD and design, in: M. Khayet, T. Matsuura (Eds.), Membrane Distillation, Elsevier, Amsterdam, 2011, pp. 17–40.
[7] P.A. Hogan, A. G. Fane Sudjito, G.L. Morrison, Desalination by solar heated membrane distillation, Desalination 81 (1) (1991) 81–90. Available from: https://doi.org/10.1016/0011-9164(91)85047-X.
[8] F. Banat, N. Jwaied, M. Rommel, J. Koschikowski, M. Wieghaus, Desalination by a "compact SMADES" autonomous solar-powered membrane distillation unit, Desalination 217 (1) (2007) 29–37. Available from: https://doi.org/10.1016/j.desal.2006.11.028.
[9] K. Murase, K. Chikamatsu, T. Kyuno, Evaluation of a solar membrane distillator hybridized with a photovoltaic cell, Desalination and Water Treatment 51 (4–6) (2013) 686–694. Available from: https://doi.org/10.1080/19443994.2012.693625.
[10] A. Krnac, M. Araiz, S. Rana, J. Velardo, A. Date, Investigation of direct contact membrane distillation coupling with a concentrated photovoltaic solar system, Energy Procedia 160 (2019) 246–252. Available from: https://doi.org/10.1016/j.egypro.2019.02.143.
[11] C.L. Ong, W. Escher, S. Paredes, A.S.G. Khalil, B. Michel, A novel concept of energy reuse from high concentration photovoltaic thermal (HCPVT) system for desalination, Desalination 295 (2012) 70–81. Available from: https://doi.org/10.1016/j.desal.2012.04.005.
[12] Q. Li, et al., An integrated, solar-driven membrane distillation system for water purification and energy generation, Applied Energy 237 (2019) 534–548. Available from: https://doi.org/10.1016/j.apenergy.2018.12.069.
[13] J. Koschikowski, M. Wieghaus, M. Rommel, Solar thermal-driven desalination plants based on membrane distillation, Desalination 156 (1) (2003) 295–304. Available from: https://doi.org/10.1016/S0011-9164(03)00360-6.
[14] H. Zwijnenberg, G. Koops, M. Wessling, Solar driven membrane pervaporation for desalination processes, Journal of Membrane Science 250 (1–2) (2005) 235–246.
[15] M. Rommel, J. Koschikowski, M. Wieghaus, Solar driven desalination systems based on membrane distillation, Solar Desalination for the 21st Century, Springer, 2007, pp. 247–257.
[16] F. Banat, N. Jwaied, Economic evaluation of desalination by small-scale autonomous solar-powered membrane distillation units, Desalination 220 (1) (2008) 566–573. Available from: https://doi.org/10.1016/j.desal.2007.01.057.
[17] H.E.S. Fath, et al., PV and thermally driven small-scale, stand-alone solar desalination systems with very low maintenance needs, Desalination 225 (1) (2008) 58–69. Available from: https://doi.org/10.1016/j.desal.2006.11.029.

[18] X. Wang, L. Zhang, H. Yang, H. Chen, Feasibility research of potable water production via solar-heated hollow fiber membrane distillation system, Desalination 247 (1) (2009) 403–411. Available from: https://doi.org/10.1016/j.desal.2008.10.008.
[19] J.-P. Mericq, S. Laborie, C. Cabassud, Evaluation of systems coupling vacuum membrane distillation and solar energy for seawater desalination, Chemical Engineering Journal 166 (2) (2011) 596–606. Available from: https://doi.org/10.1016/j.cej.2010.11.030.
[20] R.B. Saffarini, E.K. Summers, H.A. Arafat, J.H. Lienhard V, Economic evaluation of stand-alone solar powered membrane distillation systems, Desalination 299 (2012) 55–62. Available from: https://doi.org/10.1016/j.desal.2012.05.017.
[21] H. Chang, C.-L. Chang, C.-Y. Hung, T.-W. Cheng, C.-D. Ho, Optimization study of small-scale solar membrane distillation desalination systems (s-SMDDS), International Journal of Environmental Research and Public Health 11 (2014) 12064–12087. https://doi.org/0.3390/ijerph111112064.
[22] A.J. Hughes, T.S. O'Donovan, T.K. Mallick, Experimental evaluation of a membrane distillation system for integration with concentrated photovoltaic/thermal (CPV/T) energy, Energy Procedia 54 (2014) 725–733. Available from: https://doi.org/10.1016/j.egypro.2014.07.313.
[23] W.G. Shim, K. He, S. Gray, I.S. Moon, Solar energy assisted direct contact membrane distillation (DCMD) process for seawater desalination, Separation and Purification Technology 143 (2015) 94–104.
[24] H.C. Duong, L. Xia, Z. Ma, P. Cooper, W. Ela, L.D. Nghiem, Assessing the performance of solar thermal driven membrane distillation for seawater desalination by computer simulation, Journal of Membrane Science 542 (2017) 133–142.
[25] V. Karanikola, S.E. Moore, A. Deshmukh, R.G. Arnold, M. Elimelech, A.E. Sáez, Economic performance of membrane distillation configurations in optimal solar thermal desalination systems, Desalination 472 (2019) 114164. Available from: https://doi.org/10.1016/j.desal.2019.114164.
[26] S.E. Moore, S.D. Mirchandani, V. Karanikola, T.M. Nenoff, R.G. Arnold, A. Eduardo Sáez, Process modeling for economic optimization of a solar driven sweeping gas membrane distillation desalination system, Desalination 437 (2018) 108–120. Available from: https://doi.org/10.1016/j.desal.2018.03.005.
[27] K. Zarzoum, K. Zhani, H.B. Bacha, J. Koschikowski, Experimental parametric study of membrane distillation unit using solar energy, Solar Energy 188 (2019) 1274–1282.
[28] H. Deng, et al., Modeling and optimization of solar thermal-photovoltaic vacuum membrane distillation system by response surface methodology, Solar Energy 195 (2020) 230–238. Available from: https://doi.org/10.1016/j.solener.2019.11.006.
[29] G. Li, L. Lu, Modeling and performance analysis of a fully solar-powered stand-alone sweeping gas membrane distillation desalination system for island and coastal households, Energy Conversion and Management 205 (2020) 112375. Available from: https://doi.org/10.1016/j.enconman.2019.112375.
[30] M. Rabie, M.S. Salem, A.Y. Ali, A. El-Shazly, M. Elkady, S. Ookawara, Modeling of an integrated air gap membrane distillation unit utilizing a flat plate solar collector, Energy Reports 6 (2020) 1591–1596.
[31] Q. Zhao, H. Zhang, Z. Hu, S. Hou, A solar driven hybrid photovoltaic module/direct contact membrane distillation system for electricity generation and water desalination, Energy Conversion and Management 221 (2020) 113146. Available from: https://doi.org/10.1016/j.enconman.2020.113146.

[32] A. Bamasag, T. Alqahtani, S. Sinha, N. Ghaffour, P. Phelan, Solar-heated submerged vacuum membrane distillation system with agitation techniques for desalination, Separation and Purification Technology 256 (2021) 117855. Available from: https://doi.org/10.1016/j.seppur.2020.117855.

[33] G.S. Kim, T. Cao, Y. Hwang, Thermoeconomic investigation for a multi-stage solar-thermal vacuum membrane distillation system for coastal cities, Desalination 498 (2021) 114797. Available from: https://doi.org/10.1016/j.desal.2020.114797.

[34] R. Miladi, N. Frikha, S. Gabsi, Modeling and energy analysis of a solar thermal vacuum membrane distillation coupled with a liquid ring vacuum pump, Renewable Energy 164 (2021) 1395–1407. Available from: https://doi.org/10.1016/j.renene.2020.10.136.

[35] S. Al-Amshawee, M.Y.B.M. Yunus, A.A.M. Azoddein, D.G. Hassell, I.H. Dakhil, H.A. Hasan, Electrodialysis desalination for water and wastewater: a review, Chemical Engineering Journal 380 (2020) 122231. Available from: https://doi.org/10.1016/j.cej.2019.122231.

[36] A. Campione, L. Gurreri, M. Ciofalo, G. Micale, A. Tamburini, A. Cipollina, Electrodialysis for water desalination: a critical assessment of recent developments on process fundamentals, models and applications, Desalination 434 (2018) 121–160. Available from: https://doi.org/10.1016/j.desal.2017.12.044.

[37] C. Fernandez-Gonzalez, A. Dominguez-Ramos, R. Ibañez, A. Irabien, Sustainability assessment of electrodialysis powered by photovoltaic solar energy for freshwater production, Renewable and Sustainable Energy Reviews 47 (2015) 604——615. Available from: https://doi.org/10.1016/j.rser.2015.03.018.

[38] G.J. Doornbusch, M. Tedesco, J.W. Post, Z. Borneman, K. Nijmeijer, Experimental investigation of multistage electrodialysis for seawater desalination, Desalination 464 (2019) 105–114. Available from: https://doi.org/10.1016/j.desal.2019.04.025.

[39] G.J. Doornbusch, M. Bel, M. Tedesco, J.W. Post, Z. Borneman, K. Nijmeijer, Effect of membrane area and membrane properties in multistage electrodialysis on seawater desalination performance, Journal of Membrane Science 611 (2020) 118303. Available from: https://doi.org/10.1016/j.memsci.2020.118303.

[40] D.C. Buzzi, L.S. Viegas, M.A.S. Rodrigues, A.M. Bernardes, J.A.S. Tenório, Water recovery from acid mine drainage by electrodialysis, Minerals Engineering 40 (2013) 82–89. Available from: https://doi.org/10.1016/j.mineng.2012.08.005.

[41] A. Merkel, A.M. Ashrafi, M. Ondrušek, The use of electrodialysis for recovery of sodium hydroxide from the high alkaline solution as a model of mercerization wastewater, Journal of Water Process Engineering 20 (2017) 123–129. Available from: https://doi.org/10.1016/j.jwpe.2017.10.008.

[42] J. Jang, Y. Kang, J.-H. Han, K. Jang, C.-M. Kim, I.S. Kim, Developments and future prospects of reverse electrodialysis for salinity gradient power generation: influence of ion exchange membranes and electrodes, Desalination 491 (2020) 114540. Available from: https://doi.org/10.1016/j.desal.2020.114540.

[43] V. Sarapulova, et al., Transport characteristics of fujifilm ion-exchange membranes as compared to homogeneous membranes AMX and CMX and to heterogeneous membranes MK-40 and MA-41, 9, 7, 84. [Online]. Available from: <https://www.mdpi.com/2077-0375/9/7/84>, 2019.

[44] H.M.N. AlMadani, Water desalination by solar powered electrodialysis process, Renewable Energy 28 (12) (2003) 1915–1924. Available from: https://doi.org/10.1016/S0960-1481(03)00014-4.

[45] A. Albloushi, et al., Chapter 7—Renewable energy-powered membrane systems for water desalination, in: A. Basile, A. Cassano, A. Figoli (Eds.), Current Trends and Future Developments on (Bio-)Membranes, Elsevier, 2019, pp. 153–177.

[46] N.J.D. Ishimaru, Solar photovoltaic desalination of brackish water in remote areas by electrodialysis, Desalination 98 (1–3) (1994) 485–493.

[47] O. Kuroda, et al., An electrodialysis sea water desalination system powered by photovoltaic cells, Desalination 65 (1987) 161–169. Available from: https://doi.org/10.1016/0011-9164(87)90132-9.

[48] B. Peñate, F. Círez, F.J. Domínguez, V.J. Subiela, L. Vera, Design and testing of an isolated commercial EDR plant driven by solar photovoltaic energy, Desalination and Water Treatment 51 (4–6) (2013) 1254–1264.

[49] M.R. Adiga, S.K. Adhikary, P.K. Narayanan, W.P. Harkare, S.D. Gomkale, K.P. Govindan, Performance analysis of photovoltaic electrodialysis desalination plant at Tanote in Thar desert, Desalination 67 (1987) 59–66. Available from: https://doi.org/10.1016/0011-9164(87)90232-3.

[50] M.T. Ali, H.E.S. Fath, P.R. Armstrong, A comprehensive techno-economical review of indirect solar desalination, Renewable and Sustainable Energy Reviews 15 (8) (2011) 4187–4199. Available from: https://doi.org/10.1016/j.rser.2011.05.012.

[51] M. Lichtwardt, H. Remmers, Water treatment using solar powered electrodialysis reversal, in: Proceedings of the Mediterranean Conference on Renewable Energy Sources for Water Production, 1996.

[52] J.M. Ortiz, E. Expósito, F. Gallud, V. García-García, V. Montiel, A. Aldaz, Photovoltaic electrodialysis system for brackish water desalination: modeling of global process, Journal of Membrane Science 274 (1) (2006) 138–149. Available from: https://doi.org/10.1016/j.memsci.2005.08.006.

[53] J.M. Ortiz, E. Expósito, F. Gallud, V. García-García, V. Montiel, A. Aldaz, Desalination of underground brackish waters using an electrodialysis system powered directly by photovoltaic energy, Solar Energy Materials and Solar Cells 92 (12) (2008) 1677–1688. Available from: https://doi.org/10.1016/j.solmat.2008.07.020.

[54] F. Círez, J. Uche, A.A. Bayod, A. Martínez, ED Batch (Eds.), Fed by a PV Unit: A Reliable, Flexible, and Sustainable Integration, Desalination and Water Treatment 51 (2013) 4673–6685.

[55] A. Gonzalez, M. Grágeda, S. Ushak, Assessment of pilot-scale water purification module with electrodialysis technology and solar energy, Applied Energy 206 (2017) 1643––1652. Available from: https://doi.org/10.1016/j.apenergy.2017.09.101.

[56] W. He, S. Amrose, N.C. Wright, T. Buonassisi, I.M. Peters, A.G. Winter, Field demonstration of a cost-optimized solar powered electrodialysis reversal desalination system in rural India, Desalination 476 (2020) 114217. Available from: https://doi.org/10.1016/j.desal.2019.114217.

[57] A. Campione, A. Cipollina, F. Calise, A. Tamburini, M. Galluzzo, G. Micale, Coupling electrodialysis desalination with photovoltaic and wind energy systems for energy storage: dynamic simulations and control strategy, Energy Conversion and Management 216 (2020) 112940. Available from: https://doi.org/10.1016/j.enconman.2020.112940.

[58] N.C. Wright, A.G. Winter, Justification for community-scale photovoltaic-powered electrodialysis desalination systems for inland rural villages in India, Desalination 352 (2014) 82–91. Available from: https://doi.org/10.1016/j.desal.2014.07.035.

[59] T. Abraham, A. Luthra, Socio-economic & technical assessment of photovoltaic powered membrane desalination processes for India, Desalination 268 (1) (2011) 238–248. Available from: https://doi.org/10.1016/j.desal.2010.10.035.

[60] C. Li, Y. Goswami, E. Stefanakos, Solar assisted sea water desalination: a review, Renewable and Sustainable Energy Reviews 19 (2013) 136–163. Available from: https://doi.org/10.1016/j.rser.2012.04.059.

[61] A. Ali, R.A. Tufa, F. Macedonio, E. Curcio, E. Drioli, Membrane technology in renewable-energy-driven desalination, Renewable and Sustainable Energy Reviews 81 (2018) 1–21. Available from: https://doi.org/10.1016/j.rser.2017.07.047.

[62] Y.-U. Shin, J. Lim, C. Boo, S. Hong, Improving the feasibility and applicability of flow-electrode capacitive deionization (FCDI): review of process optimization and energy efficiency, Desalination 502 (2021) 114930. Available from: https://doi.org/10.1016/j.desal.2021.114930.

[63] B.B. Arnold, G.W. Murphy, Studies on the electrochemistry of carbon and chemically-modified carbon surfaces, The Journal of Physical Chemistry 65 (1) (1961) 135–138.

[64] Y. Oren, A. Soffer, Water desalting by means of electrochemical parametric pumping, Journal of Applied Electrochemistry 13 (4) (1983) 473–487.

[65] J.-B. Lee, K.-K. Park, H.-M. Eum, C.-W.J.D. Lee, Desalination of a thermal power plant wastewater by membrane capacitive deionization, Desalination 196 (1–3) (2006) 125–134.

[66] W. Tang, et al., Various cell architectures of capacitive deionization: recent advances and future trends, Water Research 150 (2019) 225–251. Available from: https://doi.org/10.1016/j.watres.2018.11.064.

[67] L. Han, K.G. Karthikeyan, M.A. Anderson, J.J. Wouters, K.B. Gregory, Mechanistic insights into the use of oxide nanoparticles coated asymmetric electrodes for capacitive deionization, Electrochimica Acta 90 (2013) 573–581. Available from: https://doi.org/10.1016/j.electacta.2012.11.069.

[68] S.-Y. Pan, A.Z. Haddad, A. Kumar, S.-W. Wang, Brackish water desalination using reverse osmosis and capacitive deionization at the water-energy nexus, Water Research 183 (2020) 116064. Available from: https://doi.org/10.1016/j.watres.2020.116064.

[69] C.C. Hu, C.F. Hsieh, Y.J. Chen, C.F. Liu, How to achieve the optimal performance of capacitive deionization and inverted-capacitive deionization" (in English) Desalination 442 (2018) 89–98. Available from: https://doi.org/10.1016/j.desal.2018.05.013.

[70] J. Gamby, P. Taberna, P. Simon, J. Fauvarque, M. Chesneau, Studies and characterisations of various activated carbons used for carbon/carbon supercapacitors, Journal of Power Sources 101 (1) (2001) 109–116.

[71] S. Porada, L. Borchardt, M. Oschatz, M. Bryjak, J.S. Atchinson, K.J. Keesman, S. Kaskel, P.M. Biesheuvel, V. Presser, Direct prediction of the desalination performance of porous carbon electrodes for capacitive deionization, Energy & Environmental Science 6 (12) (2013) 3700–3712. Available from: https://doi.org/10.1039/C3EE42209G.

[72] Y.-W. Chen, J.-F. Chen, C.-H. Lin, C.-H. Hou, Integrating a supercapacitor with capacitive deionization for direct energy recovery from the desalination of brackish water, Applied Energy 252 (2019). Available from: https://doi.org/10.1016/j.apenergy.2019.113417.

[73] R.L. Zornitta, et al., Charge and potential balancing for optimized capacitive deionization using lignin-derived, low-cost activated carbon electrodes, ChemSusChem 11 (13) (2018) 2101–2113. Available from: https://doi.org/10.1002/cssc.201800689.

[74] H. Jo, K.H. Kim, M.-J. Jung, J.H. Park, Y.-S. Lee, Fluorination effect of activated carbons on performance of asymmetric capacitive deionization, Applied Surface Science 409 (2017) 117–123. Available from: https://doi.org/10.1016/j.apsusc.2017.02.234.

[75] Y. Liu, T. Chen, T. Lu, Z. Sun, D.H.C. Chua, L. Pan, Nitrogen-doped porous carbon spheres for highly efficient capacitive deionization, Electrochimica Acta 158 (2015) 403–409. Available from: https://doi.org/10.1016/j.electacta.2015.01.179.
[76] X. Xu, et al., Capacitive deionization using nitrogen-doped mesostructured carbons for highly efficient brackish water desalination, Chemical Engineering Journal 362 (2019) 887–896. Available from: https://doi.org/10.1016/j.cej.2019.01.098.
[77] S. Santangelo, et al., Evaluation of the electrochemical performance of electrospun transition metal oxide-based electrode nanomaterials for water CDI applications, Electrochimica Acta 309 (2019) 125–139. Available from: https://doi.org/10.1016/j.electacta.2019.04.075.
[78] G. Zhu, H.Y. Wang, H.F. Xu, L. Zhang, Enhanced capacitive deionization by nitrogen-doped porous carbon nanofiber aerogel derived from bacterial-cellulose" (in English)Journal of Electroanalytical Chemistry 822 (2018) 81–88. Available from: https://doi.org/10.1016/j.jelechem.2018.05.024.
[79] Y. Liu, J. Ma, T. Lu, L. Pan, Electrospun carbon nanofibers reinforced 3D porous carbon polyhedra network derived from metal-organic frameworks for capacitive deionization, Scientific Reports 6 (2016) 32784. Available from: https://doi.org/10.1038/srep32784.
[80] W. Dianbudiyanto, S.-H. Liu, Outstanding performance of capacitive deionization by a hierarchically porous 3D architectural graphene, Desalination 468 (2019) 114069. Available from: https://doi.org/10.1016/j.desal.2019.07.009.
[81] W. Shi, et al., Ultrahigh performance of novel capacitive deionization electrodes based on a three-dimensional graphene architecture with nanopores, Scientific Reports 6 (1) (2016) 18966. Available from: https://doi.org/10.1038/srep18966.
[82] A.S. Yasin, H.O. Mohamed, I.M.A. Mohamed, H.M. Mousa, N.A.M. Barakat, Enhanced desalination performance of capacitive deionization using zirconium oxide nanoparticles-doped graphene oxide as a novel and effective electrode, Separation and Purification Technology 171 (2016) 34–43. Available from: https://doi.org/10.1016/j.seppur.2016.07.014.
[83] X. Xu, L. Pan, Y. Liu, T. Lu, Z. Sun, D.H. Chua, Facile synthesis of novel graphene sponge for high performance capacitive deionization, Scientific Reports 5 (2015) 8458. Available from: https://doi.org/10.1038/srep08458.
[84] J. Ma, L. Wang, F. Yu, Water-enhanced performance in capacitive deionization for desalination based on graphene gel as electrode material, Electrochimica Acta 263 (2018) 40–46. Available from: https://doi.org/10.1016/j.electacta.2018.01.041.
[85] Y. Wang, et al., Graphite felt 3D framework composites as an easy to scale capacitive deionization electrode for brackish water desalination, Chemical Engineering Journal (2019). Available from: https://doi.org/10.1016/j.cej.2019.123698.
[86] K. Wei, et al., A novel capacitive electrode based on TiO_2-NTs array with carbon embedded for water deionization: fabrication, characterization and application study, Desalination 420 (2017) 70–78. Available from: https://doi.org/10.1016/j.desal.2017.07.001.
[87] A.S. Yasin, I.M.A. Mohamed, H.M. Mousa, C.H. Park, C.S. Kim, Facile synthesis of TiO_2/ZrO_2 nanofibers/nitrogen co-doped activated carbon to enhance the desalination and bacterial inactivation via capacitive deionization, Scientific Reports 8 (1) (2018) 541. Available from: https://doi.org/10.1038/s41598-017-19027-w.
[88] C. Hu, T. Wang, J. Dong, R. Liu, H. Liu, J. Qu, Capacitive deionization from reconstruction of NiCoAl-mixed metal oxide film electrode based on the "memory effect,", Applied Surface Science 459 (2018) 767–773. Available from: https://doi.org/10.1016/j.apsusc.2018.08.055.

[89] J. Shen, et al., Hollow ZIFs-derived nanoporous carbon for efficient capacitive deionization, Electrochimica Acta 273 (2018) 34—42. Available from: https://doi.org/10.1016/j.electacta.2018.04.004.
[90] T.N. Baroud, E.P. Giannelis, High salt capacity and high removal rate capacitive deionization enabled by hierarchical porous carbons, Carbon 139 (2018) 614—625. Available from: https://doi.org/10.1016/j.carbon.2018.05.053.
[91] M. Li, H.G. Park, Pseudocapacitive coating for effective capacitive deionization, ACS Applied Materials & Interfaces 10 (3) (2018) 2442—2450. Available from: https://doi.org/10.1021/acsami.7b14643.
[92] T. Alencherry, A.R. Naveen, S. Ghosh, J. Daniel, R. Venkataraghavan, Effect of increasing electrical conductivity and hydrophilicity on the electrosorption capacity of activated carbon electrodes for capacitive deionization, Desalination 415 (2017) 14—19. Available from: https://doi.org/10.1016/j.desal.2017.04.001.
[93] B.Q. Qian, et al., Sulfonated graphene as cation-selective coating: a new strategy for high-performance membrane capacitive deionization" (in English)Advanced Materials Interfaces 2 (16) (2015). Available from: https://doi.org/10.1002/admi.201500372. ARTN1500372.
[94] A. Jain, et al., Aqueous-processed, high-capacity electrodes for membrane capacitive deionization, Environmental Science & Technology 52 (10) (2018) 5859——5867. Available from: https://doi.org/10.1021/acs.est.7b05874.
[95] C. Zhang, L. Wu, J. Ma, M. Wang, J. Sun, T.D. Waite, Evaluation of long-term performance of a continuously operated flow-electrode CDI system for salt removal from brackish waters, Water Research 173 (2020) 115580. Available from: https://doi.org/10.1016/j.watres.2020.115580.
[96] J. Lee, K. Jo, J. Lee, S.P. Hong, S. Kim, J. Yoon, Rocking-chair capacitive deionization for continuous brackish water desalination" (in English)ACS Sustainable Chemistry & Engineering 6 (8) (2018) 10815—10822. Available from: https://doi.org/10.1021/acssuschemeng.8b02123.
[97] X. Gao, A. Omosebi, J. Landon, K. Liu, Enhanced salt removal in an inverted capacitive deionization cell using amine modified microporous carbon cathodes, Environmental Science & Technology 49 (18) (2015) 10920—10926. Available from: https://doi.org/10.1021/acs.est.5b02320.
[98] J. Ma, et al., Development of redox-active flow electrodes for high-performance capacitive deionization, Environmental Science & Technology 50 (24) (2016) 13495—13501. Available from: https://doi.org/10.1021/acs.est.6b03424.
[99] J. Ma, C. He, D. He, C. Zhang, T.D. Waite, Analysis of capacitive and electrodialytic contributions to water desalination by flow-electrode CDI, Water Research 144 (2018) 296—303. Available from: https://doi.org/10.1016/j.watres.2018.07.049.
[100] Y. Cho, K.S. Lee, S. Yang, J. Choi, H.-r Park, D.K. Kim, A novel three-dimensional desalination system utilizing honeycomb-shaped lattice structures for flow-electrode capacitive deionization, Energy & Environmental Science 10 (8) (2017) 1746—1750. Available from: https://doi.org/10.1039/C7EE00698E.
[101] C. Zhang, L. Wu, J. Ma, A.N. Pham, M. Wang, T.D. Waite, Integrated flow-electrode capacitive deionization and microfiltration system for continuous and energy-efficient brackish water desalination, Environmental Science & Technology 53 (22) (2019) 13364—13373. Available from: https://doi.org/10.1021/acs.est.9b04436.
[102] C. Tan, C. He, W. Tang, P. Kovalsky, J. Fletcher, T.D. Waite, Integration of photovoltaic energy supply with membrane capacitive deionization (MCDI) for salt removal from brackish waters, Water Research 147 (2018) 276—286. Available from: https://doi.org/10.1016/j.watres.2018.09.056.

[103] C. Tan, C. He, J. Fletcher, T.D. Waite, Energy recovery in pilot scale membrane CDI treatment of brackish waters, Water Research 168 (2020) 115146. Available from: https://doi.org/10.1016/j.watres.2019.115146.
[104] C. Bales, B. Lian, J. Fletcher, Y. Wang, T.D. Waite, Site specific assessment of the viability of membrane capacitive deionization (mCDI) in desalination of brackish groundwaters for selected crop watering, Desalination 502 (2021) 114913. Available from: https://doi.org/10.1016/j.desal.2020.114913.

CHAPTER 6

Unlocking nanomaterials for next generation solar-driven water treatment technologies

Todd P. Otanicar and Nipun Goel
Mechanical and Biomedical Engineering, Boise State University, Boise, ID, United States

6.1 The potential of nanomaterials for solar-driven water treatment

A significant body of literature has been devoted to exploring the use of nanoparticles suspended in liquids for direct absorption solar collectors (DASCs) [1], vapor generation for sterilization [2], nanoparticles to enhance the properties PV cells [3], and nanoparticles for advanced window thermal control [4]. Table 6.1 provides a high-level overview of the role of nanomaterials in various solar technologies, the vast majority of these are used for enhancing the absorption of solar energy collection. In PV materials, integrating plasmonic nanoparticles can vastly increase the solar absorption [5], while semiconducting quantum dots can be used to downshift high energy photons to lower energy levels where the cells have higher quantum efficiency [6].

With regards to water treatment, not only do nanomaterials have unique properties due to their size, they can be designed in ways that incorporate feature sizes (pores, channels, etc.) which can target the sized-based removal of elements during a separation process. Further, multifunctional materials [7] (e.g., combining high optical absorption with high physical/chemical absorption) may open up new areas to remove water-borne pollutants (as shown in Table 6.2). While all of the previous references and studies discussed through this chapter demonstrate that nanotechnology holds significant promise, there is a lack of government and regulatory oversight in these materials. In addition, scaling up many of these technologies represents a limiting factor for incorporation into industrial wastewater treatment facilities [8]. This chapter will step through the nanomaterial classes that could be used in such applications: nanoparticles, nanofibers, and nanomembranes,

Solar-Driven Water Treatment
DOI: https://doi.org/10.1016/B978-0-323-90991-4.00002-5

Table 6.1 The role of nanomaterials in enhancing solar energy harvesting.

Level of difficulty	Nanotechnology type	Nanotechnology-enhanced property
Basic -> Advanced	Nanoparticles	Thermal conductivity, solar absorptivity, photocatalysis
	Nanorods	Thermal conductivity, solar absorptivity, photocatalysis
	Nanofibers	Solar absorptivity, photocatalysis
	Nanomembranes	Energy transfer

Table 6.2 Water treatment objectives and the opportunities for nanotechnology to have an impact.

Level of difficulty	Target pollutant	Nanotechnology-enhanced process
Basic -> Advanced	Pathogens	Photocatalysis
	Turbidity, and taste	Selective adsorbents
	Heavy Metals	Selective adsorbents, membranes, filtration
	Hardness	Electrosorption, nucleation
	Pesticides	Sorbents, photocatalysis, membranes
	Pharmaceuticals	Filtration, membranes, sorbents, photocatalysis
	Salts	Membranes, electrosorption, distillation

Recreated from P.J.J. Alvarez, C.K. Chan, M. Elimelech, N.J. Halas, D. Villagrán, Emerging opportunities for nanotechnology to enhance water security. Nature Nanotechnology 13 (8) (2018) 634–641. https://doi.org/10.1038/s41565-018-0203-2.

while also providing commentary on the resulting environmental implications of employing these materials. The discussion will focus on how nanomaterials can enhance solar energy collection, enhance water treatment processes, and, importantly, how these enhancements can be synergistically combined to create the next generation of solar-driven water treatment technologies.

6.2 Nanoparticles

A nanoparticle is typically defined as a particle that has all its three dimensions measuring in at 100 nm or less [9]. They can be considered as a subset within the broader category of *nanomaterials*, which includes nanofibers, nanorods, nanotubes, nanoribbons, nanomembranes, etc. The small size of nanoparticles gives them the distinct advantage of an increased surface area compared to materials at macroscale, and also enhanced chemical reactivity, and mechanical, electrical, optical, and magnetic properties. When dispersed in a bulk medium, the alterations in these properties enable enhanced bulk properties, making them a promising additive in a wide range of applications such as electronics, healthcare, chemicals, cosmetics, composites, energy, etc.

The list of known nanoparticles is fairly inexhaustible resulting in classification systems not only based on the material used (i.e., metallic, nonmetallic, composites, and polymers), but also based on their preparation methodology and fundamental properties such as size distribution, crystal structure, surface area, porosity, surface chemistry, surface charge, etc. [10]. Since the first demonstrated enhancement in thermal conductivity of heat transfer fluids from the use of nanoparticles by Choi and Eastmen [11], a wide array of nanoparticle synthesis techniques have been developed that allow for control of nanoparticle size distribution, morphology, composition, crystallinity, etc. The use of material and technology of preparation of nanoparticles is ultimately determined by their field of application and the need for sustainability and environmental effects from their use. Coprecipitation, for example, is a simple and rapid method of preparation using chemical reactions, which provides easy control of particle size, homogeneity, and composition at low temperatures. However, it faces problems with reproducibility between batches and can only be applied to charged particles [12]. Sputtering is another prominent synthesis method in which ions of the desired target material are bombarded off the cathode using energetic particles and deposited on a substrate to form a layer or film. The biggest advantage of this technique, in addition to the fact that it can be applied to almost any material, is that the films developed this way have the same composition as the target material [13]. For a more detailed explanation on these nanoparticle synthesis techniques, as well as other commonly used techniques such as hydrothermal synthesis, ball milling, micromechanical synthesis, laser ablation, sol gel, chemical and physical vapor deposition, etc. [14], readers are encouraged to read the review articles by Jamkhande et al. [15] and Rane et al. [16].

Some examples of commonly used nanoparticles in the field of energy generation and water purification include, but are not limited to, TiO_2, Fe_2O_3, ZnO, Al_2O_3, Ag, Au, CNT. Table 6.3 attempts at categorizing these nanoparticles by the specific properties enhanced, and a detailed discussion of these nanoparticles is presented in the following sections.

Table 6.3 Nanoparticles for specific application driven property enhancement.

Material property	Nanoparticles
Thermal conductivity	CuO, TiO_2, CNT, Al_2O_3, Au, Ag, Fe_3O_4, Fe_2O_3, ZnO, Graphene
Optical absorption	Ag, Au, ZnO, Te, ZnS, TiO_2
Electrical conductivity	Ag, Al_2O_3, SiO_2, ZnO
Adsorption	CNT, TiO_2, Fe_3O_4, CuO, ZrO_2
Chemical reactivity	TiO_2ZnO, CuS, Fe_2O_3

6.2.1 Nanoparticles for enhancing solar energy collection

Improvements in the absorption of the short-wavelength, or optical, part of the solar spectrum are of interest in photovoltaic applications, whereas absorption of both the longer-wavelength, or thermal, as well as the short-wavelength part of the solar spectrum are of interest in solar thermal applications such as solar steam generation, concentrated solar power, solar thermal-driven desalination.

In general, the efficiency of a typical solar material is limited by how much of the solar spectrum it can absorb. For example, the addition of nanoparticles to phase change materials can significantly increase their conductive and convective heat transfer properties [17]. Nanoparticles can also increase the plasmonic resonance when integrated into the materials for solar cells. While some of these effects are achieved through an increase in effective absorption of the solar spectrum, other can be attributed to the enhanced scattering of incident solar energy [18–21]. Absorption is the process by which incident solar radiation is converted to heat, either directly from plasmon decay into phonons or from ohmic losses. The ability of the nanoparticles to absorb incident radiation is determined in terms of absorption cross-section, and can be calculated from wave number in the media (k), and the dipole polarizability of the particle (α) as:

$$\sigma_{Abs} = 4\pi k \mathrm{Im}\{\alpha\} \tag{6.1}$$

Scattering, on the other hand, can be defined as the process that results in a change in direction of propagation of light while conserving the total energy of incident light. Since the size of nanoparticles is of the order of the incident wavelength of light, the individual nanoparticles can be treated as a point dipole with the dipole oscillation in the direction of light polarization. In other words, the nanoparticles re-radiate light that was converted to a plasmon oscillation in the pattern of dipole radiation. The scattering cross-section for the nanoparticle can be computed using the wave number in the media (k), and the dipole polarizability of the particle (α) as:

$$\sigma_{sca,dipole} = \frac{8}{3}\pi k^4 |\alpha|^2 \tag{6.2}$$

Enhanced scattering is particularly beneficial in plasmonic solar cells as it not only provides a spectral shift, but the scattered light also takes a longer path through the semiconductor material. The plasmon scattering and absorption of nanoparticles is typically governed by size, shape and composition of the nanoparticles, the surrounding material (solid or fluid), and volume fraction of nanoparticles which influence the proximity between nanoparticles. For example, change in shape and size can alter the electron oscillation pattern on the surface. More specifically, an increase in size of the particle or aspect ratio of the particle redshifts the plasmon resonance to longer wavelengths and an increased scattering is observed [22]. Similarly, changes in nanoparticle composition can alter the free carrier density and plasmon decay through losses in the material. For example, metallic nanoparticles have a set carrier density, however with addition of dopants or use of metal oxides can cause a variation in plasmon resonance. This free charge carrier concentration can be further modulated by adjusting the oxide concentration [23,24]. The choice and behavior of the surrounding media can alter the refractive index around a particle, which in turn changes the number of molecules capable of producing induced molecular dipoles in the media from electronic oscillations. This can result in shifts and amplitude changes in plasmonic behavior.

Lastly, nanoparticles in proximity to each other can interact and modify the resonance location. Although direct interactions from neighboring nanoparticles suspended in the surrounding media in low enough concentrations can be ignored, nanoparticle agglomeration from settling over time or due to chemical interactions with the surfactant can effectively increase the size of the nanoparticle, resulting in a redshifted resonance and an increase in scattering [25]. These observations have been the primary target of several studies to achieve higher thermal efficiencies in solar collectors.

6.2.1.1 Nanoparticles for photovoltaics

Depending on the application, the goals from the use of nanoparticles in solar energy are very specific. Photovoltaic applications typically desire a higher scattering of light, whereas solar thermal applications aim for higher solar absorption. The different mechanisms that could be utilized for photovoltaic applications can be broadly broken down into three types: (1) far field effect, where metal nanoparticles can be used to enhance scattering of incident radiation, (2) the near field enhancement, where an increased number of photogenerated charge carriers is obtained due to enhancement in electric field, and (3) direct generation of charge carriers in the semiconductor substrate [26]. Stuart and Hall are often regarded as pioneers for their demonstration of a nearly eightfold enhancement in photocurrent of silicon-on-insulator photodetectors from plasmonic scattering using Ag nanoparticles for light of wavelength 800 nm [27]. The technique has since been advanced for use in photovoltaics and solar cells as well. Derkacs et al. used Au nanoparticles deposited on the surface of thin-film amorphous silicon p–i–n solar cells and observed an 8.1% and 8.3% increase in short-circuit current density and energy conversion efficiency, respectively [28]. Rand et al. investigated the near field enhancement in organic tandem thin-film solar cells containing Ag nanoparticles. They created organic thin films by thermal evaporation at room temperature in high vacuum and subsequently deposited a nanocluster layer of Ag nanoparticles to a thickness of 1 nm. They recorded an increase in the optical-field intensity in the near field of the chain of Ag nanoclusters by upto a factor of 100 compared to the incident light intensity, and subsequently an increased absorption in thin organic films placed in contact with the clusters [29].

A scarcity of semiconductor materials and the need to lower the cost have driven investigators to reduce the thickness of photovoltaic cells. This has further reinforced the need for light trapping so that the efficiency of solar cells is not hindered with reduced cell thickness. In a detailed review done by Atwater and Polman on the use of plasmonics in solar cells, they showcased that metal nanoparticles allow for light trapping by selective scattering of light, coupling the plasmonic near field of the nanoparticle to the semiconductor, and excitation of surface plasmon polaritons at the metal/semiconductor interface as illustrated in Fig. 6.1 [30].

6.2.1.2 Nanoparticles for solar thermal

Enhancement of thermal properties with nanoparticles can occur due to their small size which results in an increased surface area, making them extremely promising for solar thermal applications, including but not limited to direct absorption solar collectors, phase change materials for thermal energy storage, solar-driven space, water heaters, etc.

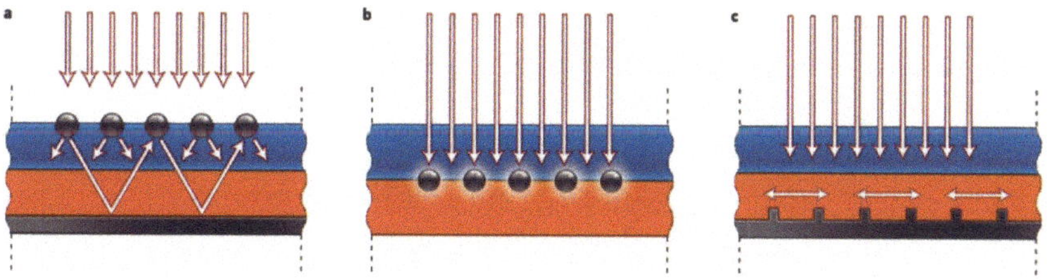

Figure 6.1 Light trapping by (A) scattering from metal nanoparticles at the surface of the solar cell, (B) the excitation of localized surface plasmons in metal nanoparticles embedded in the semiconductor, and (C) the excitation of surface plasmon polaritons at the metal/semiconductor interface. *Reproduced with permission from H.A. Atwater, A. Polman, Plasmonics for improved photovoltaic devices. Nature Materials 9 (2010) 205–213. https://doi.org/10.1038/nmat2629.*

The use of nanoparticles in phase change materials (PCMs) for thermal energy storage has garnered much interest due to the promising increase in thermal conductivity of the PCMs [31–33]. Kumar et al. studied the changes in thermal conductivity of commercial grade paraffin wax PCM when dispersed with different weight percentages of Si_3N_4 nanoparticles. They noted a nonlinear improvement in the thermal conductivity of paraffin wax with the increase in weight percentage of nano-Si_3N_4 particles. They argued that the increase in weight percentage of nanoparticles also resulted in localized agglomeration of the particles resulting in decline in the rate of thermal conductivity improvement for the paraffin wax PCM [32].

In another study, Saranprabhu and Rajan investigated the changes in solid phase specific heat, and thermal conductivity of solar salt mixed with lower concentrations of MgO nanoparticles. They noted that the addition of 0.25 wt.% of MgO nanoparticles resulted in 17.5% increase in solid phase thermal conductivity of solar salt. They also noted that the specific heat for the MgO solar salt composite increased nonlinearly with the increase in MgO wt.%, whereas the solidification time during the discharge cycle reduced by 30% due to higher thermal diffusivity for the MgO solar salt composite [34].

Improvements in thermal conductivity of nanofluids has also been an active field of research [35,36]. Moghadassi et al. studied the effects of CuO nanoparticle size and concentration on the thermal conductivity of paraffin and mono-ethylene glycol fluids. They noted a consistently higher thermal conductivity for CuO nanofluids prepared with nanoparticles of size 30 nm for all nanofluid concentrations tested compared to those prepared with CuO nanoparticles of size 40 nm. More surprisingly, they observed a 50% increase in thermal conductivity of both the mono-ethylene glycol and paraffin nanofluids when mixed with 5% volume fraction of CuO nanoparticles [37].

Devarajan et al. tested for improvements in efficiency of solar water heating system using CuO, Al_2O_3, and ZrO_2 nanoparticles [38]. They designed a double loop solar water heating system with nanoparticle laden water in the primary loop to absorb solar energy, and a secondary loop to transfer the absorbed heat to the water in the storage tank. They observed that the use of nanoparticles resulted in a stratification of temperature in the working fluid, resulting in an overall rise in the temperature of water in the storage tank. They also noted that of although efficiency of the solar heater increased with the increase in the weight fraction of the nanoparticles for all three nanoparticle materials tested, the increase in efficiency from the use of Al_2O_3 nanoparticles was consistently higher than those from ZrO_2 and CuO.

Menbari et al. investigated the effect of CuO/water nanofluid on the efficiency of a direct absorption solar collector. They used CuO nanoparticles of $>$ 99.9% purity and dispersed in the base heat transfer fluid using ultrasonic instigation and maintained a pH level of about 8–9. For the different nanofluid flow rates tested, they noted that due to an increased scattering and absorption of incoming solar irradiance by the nanoparticles, the nanofluid temperature was highest near the receiver tube wall and subsequently decreased with the increase in distance from the wall. Additionally, they noted that the variation in temperature and reduced energy absorption with increase in distance from the tube wall could be offset by increasing the nanofluid flow rate, and that an increase in the volume fraction of CuO nanofluid from 0.004% to −0.008% resulted in an increase in thermal efficiency of the solar collector by 18%–52% [14].

Liu et al. studied the performance of high-temperature direct absorption solar collector using graphene/ionic liquid nanofluids. They varied graphene concentration from 0.0005 to 0.01 wt.% and then dispersed it in $[HMIM]BF_4$ after 30 minutes of sonification without any surfactant. The use of graphene/$[HMIM]BF_4$ resulted in a 40% reduction in transmittance through the heat transfer fluid, and they were able to achieve upto 100% solar radiation absorption by increasing graphene concentration [39] (see Fig. 6.2). An increased concentration in graphene, however, also resulted in a localization of the absorbed energy as graphene particles were mainly concentrated in a thin layer. As such, they noted that although the

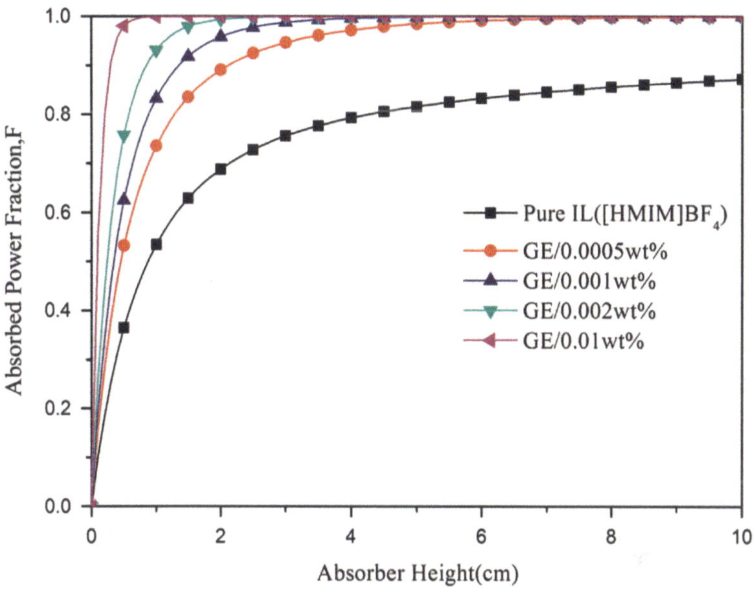

Figure 6.2 Absorbed power fraction versus graphene concentration and height of receiver. *Reproduced with permission from J. Liu, Z. Ye, L. Zhang, X. Fang, Z. Zhang, A combined numerical and experimental study on graphene/ionic liquid nanofluid-based direct absorption solar collector. Solar Energy Materials and Solar Cells 136 (2015) 177–186. https://doi.org/10.1016/j.solmat.2015.01.013.*

receiver's efficiency increased with increase in solar concentration and receiver's height, the receiver efficiency decreased from 0.72 to 0.23 at 550 K as graphene concentration increased from 0.0005 to 0.01 wt.%.

Nanoparticles have been tested for use in combined PV/thermal systems as well. Taylor et al. developed optimized nanofluid-based filters from specialized nanoparticle suspensions to match with several PV cell options: InGaP, CdTe, InGaAs, Si, and Ge. They optimized a hybrid solar photovoltaic/thermal model for each of these PV cell option based on an "ideal" approximations of PV material response curves and tested several optical filters created using specialized nanoparticle suspensions. They noted that the strong absorptivity of the free electrons at the surface of the nanoparticles under specific wavelength, polarization and incident angle conditions, resultedin a superior solar-weighted efficiency compared to pure fluids [40]. This observation was particularly significant as liquid filters are not only able to be controlled dynamically with pumps, but, the resulting nanofluid filters are also considerably more compact than a pure fluid filter or a conventional filter surrounded by a pure fluid.

6.2.2 Nanoparticles for enhancing water treatment

Nanoparticles have also been showcased for use in water treatment either as reaction catalysts or as adsorbents. When used as catalysts, nanoparticles tend to enhance the degradation, oxidation, and reduction of pollutant compounds. When used as nano-adsorbents, nanoparticles form strong bonds with specific compounds in a reversible or nonreversible way. For both methods of nanoparticle water treatment, the high surface area, high surface charge density, high electron charge transfer rates, and the selective chemical affinity for the specific targeted compounds present in water are utilized [41]. While both catalysis and adsorption can be used individually, these techniques are often superimposed with other techniques such as filtration and distillation to ensure high water production. An alternative way of describing the methodology of using nanoparticle for water treatment would be based on whether the nanoparticles are freely suspended in the fluid or are attached to a fixed structure or membrane.

6.2.2.1 Freely suspended nanoparticle adsorbents

Adsorption is perhaps the simplest examples of water treatment technology where nanoparticles are freely suspended in the fluid. Adsorption is one of the most widely used methods for water purification due to its effectiveness and scalability. Among the commercially available adsorbents, activated carbon is perhaps the only industrially used adsorbent for water treatment, albeit in micron sizes only. Nanoparticles such as graphene, carbon nanotubes (CNTs), siliceous adsorbents, and metal oxides have also gained traction due to their higher aspect ratio and enhanced reactivity which in turn translate to higher adsorption capacity. For example, oxidized CNTs offer high adsorption capacity for metal ions through electrostatic attraction and chemical bonding at their surface functional groups (e.g., carboxyl, hydroxyl, and phenol) [42]. More specifically, the dissolved metal ions in aqueous solution replace the protons in the carboxylic and phenolic groups of CNTs, resulting in a drop in the pH of the solution. Li et al. studied the performance of CNTs as adsorbents for heavy metals like Cu^{2+}, Pb^{2+}, and Cd^{2+}. They prepared multiwalled CNTs by dissociating methane in a hydrogen flow at 900 K using Ni nanoparticles on diatomites as catalysts. The catalysts and nanoparticles were subsequently removed by immersing CNTs in hydrofluoric acid and concentrated nitric acid, followed by repeated washing of CNTs using hot distilled water. These CNTs were added to metal ion solutions and shaken for 4 hours before analyzing on an atomic absorption spectrometer. They demonstrated that CNTs are better adsorbents than activated carbon for heavy metals like $Cu2^+$, $Pb2^+$, and $Cd2^+$ due to highly accessible adsorption sites and the short intraparticle diffusion distance on CNTs [43].

Among all contaminants, arsenic and fluoride removal has garnered much attention. Deliyanni et al. used metal oxide nanoparticles such as nanosized magnetite and $TiO_2$and observed a superior adsorption performance compared to activated charcoal. They prepared β-FeO(OH), also known as Akaganéite, from an aqueous solution of iron(III) chloride salt using ammonium carbonate as precipitating agent at room temperature. The solution was stirred until its pH dropped to 8 and the precipitate was decanted. The nanomaterial was obtained from the precipitate by osmosis through a cellulose membrane. The performance of Akaganéite was then tested in an arsenate stock solution of known pH value for 24 hours in a reciprocal shaker at room temperature. The residual arsenic in the solute was separated from the solid/liquid mixture using membrane filtration and was chemically analyzed. They noted that the amount of arsenate adsorption increased by lowering the pH of the system and increasing the amount of the sorbent, and a maximum load capacity of 100–120 mg As(V) per gram of Akaganéite [44]. A combination of activated charcoal and metal oxide nanoparticle through skeletal impregnation has opened avenues to achieve simultaneous removal of arsenic and organic contaminants, thus finding application in point-of-use water treatment systems [45,46]. Table 6.4 lists some of the state of the art nano-adsorbents and

Table 6.4 State of the art nanoparticles for removal of pollutants by adsorption.

Adsorbent nanoparticle	Pollutant	Adsorption capacity (mg/g)	Refs
Akaganéite	As(V)	100–120	[44]
Magnetic chitosan	Co(II)	27.5	[47]
$MnFe_2O_4$	Cr(VI)	31.5	[48]
δ-FeOOH coated maghemiteγ-Fe_2O_3	Cr(VI)	25.8	[49]
Gum Arabic magnetic polymer	Cu(II)	38.5	[50]
Fe_3O_4	Hg(II)	125.0	[51]
hexadecyl functionalized magnetic silica	Rhodamine 6G	35.6	[52]
CNTs	Pb(II)	97.1	[43]
CNTs	Cu(II)	28.5	[43]
CNTs	Cd(II)	10.9	[43]

their absorption capacities as well as the target pollutant they are designed to remove from water.

6.2.2.2 Nanoparticles attached to fixed structure or membranes

While adsorption using freely suspended nanoparticles has been proven to be extremely effective, it is often times more convenient to have the nanoparticles affixed to surfaces for ease of nanoparticle recovery and cleaning for reuse. The use of filtration systems using membranes has been in use for decades. It is, however, only recently that the advantages brought by nanoparticles have been applied to the membrane distillation technology for increased treated water production. Membrane distillation is driven by the vapor pressure difference between the membrane surfaces that can be achieved by creating a thermal gradient on across the membrane [53,54]. Membrane distillation systems can therefore be paired with low grade heat and renewable energy sources such as solar, wind, or geothermal for more ecological and cost-effective systems [55]. However, the process of phase transition in membrane distillation reduces the supply of heat for evaporation to the feed-membrane interface causing a polarization in temperature at the membrane and thus reduce the distillation efficiency of the membranes over time due to the loss of driving force [56,57]. A significant number of studies on membrane nanotechnology have focused on adding nanomaterials into polymeric or inorganic membranes to enhance removal of heavy metals, membrane permeability, and reduce fouling.

Maximous et al. studied the effects of using Al_2O_3 nanoparticles on performance of polymeric ultrafiltration membranes. They dissolved Al_2O_3 nanoparticles with average particle size of 48 nm and a surface area of 34 m^2/g into the N-methyl pyrrolidone solution and sonicated the solution at 60°C for 72 hours to obtain a uniform and homogeneous casting suspension. They then added 18 wt.% polyethersulfone polymer and sonicated the mixture again for a week before casting the membranes with a 100 μm casting knife onto a glass plate at room temperature. The membranes were coagulated and cleaned in deionized water bath before use in ultrafiltration of activated sludge. They noted that membrane deionized water permeation increased with the increase in nanoparticles concentration in the casting solution upto 0.05 Al_2O_3/polyethersulfone because of the higher affinity for water to Al_2O_3 than polyethersulfone polymer. The decreased interaction between polymer and solvent molecules due to the hindrance of nanoparticles resulted in an increase in penetration velocity of water into nascent membrane during the phase inversion. Any further increase in concentration, however, resulted in a drastic decline declined in deionized water permeation from clogging of membrane pores during the phase inversion leading to decrease in the water flux [58].

In a similar experiment, Bae and Tak studied the effects of TiO_2 nanoparticles on fouling mitigation in ultrafiltration membranes [59]. They prepared TiO_2 entrapped membranes by phase inversion methods using PSf, PVDF, and PAN as membrane materials and casting solutions consisting of polymer, *n*-methyl-2-pyrrolidone, and TiO_2 nanoparticles. The membranes were cast with 200 μm casting knife onto polyester non-woven fabric, evaporated at 25°C ± 1°C, 65% ± 5% relative humidity and then immersed in deionized water coagulation bath. Lastly, TiO_2 deposited membrane was prepared by dipping the PSf into 1% TiO_2 aqueous suspension then pressurizing at 400 kPa for 2 hours using compressed nitrogen gas. They noted that the addition of TiO_2 nanoparticles resulted in an increase in hydrophilicity of the all the membrane materials tested. This resulted in lower flux decline compared to neat polymeric membrane and an increase in fouling mitigation regardless of polymer material. They also noted that TiO_2 deposited membranes showed greater fouling mitigation effect due to the presence of a larger number of nanoparticles on membrane surface compared to TiO_2 entrapped membranes.

Lind et al. investigated the use of nano-zeolites in thin-film nanocomposite membranes on membrane permeability [60]. They coated zeolite-polyamide thin-film nanocomposite membranes onto polysulfone ultrafiltration membranes by interfacial polymerization of amine and acid chloride monomers in the presence of zeolite nanocrystals. By varying size of the zeolite crystal used and using different interfacial polymerization chemistries, they were able to produce nanocomposite thin films with widely varying structure, morphology, charge, hydrophilicity, and separation performance. They noted that the addition of nano-zeolites not only lead to more permeable, negatively charged, and thicker polyamide active layer, doping nano-zeolites at 0.2 wt.% achieved >99.4%better salt rejection than commercial RO membranes due to the creation of preferential paths for water from the small, hydrophilic pores of nano-zeolites [61]. Besides an increased permeability, they also demonstrated that nano-zeolites could be used as carriers for Ag^+ antimicrobial agents to inhibit adhesion of bacteria cells to the membranes and thus increasing antifouling property to the membrane [62].

6.2.3 Nanoparticles for enhancing solar-driven water treatment

Nanoparticles that can demonstrate a combination of characteristics such as high surface area, surface charge density, and selective chemical affinity for the specific targeted compounds present in water, as well as an enhanced absorption and conversion of incident solar irradiation into thermal energy within the heat transfer fluid can prove to be perfect fit for solar water treatment systems like Solar Stills, where performance of the still depends on the thermal characteristics of basin water [63,64]. The addition of high thermal conductivity nanofluids to water increases the solar absorptance of the still compared to pure water resulting in an increase in the temperature differential between the basin fluid and the glass cover. Elango et al. demonstrated this principle using a single slope solar still fabricated from 0.01 m GI sheet and 0.5×0.5 m absorber area. They evaluated the performance of water nanofluids of Aluminum Oxide (Al_2O_3), Zinc Oxide (ZnO), Iron Oxide (Fe_2O_3), and Tin Oxide (SnO_2) of different concentrations. They recorded a 29.95% higher production from the still with Al_2O_3 nanofluid than the still with water. The corresponding production from the still with ZnO and SnO_2 nanofluids was 12.67% and 18.63% more than the still with water [65].

Wang et al. explored a combination of photodegradation and solar distillation using a novel trifunctional membrane made of mixed cellulose ester, ZnO particles, and Gold nanoparticles [66]. The trifunctional membrane was prepared by a simple vacuum filtration method, where mixed cellulose ester was used as a porous supporting layer, and then a layer-by-layer strategy was applied between the positive charged hedgehog-like hierarchical ZnO particles and negatively charged Au nanoparticles.

Lastly, they used 1H,1H,2H,2H-perfluorooctyl triethoxy silane to modify the surface properties of the membrane to obtain a superhydrophobic surface to ensure that the solar absorber can float for long periods to allow for a long-term localized solar heating at the water–air interface. By floating this membrane over surface of water dispersed with Rhodamine water, they were able to induce localized plasmon resonance resulting in a 300% enhancement in the steam generation for clean water production and a reduction in Rhodamine concentration to 30% of the initial concentration.

Polymeric membrane distillation has inspired additional materials in the form of nano-porous foams for use in direct solar distillation processes. Foams made from metal oxides such as aluminum oxide [67], titania [68], and graphite [69] have been used with nanoparticles to combine the enhanced heating effects from nanoparticles with the capillary action and adherence to water of the foam materials, thus achieving enhanced water desalination. Zhu et al. prepared nano-porous foam like cages from black titania at different preparation temperatures that indicated a varying efficacy of absorbing the solar irradiation and a localized increase in temperature at the water–air interface rather than uniformly heating the bulk (see Fig. 6.3) [68]. Among the key findings, they noted that the darker colored titania films had a higher light to thermal conversion efficiencies, which resulted in a higher evaporation rate. Of the different Titania cages tested, the black titania nanocages (210°C) recorded the fastest water evaporation rate of 1.13 kg/m^2/hour and the highest conversion efficiency of 70.9% under 1-sun illumination. Furthermore, they also noted that the concentrations of Na^+, K^+, Mg^{2+}, Ca^{2+}, and B^{3+} in the seawater were all significantly reduced after desalination. Lastly, they demonstrated that the titania nanocages were able to maintain performance for more than 10 cycles of more than 5 hours duration each.

Nanoparticle membranes have also found a unique application for wastewater treatment in the oil industry where the typical output of produced water to oil is at a volumetric ratio of 3:1. Ibrahim et al. used membrane distillation techniques with nano-photonics-enhanced materials to achieve an overall dissolved organic carbon rejection rate between 72.5% and 96%, and more than 99% rejection rate of dissolved solids [70]. They used commercially available Polytetrafluoroethylene (PTFE) membranes containing polypropylene support layer and applied a polydopamine adhesion layer to the polypropylene side of the membrane, followed by a second coating of carbon black nanoparticles. The performance of the membrane was tested against two different samples of produced water under 1-sun illumination. They noted an excellent rejection of both dissolved organic carbon and other dissolved solids in oil produced waters. Furthermore, the hydrophilicity of the membrane could be recovered by washing with deionized water resulting in a mitigation of wetting of foulants into hydrophobic PTFE layer.

The introduction of nanoparticles in desalination systems also brings the need for additional processes to separate the nanoparticles from the solutions with suspended solids or precipitated salt in order not to waste the nanoparticles. Wang et al. developed a solar thermal evaporation system that used Fe_3O_4 magnetic nanoparticles on reduced graphene oxide. Besides an evaporation efficiency of upto 70% under 1 kW/m^2

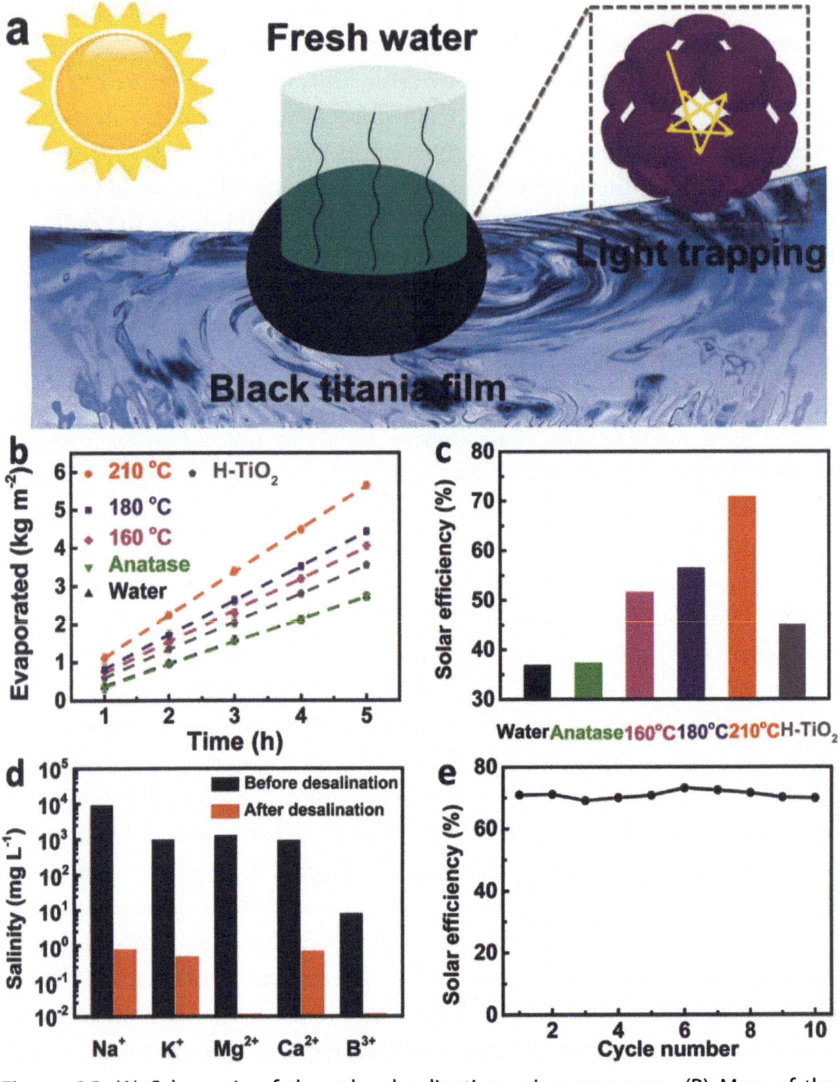

Figure 6.3 (A) Schematic of the solar desalination using nanocage. (B) Mass of the evaporated water as a function of the irradiation time, and (C) corresponding solar efficiency in the absence of the titania film and in the presence of anatase film, colored titania synthesized at 160, 180, and 210°C films. (D) Measured concentrations of four primary ions in an actual seawater sample before and after desalination. (E) Cycle performance of the black titania nanocages synthesized at 210°C film. *Reproduced with permission from G. Zhu, J. Xu, W. Zhao, F. Huang, Constructing black titania with unique nanocage structure for solar desalination. ACS Applied Materias & Interfaces 8 (46) (2016) 31716–31721. https://doi.org/10.1021/acsami.6b11466.*

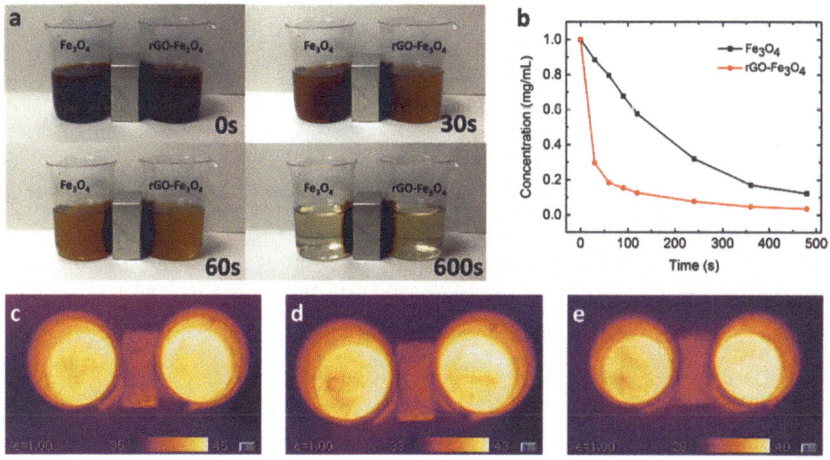

Figure 6.4 Recycling of Fe_3O_4 MNPs and rGO-Fe_3O_4. (A) Optical photos of MNPs separating from saline solution by using a magnet at 0, 30, 60, and 600 s. (B) Concentration change as a function of time estimated from the absorbance data. (C–E) IR images of the separation process. *Reproduced with permission from X. Wang, G. Ou, N. Wang, H. Wu, Graphene-based recyclable photo-absorbers for high-efficiency seawater desalination. ACS Applied Materials and Interfaces 8 (14) (2016) 9194–9199. https://doi.org/10.1021/acsami.6b02071.*

irradiation, the presence of magnetic nanoparticles allowed for the extraction and subsequent recycling of reduced graphene oxide using magnetic force [71]. Fig. 6.4 shows the separation process of different MNPs by using a magnet.

6.2.3.1 Photocatalytic nanoparticles

The use of photocatalysis is a relatively new technique for water treatment and primarily finds application in removal of heavy metals from water and has been determined to be driven by electro-optical properties and specific surface area of the photocatalyst. Mungondhori et al. evaluated the performance of asymmetric polymeric membranes of PAN, PVDF and PMAA-g-PVDF/PAN blended with Nitrogen-doped titanium dioxide nanoparticles to remove Pb^{2+} and Fe^{3+}ions from synthetic wastewater [72]. They used titanium tetrachloride and ammonia solution to prepare amorphous $TiO_{2-x}NH_x$, and used dry–wet phase inversion technique to prepare the three sets of membranes, each with 0, 1, or 3% N-TiO_2 (w/w) photocatalyst. They prepared two separate solutions of Pb^{2+} and Fe^{3+}were prepared with concentration of 10 mg/L each and tested for photoreduction under solar irradiation. Using atomic absorption spectroscopy to monitor changes in the concentration of the heavy metals, they noted that although the photoreduction of Fe^{3+} occurred at very low rates irrespective of the loading percentage of nitrogen-doped titanium oxide, photoreduction efficiency as high as 76.2% could be achieved using the novel 1% N-TiO2-PMAA-g-PVDF/PAN asymmetric membrane. However, any further increase in the doping percentage resulted in a 10%–17% decline in the photoreduction efficiency potentially due to agglomeration of the fine powders, resulting in reduced surface area for photoreduction to occur. Similar test with Pb^{2+}solution returned upto 90.5% reduction efficiency for 1% N-TiO_2 loading, though any further increase in loading resulted in only a slight increase in reduction efficiency.

Ibrahim and Asal used precipitation methods to dope lanthanides in ZnO nanoparticle to increase the surface area of the nanoparticles and also an improvement in the crystal, porosity, morphology, as well as the optical adsorption and emission of UV light properties of the prepared photocatalysts. The nanoparticles were then tested for photobactericidal ability under UV light exposure. They noted that compared to pure ZnO, the survival ratio for *Escherichia coli* on surface of lanthanide-doped ZnO nano-powders decreased to its minimum value within 120 minutes. They deduced that the increased bactericidal effect of UV/Ln^{3+}-ZnO photocatalyst was due to the generation of HO oxidizing radical that attacks the outer membrane of bacteria. Furthermore, the enhanced porosity and surface area from Ln^{3+}lead to an increased adsorption property of bacteria cell on the photocatalyst, resulting in an increased water disinfection rate [73].

In another work, Sreeja and Shetty used TiO_2 coated with silver nanoparticles to increase the absorption spectrum which enhances the rate of production of reactive oxygen responsible to destroy cell wall of bacteria. This results in a more effective disinfection process compared to using TiO_2 alone [74]. They observed a complete disinfection of *E. coli* infected water within 40 minutes owing to the photocatalytic activity of Ag@TiO_2 nanoparticles, whereas in the absence of catalyst and presence of light, 78% disinfection was observed at the end of 120 minutes, as shown in Fig. 6.5. They also noted that not only the regrowth of cells after photocatalytic disinfection was below the detectable limits, but the nanoparticles were also found to be photo catalytically active in endotoxin degradation during photocatalytic disinfection. With a simultaneous disinfection of microbes and degradation of endotoxins, the end product was a safe drinking water for direct use.

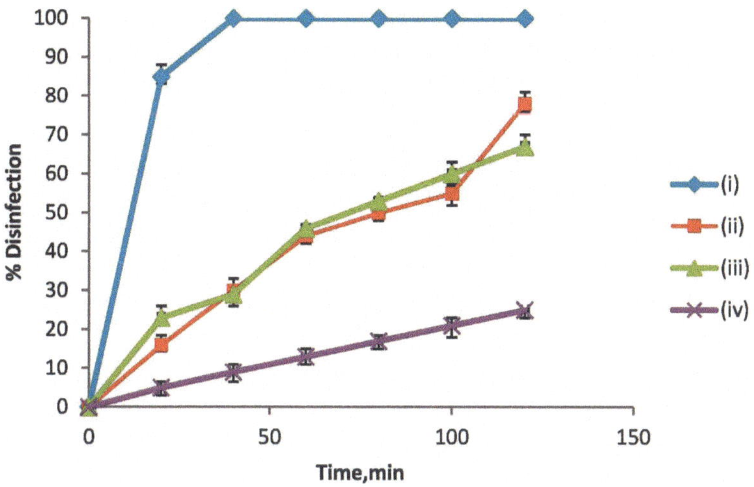

Figure 6.5 Disinfection studies (i) in the presence of light and catalyst (catalyst loading = 0.3 g/L), (ii) in the presence of light and absence of catalyst, (iii) in the presence of catalyst (catalyst loading = 0.3 g/L) and absence of light, and (iv) in the absence of catalyst and light. Average initial viable cell concentration = 40*108 CFU/mL. *Reproduced with permission from S. Sreeja, K.V. Shetty, Photocatalytic water disinfection under solar irradiation by Ag@TiO₂ core-shell structured nanoparticles. Solar Energy 157 (2017) 236–243. https://doi.org/10.1016/j.solener.2017.07.057.*

Although TiO_2 has come to be the most predominant photocatalytic material due to its high adsorption in the UV region and high chemical stability, metal oxides such as WO_3 [75], Fe_2O_3 [76], ZnO [77] as well as bismuth compounds [78] and CuS [79] have also been investigated for use as photocatalytic materials for water treatment. There is also an increasing emphasis on developing a new generation of photocatalysts that can utilize the entire visible spectrum through an increase in electron band gap. For more details on advances in photocatalysis, readers are encouraged to the read the review article by Anis et al. [80].

6.3 Nanomembranes

Nanomembranes are a relatively ill-defined class of nanotechnology as it can mean a variety of things. The key aspect is that at least one critical dimension is of nanoscale length. Here, we focus specifically on materials with nanoscale thickness that are used for selective transport across a medium versus more traditional thin films that are primarily used for enhancing solar absorption. Of particular interest in many applications are carbon or graphene nanomembranes, which are 2D sheets with controlled physical and/or chemical properties [81]. Such tailoring creates "interfaces without bulk" due to the molecular thickness [81]. Three primary methods exist for fabrication of nanomembranes: exfoliation, chemical vapor deposition, and self-assembly. While the first two methods are commonly used to create perfect single layers of graphene, they do not lend themselves well to chemical surface modification, and hence the use of self-assembly. The self-assembly route is shown in Fig. 6.6, where a self-assembled molecular monolayer is built followed by crosslinking via exposure to low energy electron or photon radiation.

Importantly by controlling the size of the precursor molecules one has extraordinary control over the membrane structure. As seen in Fig. 6.7, it is possible to control both the thickness of the nanomembrane (from 0.6 to 2.2 nm) and the resulting porosity is controlled by the size of precursor molecule as well as the self-assembly packing density. Further control of porosity can be achieved through perforation using highly charged ions, UV crosslinking, and conventional lithography techniques [81].

Importantly the nanomembrane process allows for a wide range of control for functionalization, size, and porosity that allows detailed tuning for specific applications.

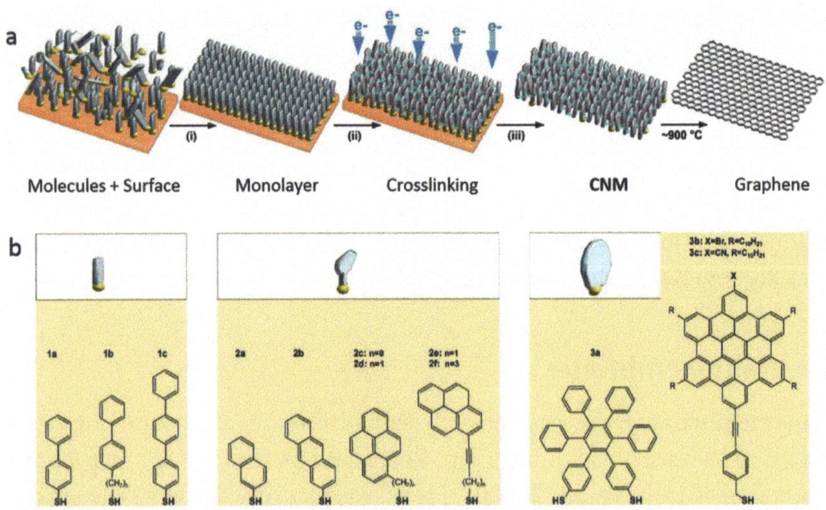

Figure 6.6 Self-assembly process of single layer graphene. *Reproduced with permission from A. Turchanin, A. Gölzhäuser, Carbon nanomembranes. Advanced Materials 28 (29) (2016) 6075–6103. https://doi.org/10.1002/adma.201506058.*

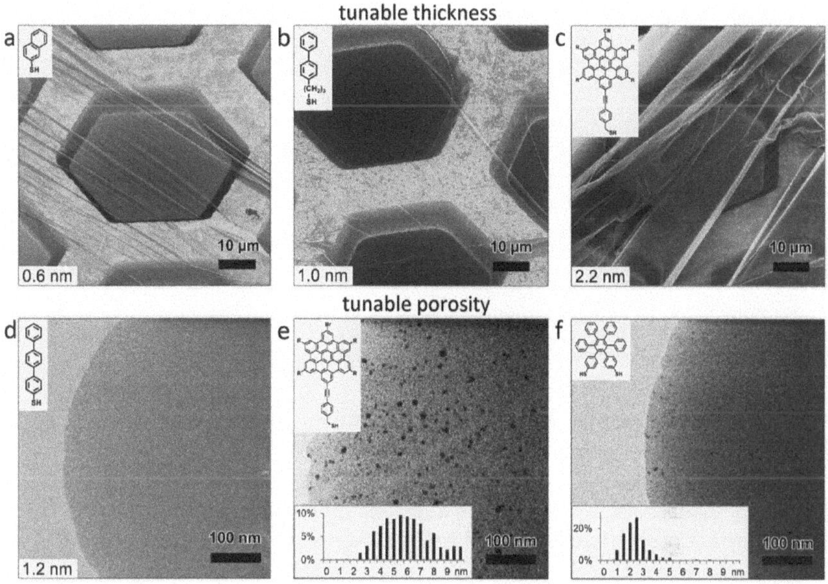

Figure 6.7 Control of nanomembrane porosity through self-assembly: (A) controlled thickness and (B) controlled porosity. *Reproduced with permission from A. Turchanin, A. Gölzhäuser, Carbon nanomembranes. Advanced Materials 28 (29) 2016 6075–6103. https://doi.org/10.1002/adma.201506058.*

6.3.1 Nanomembranes for solar energy harvesting

Nanomembranes within solar energy harvesting is significantly less heavily researched than nanoparticles but there are a few applications where a true nanomembrane approach is being applied. One interesting application is the use of thin silicon nanomembranes that enable energy transfer (ET) across the interface with highly absorbing nanostructures [82]. These membranes also allow for high charge carrier mobility, which could increase the efficiency of thin-film PV devices [82]. A similar approach has been developed for developing ultrathin silicon nanomembranes that function as highly sensitive photodetectors (performance comparable to bulk materials) but in ultrathin materials [83]. Another recent approach focused on enhancing the UV absorption of a graphene nanomembrane [84]. A challenge with graphene is the relatively low absorption in the UV wavelengths due graphene lacking a bandgap, while surface modifications with metal layers have resulted in increased UV absorption, achieving increased absorption with only modifications to the graphene layers is desired. In one approach a nanomembrane of graphene is created as a

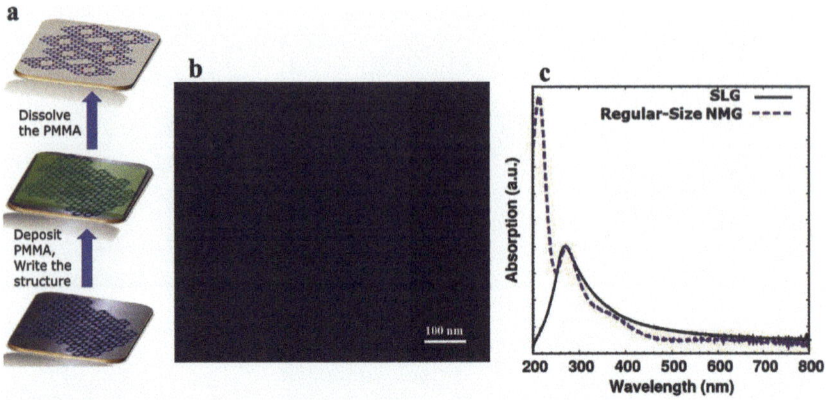

Figure 6.8 Nanomembrane fabrication via EBL (A), the resulting porous structure (B), and absorption (C). *Reproduced with permission from A. Kasry, M.M. Fadlallah, N.H. Voelcker, A.A. Maarouf, Experimental and theoretical demonstrations of ultraviolet absorption enhancement in porous nanomembrane graphene. Carbon N. Y. 155 (2019) 65–70. https://doi.org/10.1016/j.carbon.2019.08.038.*

nanomesh that results in a super lattice of pores and the establishment of an electronic gap. The nanomembrane graphene is created using electron beam lithography (EBL), see Fig. 6.8, such that it is possible to create a controlled pore structure. The authors demonstrated that this controlled and regular pore size and structure was the key to enhancing UV absorption as a random size distribution of pores on a layer of graphene had similar optical properties to that of a single layer [84].

It remains to be seen if further applications of true nanomembranes will arise for solar energy harvesting outside of photovoltaics but it appears that traditional thin-film structures would provide the tailored light absorption needed. Additionally, the recent work demonstrating the enhancement of absorption within the UV spectrum could provide a guide for designing tailored structures for further absorption enhancement.

6.3.2 Nanomembranes for water treatment

Often nanomembranes serve as a thin layer on top of an existing membrane structure to enhance a specific property and add multifunctional capability [85]. In many cases the ultra-small length scales of the pores formed in nanomembranes result in ultra-selective filtration useful for the nanofiltration. In one approach, carboxyl multiwalled carbon nanotubes were added to a calcium alginate hydrogel to increase the mechanical strength associated with nanofiltration [86]. While this application is not a pure nanomembrane it is indicative of many of the works classified under nanomembranes.

Traditional membranes for reverse osmosis rely on the solution-diffusion process, which creates a tradeoff between selectivity and membrane flux. One of the strongest arguments for the use of nanomembranes is the switch from the solution-diffusion mechanism seen in "thick" membranes to one of ballistic transport in nanomembranes [81]. The solution-diffusion process is the limiting factor in most membrane transport processes as the diffusion can be quite slow, hence the need for higher pressures (and increasing energy) to achieve high degrees of flux. As the membrane becomes thinner the pressure needed for the same level of flux decreases, and a ballistic limit is approach for ultrathin nanomembranes. In this limit the membranes act like physical barriers such that molecules cross the ultrathin membrane through a pore or are blocked by the surface, and hence don't rely on diffusion through the membrane. Because of this, nanomembranes hold promise to be energy efficient membranes [81] (Fig. 6.9).

Recent applications of porous membranes in carbon and graphene where the pore size can be controlled have been shown to be highly attractive for water desalination. In these cases, the membranes are true nanomembranes in thickness. One such approach demonstrates that nanometer scale pores in a single layer of graphene are effective filters of NaCl ions in water [87]. Through computer simulations it was demonstrated that smaller

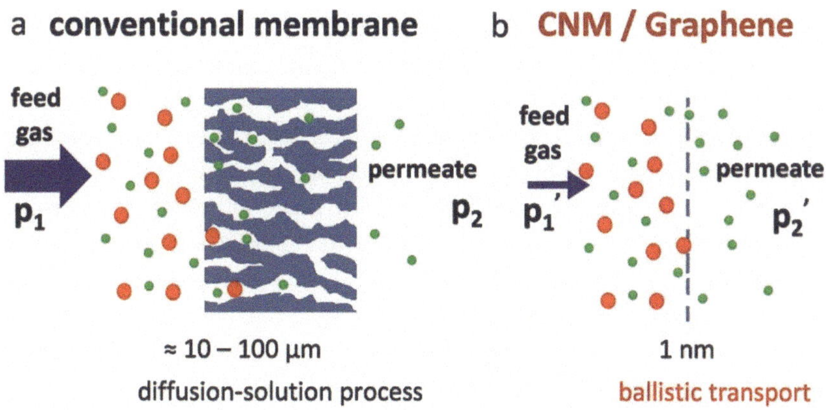

Figure 6.9 Schematic of solution-diffusion versus ballistic transport. *Reproduced with permission from A. Turchanin, A. Gölzhäuser, Carbon nanomembranes. Advanced Materials 28 (29) (2016) 6075–6103. https://doi.org/10.1002/adma.201506058.*

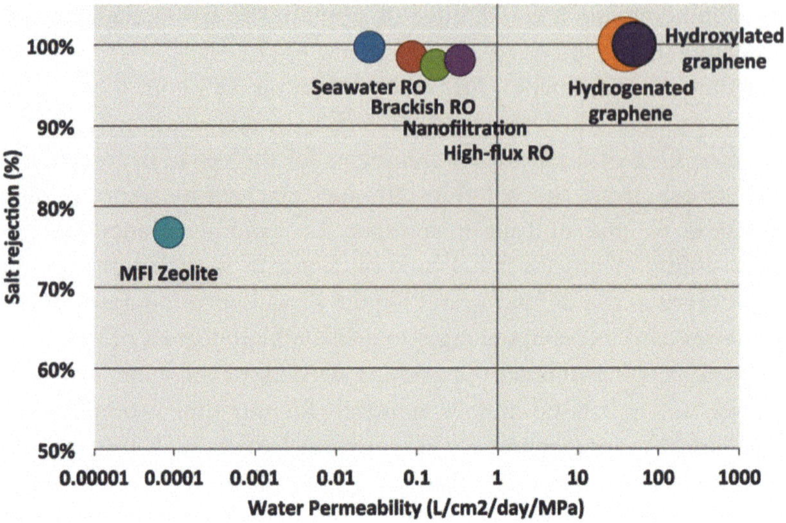

Figure 6.10 Salt rejection versus permeability for various RO membranes in comparison to ultrathin porous graphene nanomembrane. *Reproduced with permission from D. Cohen-Tanugi, J.C. Grossman, Water desalination across nanoporous graphene. Nano Letters 12 (7) (2012) 3602–3608. https://doi.org/10.1021/nl3012853.*

pore sizes were more capable of rejecting salt, particularly when the pores were hydrogenated, and the resulting water permeability was two to three orders of magnitude larger than existing membrane technologies, see Fig. 6.10 [87]. This is driven by the transition from a solution-diffusion limited process to a ballistic limited process, which allows for higher permeate flux for a given process and allowing separation efficiency based on size selectivity [81]. While the potential for exciting opportunities exists with graphene and graphene oxide, particularly due to the easy fabrication, there are challenges. Graphene grown via chemical vapor deposition and then placed on a polymer support, often results in subnanometer defects and nanometer scale tears that affect the performance, although recent routes for sealing these defects have been discovered [88]. Another challenge, is the swelling of graphene oxide membranes when soaked or fabricated in an aqueous solution. This swelling results in a significant increase in the thickness of interlayer spacing of a staked membrane, resulting in significantly reduced water mobility [89].

Due to swelling issue and challenges in manufacturing graphene membranes free of defects other approaches are being investigated for the fabrication of highly selective nanomembranes. One promising approach is

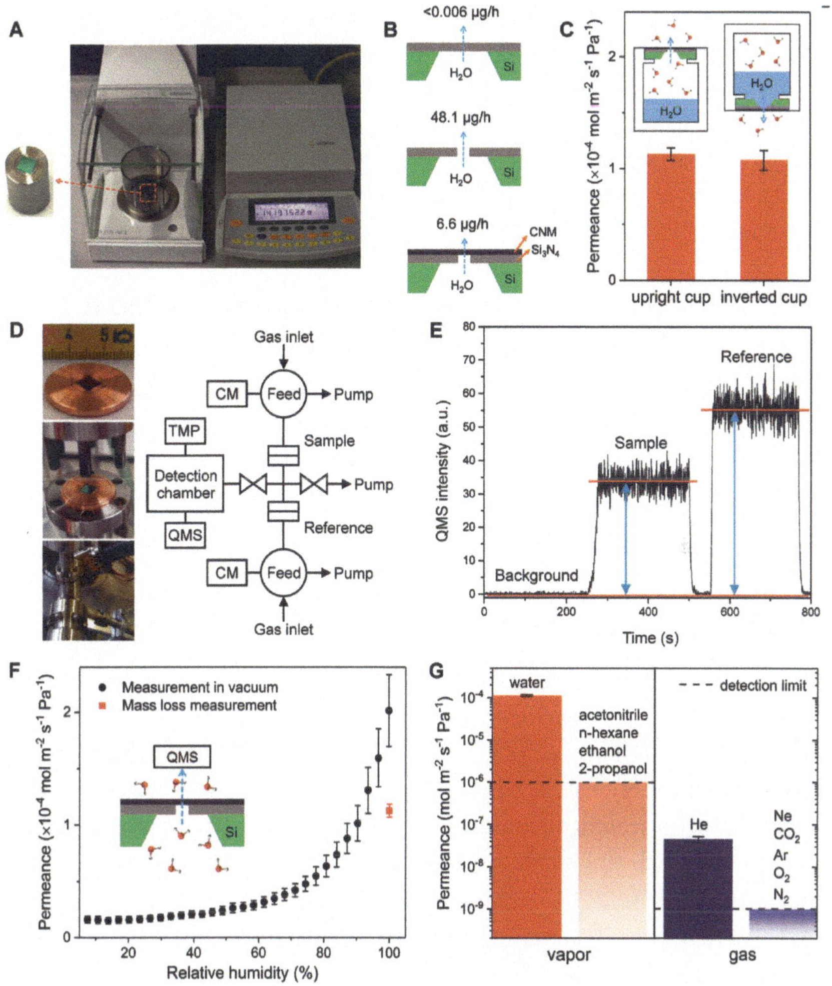

Figure 6.11 Demonstrated molecular transport across carbon nanomembranes. *Reproduced with permission from Y. Yang, P. Dementyev, N. Biere, D. Emmrich, P. Stohmann, R. Korzetz, et al., Rapid water permeation through carbon nanomembranes with subnanometer channels. ACS Nano 12 (5) (2018) 4695–4701. https://doi.org/10.1021/acsnano.8b01266.*

through the fabrication of ~1.2 nm thick carbon nanomembranes which were created by irradiation induced cross linking of self-assembled monolayers [90]. The resulting analysis demonstrated that the dense network of subnanometer channels formed in the fabrication blocked the transport of most gases and liquids, but allowed for highly effective transport of water, see Fig. 6.11 [90].

In a recent work a 2-D graphene nanomembrane fabricated by layer-by-layer self-assembly was utilized to demonstrate control of thickness, morphology, charge density, and internal structure [91]. They were able to demonstrate control ion transport by controlling the lateral size of the graphene oxide as well as the thickness, and that functional groups can be used to create selective ionization through the pH of the solution during self-assembly [91].

6.3.3 Nanomembranes for solar-driven water treatment

Overall, the future of nanomembranes is very bright as a potential for creating highly selective, and easy to fabricate membrane structures. While these membranes have shown potential for high performance, integration with solar energy for the driving force has not yet been demonstrated as of writing in 2021, to our knowledge. This area represents a potential area for significant investigation.

6.4 Nanofibers

Nanofibers are typically considered a nanostructure where the diameter is on the nanoscale and the length can range from nanoscale to significantly longer, although there is no hard and fast limit. Arguably, there is very little difference between a nanorod and a nanofiber, but here we focus specifically on research devoted to nanofibers where the length is significantly greater than the diameter (i.e., extremely high aspect ratio).

6.4.1 Nanofibers for enhanced solar absorption

Nanofibers, like nanoparticles and nanorods, have the potential for controlled optical properties that can be leveraged for high solar absorption with a relative low loading of nanomaterial. Nanofibers can be used to enhance the solar absorption of selective absorber coatings for solar thermal applications, to enhance solar absorption in dye-sensitized solar cells, and to increase scattering in solar cell applications as well.

Solar selective absorbers are the industry standard coatings for solar thermal energy conversion that maximize solar energy absorption while simultaneously minimizing thermal emission. A recent demonstration of this was achieved utilizing copper oxide (CuO) nanofiber-based thin films [92]. The nanostructured coating was prepared via a chemical oxidation route where a copper plate is soaked in an aqueous solution of sodium hydroxide (NaOH) for various pH levels. After the reaction the copper foil has an apparent black film on the surface indicating the formation of CuO. As can be seen in Fig. 6.12 elongated fibers are formed on the surface of the copper foil, which are subsequently characterized for the solar and infrared properties. Results reveal that the nanofibers have exceptionally high solar-weighted absorptance of 95% with exceptionally low values of thermal emittance (<7%) when measured at room temperature.

Nanofibers for enhancing dye-sensitized solar cells have also been demonstrated utilizing tin oxide (SnO_2) as the photoanode [93]. Typical utilizing tin oxide results in poor photovoltaic performance due to low Fermi energy but by incorporating TiO_2 nanoparticles into the tin oxide nanofibers the performance can be nearly doubled [93].

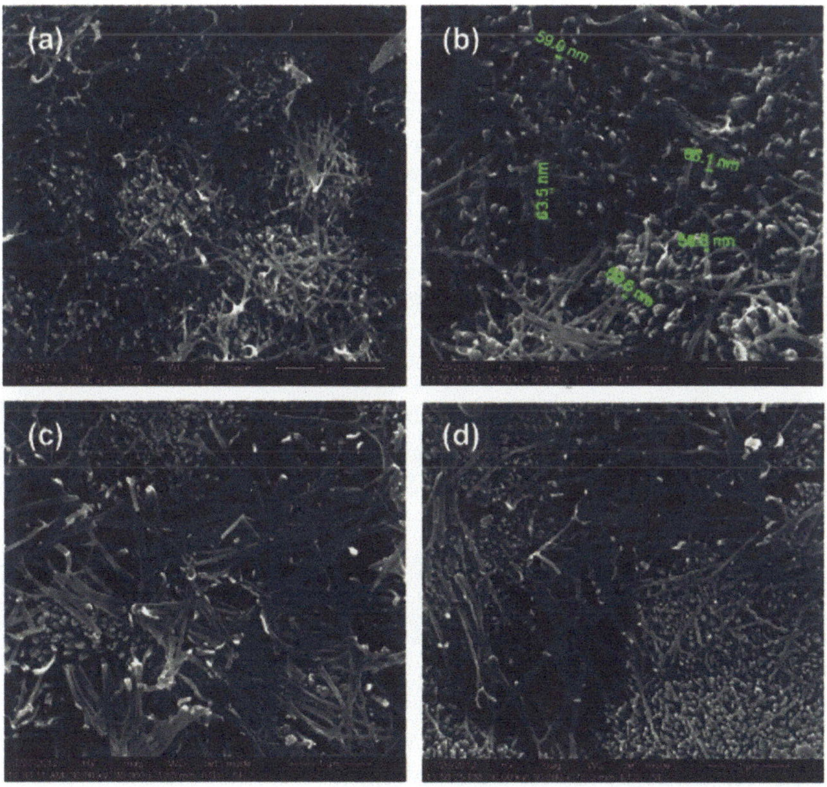

Figure 6.12 SEM images of CuO nanostructures (A and B) correspond to nanofibers prepared at a pH of 11.4 while (C and D) are for a pH of 12. *Reproduced with permission from S. Karthick Kumar, S. Suresh, S. Murugesan, S.P. Raj, CuO thin films made of nanofibers for solar selective absorber applications. Solar Energy 94 (2013) 299–304. https://doi.org/10.1016/j.solener.2013.05.018.*

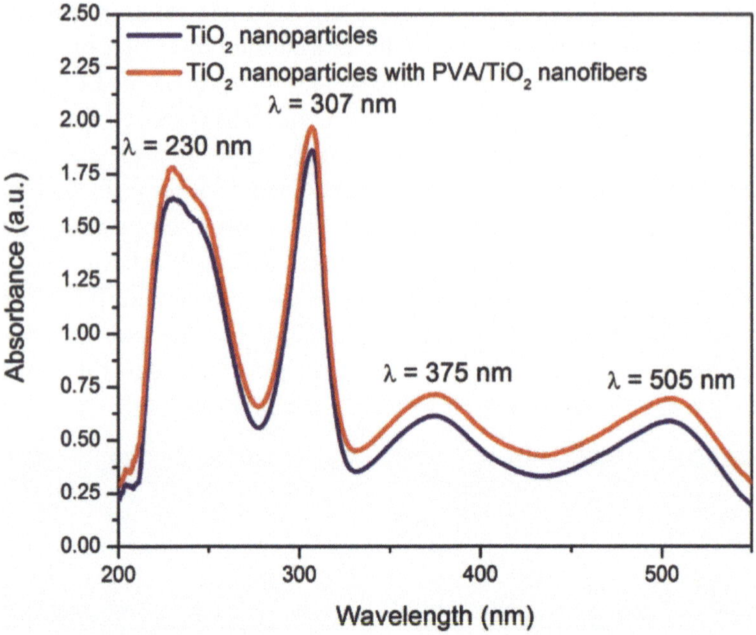

Figure 6.13 Absorption spectrum of TiO_2 nanoparticles and TiO_2 nanoparticles with a layer of PVA/TiO_2 nanofibers. *Reproduced with permission from M.N. Mustafa, S. Shafie, M.H. Wahid, Y. Sulaiman, Light scattering effect of polyvinyl-alcohol/titanium dioxide nanofibers in the dye-sensitized solar cell. Scientific Reports 9 (1) 2019 14952. https://doi.org/10.1038/s41598-019-50292-z.*

In another application of nanofibers to dye-sensitized solar cells, the nanofibers are used to create a light scattering layer that reduce the amount of light loss in these types of solar cells [94]. In this application the nanofibers as the light scattering layer were created via electrospinning of 10 wt.% polyvinyl acetate dissolved in water and then to which 0.06 M titanium isopropoxide was added. As can be seen in Fig. 6.13 the addition of the nanofiber layer on the top of the pure nanoparticles increases the absorbance of the composite due to the scattering of light, which is also shown to increase the maximum power density of the solar cell [94].

6.4.2 Nanofibers for water treatment

Arguably one of the most heavily researched areas of nanotechnology for membranes is the area of nanofibers. One of the primary drivers of this explosion is the availability of electrospinning technique for facile fabrication of nanofibers, particularly polymer-based nanofibers which form the backbone of most existing commercial membrane technologies. Electrospun nanofibers result in high specific surface area, high porosity, and high orientation or alignment of the nanofibers [95]. While the technique is simple and allows for the easy incorporation of other additives, the relatively slow yield speed has limited commercial deployment of polymer membranes which is still typically done with phase inversion processes [95]. While polymer nanofibers are ubiquitous for membrane processing research, we focus here on a few novel approaches for different uses of nanofibers in water treatment.

One interesting application of nanofibers is the use of a controlled composition nanofibers for use as anodes in electrocatalysis. In this approach, ruthenium oxide (RO) is composited with antimony doped tin oxide (ATO) for enhanced chemical stability, enhanced electrical conductivity, and high oxygen evolution reaction (OER) [96]. Electrospinning was used to create ATO/RO composite nanofiber anodes for comparison, to commercial RO anodes. It was demonstrated that the controlled fabrication of material content achieved in the electrospinning technique resulted in superior electrical conductivity, faster organic oxidation of BPA, and a higher degradation efficiency when compared to commercial RO anodes [96].

Another unique application is the use of nanofibers acting as sorbents, in this case the enhancement of graphite felts through controlled growth of carbon nanofibers [97]. This approach mitigates some of the issues associated with using carbon powders such as porosity control, toxicity, and shaping by hierarchically constructing low-dimensional nanomaterials. This is done by continuous deposition of carbon nanofibers on the graphite felt, which continuously increased the contact angle and adsorption capacity of the mat. Fig. 6.14 demonstrates the application of the felt as a sorbent for a variety of different organic liquids and the potential for regeneration using combustion [97].

Another interesting approach is the fabrication of nanofibers from widely available and environmentally benign chitosan. The unique aspect is that the nanofibers can easily be fabricated, particularly when using electrospinning, with modification through the addition of other functional materials such as silver nanoparticles to include an antibacterial effect [98]. One of the downsides of utilizing chitosan is the high degree of water uptake, poor mechanical and thermal properties. Recently, it has been demonstrated that these properties can be highly improved with functionalized multiwalled carbon nanotubes [99].

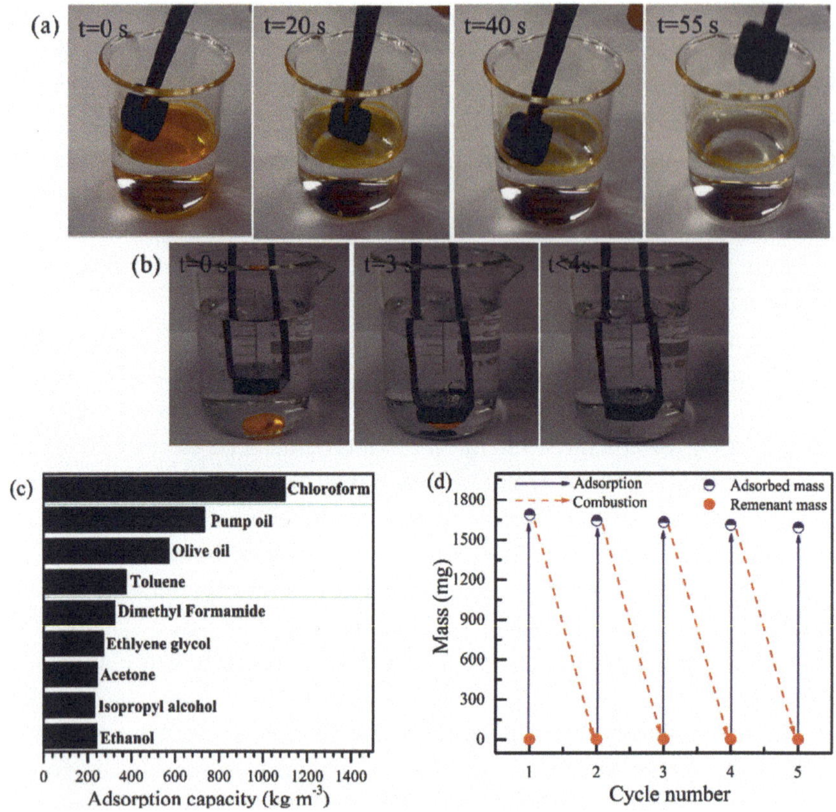

Figure 6.14 (A, B) Photographs showing sorption process for different materials, (C) sorption capacities for various organic liquids, and (D) demonstration of cyclic sorption and regeneration via combustion. *Reproduced with permission from Y. Shen, L. Li, K. Xiao, J. Xi, constructing three-dimensional hierarchical architectures by integrating carbon nanofibers into graphite felts for water purification. ACS Sustainable Chemistry & Engineering 4 (4) (2016) 2351–2358. https://doi.org/10.1021/acssuschemeng.6b00030.*

6.4.3 Nanofibers for solar-enhanced water treatment

The bulk of the current studies where solar (or light energy) is being utilized to enhance water treatment is based upon photocatalysis. One novel approach is to combine the photocatalytic properties of titanium dioxide (TiO_2) with the antibacterial properties of carbon quantum dots (CQDs) to create a multifunctional nanofiber [100]. The authors of this study then tested the nanofibers for photocatalytic behavior under outdoor sun, see Fig. 6.15, which demonstrated significantly improved photodegradation of methylene blue relative to pure TiO_2 nanofiber. Additionally, they

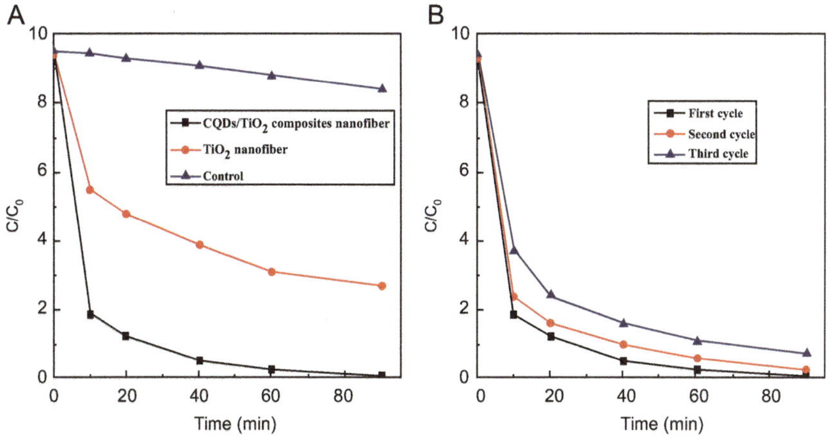

Figure 6.15 CQDs/TiO_2 Nanofibers for the catalytic removal on methylene blue. *Reproduced with permission from P.S. Saud, B. Pant, A.-M. Alam, Z.K. Ghouri, M. Park, H.-Y. Kim, Carbon quantum dots anchored TiO2 nanofibers: effective photocatalyst for waste water treatment. Ceramics International 41 (9) (2015) 11953–11959. https://doi.org/10.1016/j.ceramint.2015.06.007.*

demonstrated that the CQDs/TiO_2 nanofibers inhibited the growth of *E. coli* bacteria [100].

Another application of nanofibers for advanced photocatalysis is with TiO_2 fibers wrapped with a layer of graphene [101]. The ability to wrap the nanofibers with graphene reduces the agglomeration typically seen when trying to integrate TiO_2 into graphene, but all provides efficient charge separation in the materials. The work demonstrated that the photocatalytic degradation of methylene oxide was 90% for the graphene wrapped nanofibers in relation to 59% for pure TiO_2 nanofibers. Both of the prior two papers represent the multifunctional potential afforded by integrating nanomaterials into water treatment.

In addition to photocatalysis applications, nanofibers have also been used for enhanced solar steam generation. Solar steam generation has attracted a lot of interest as a means for high-temperature sanitation of medical instruments using nanoparticles [2], and has recently seen an explosion in research for potential desalination applications. In most of these applications a highly solar absorbing (typically carbon) structure is floated on top of the water surface, the structure provides both capillary transport of water as well as photothermal heating [102]. Here, we will focus on discussing recent research that has specifically utilized nanofibers as the means for driving the solar steam generation process.

The efficiency of solar steam generation can be quantified with the following formula:

$$\eta = \frac{\dot{m} h_{fg}}{G} \tag{6.3}$$

where $\dot{m}$ is the evaporation rate per unit surface area, h_{fg} is the latent heat of vaporization, and G is the solar flux on the surface of the device [103]. For 1-sun conditions, 1000 W/m^2, the efficiency of solar steam generation on a pure surface of water is roughly 15% [103]. An important consideration that this efficiency doesn't capture is the energy required to complete the condensation process, resulting in a much lower efficiency for the full water capture cycle.

An early approach for utilizing nanofibers in solar steam generation focused on a bilayer system comprised of cellulose naonfibrils (CNF) with a thin layer of carbon nanotubes on the top surface to enhance solar absorption [103]. In this concept the CNF is used to form an aerogel-like porous structure with vertical channels (200–300 μm) which actively pump water to the top surface, due to its high hydrophilicity. Additionally, the CNF aerogel structure serves as a thermal insulator from the hot top surface and the bulk water. The carbon nanotubes on the top surface layer, see Fig. 6.16, increases the solar absorptivity to over 98% in the wavelength range from 250–1200 nm. The assemble system had a bulk density of 0.0098 g/cm^3 with a porosity of 94%. At 1 sun, this structure was capable of a steam generation efficiency of 76%, with higher efficiencies and higher steam generation rates under moderately concentrated sunlight [103].

In an approach similar to what is seen in conventional membranes, another group demonstrated solar steam generation with a highly absorbing photothermal "cloth" [104]. The cloth was created through the electrospinning of polymer nanofibers which integrated carbon black nanoparticles into the carbon matrix. The cloth was demonstrated to be hydrophilic as well as highly solar absorbing (~94%), which resulted in a 1-sun solar steam generation efficiency of 83%. Importantly, the work also demonstrated a highly mechanically strong textile, that could be handwashed for more than 100 times without degradation as a means for removing foulants [104].

Furthering this concept, a combination of the two previous approaches was investigated by combining a layer of electrospun hydrophobic PVDF nanofibers with a secondary layer of hydrophilic carbon black nanofiber layer [105]. The hydrophobic layer serves as the floating support that also

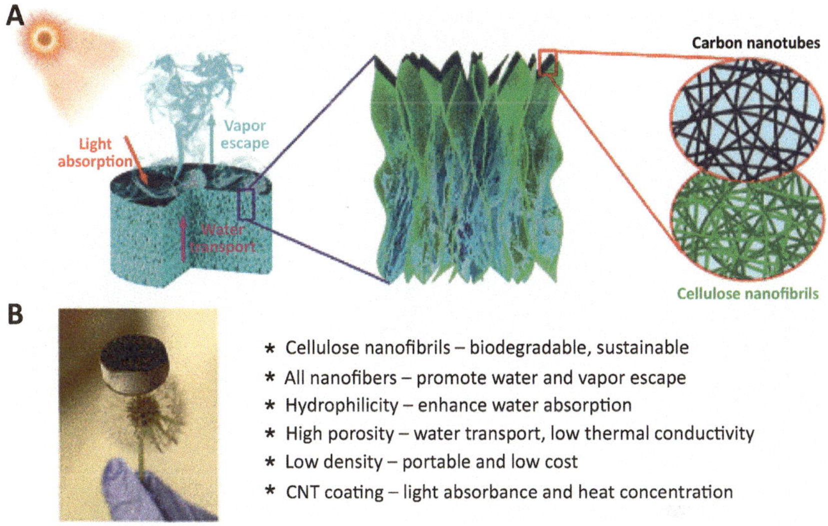

Figure 6.16 Working structure of solar steam generation utilizing nanofiber-based aerogel and carbon nanotube absorber. *Reproduced with permission from F. Jiang, H. Liu, Y. Li, Y. Kuang, X. Xu, C. Chen, et al., Lightweight, mesoporous, and highly absorptive all-nanofiber aerogel for efficient solar steam generation. ACS Applied Materials and Interfaces 10 (1) (2018) 1104–1112. https://doi.org/10.1021/acsami.7b15125.*

provides an insulative barrier. The carbon black and polyacrylonitrile nanofiber layer acts as a hydrophilic broadband solar absorber with over 98% solar absorptance. This bilayer nanofiber-based system demonstrated 1-sun solar thermal conversion efficiency of 82% [105].

In a recent work, a loofah, which is a naturally occuring polymer that has a natural vascular architecture but inherently isn't solar absorbing, was carbonized and integrated with carbon nanotubes to enhance solar absorptance, through the integration of carbon nanotubes [106]. Because the loofah is inherently not solar absorbing, see Fig. 6.17, the loofah is put through a carbonization process by heating in a tube furnace at 500°C under a nitrogen atmosphere for 2 hours. This carbonization in conjunction with a coating of carbon nanotubes results in a high level of solar absorbance. Notably, this system uses a layer of ethylene-vinyl acetate foam to act as a float due its hydrophobic nature and also as a thermal insulator, it also adds a mid-layer cotton pad to act as a water transport layer. This system demonstrated a 1-sun efficiency of 92.5%. In addition to the traditional evaporation tests, experiments were conducted relative to seawater and simulated wastewater. One thing of note, is that the seawater evaporation rate is reduced relative to freshwater which is a

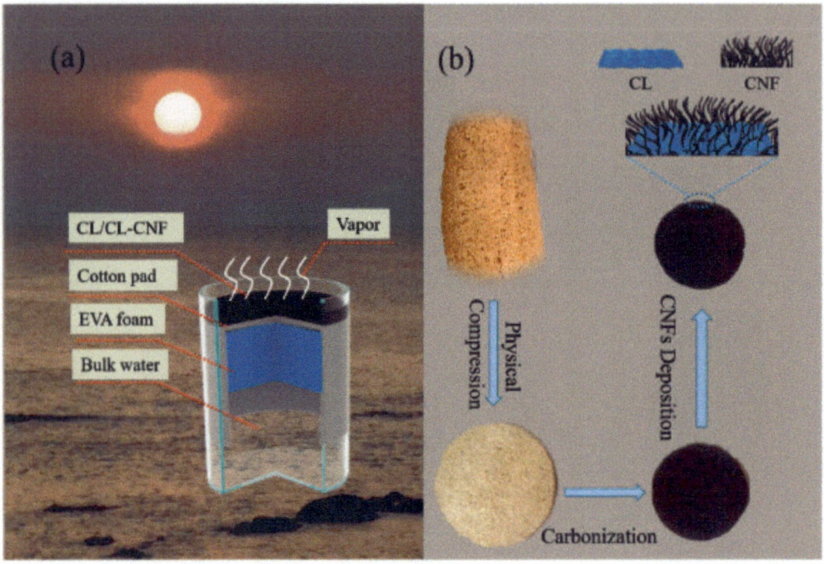

Figure 6.17 Steam generation using a carbonized loofah. *Reproduced with permission from C. Zhang, B. Yuan, Y. Liang, L. Yang, L. Bai, H. Yang, et al., Carbon nanofibers enhanced solar steam generation device based on loofah biomass for water purification. Materials Chemistry and Physics 258 (2021) 123998. https://doi.org/10.1016/j.matchemphys.2020.123998.*

notable and relevant concern. In the test on simulated wastewater the condensed water which showed well over 99% ion removal. Additional testing was performed on performance after cycling the system on and off sun 20 times with little degradation in performance [106].

6.4.4 Environmental impacts

A significant effort has been devoted to understanding the potential health and environmental impacts of nanomaterials, most of the work is focused on transport, toxicity, and detectability [107–109]. Nanomaterials can be created both naturally through the conventional breakdown of larger materials through mechanisms such as particle attrition or can "engineered" through a variety of synthesis and manufacturing methods. Through our literature review no studies were found, to our knowledge, that investigated the release of engineered nanoparticles from either nano-enhanced solar or water treatment systems, most of the focus was on the fate and transport of single material systems. For instance, fullerenes (an engineered nanoparticle) have been shown to induce lipid peroxidation in smallmouth bass, Fe_2O_3 causes serious delays in zebrafish embryo hatching, and silver nanoparticles proving to be lethal to zebrafish [110]. While these studies are useful, they are typically devoted to a single particle or set of particles and the potential impact. On a positive note, it is expected that nanoparticles are expected to be present in water environments at extremely low levels (mg/L to μg/L) based on modeling work [110].

For instance, Zhang et al. investigated the stability of commercial metal oxide nanoparticles in water [111]. During this investigation they discovered that commercially available nanoparticles (TiO_2, Fe_2O_3, ZnO, NiO, and silica) were nearly impossible to disperse into tap water, but were dispersible in nanopore water. This indicates that the aggregation is highly influenced by the presence of electrolytes. Interestingly, although the particles aggregated quickly the particles did not settle out of water, with 20%–60% remaining in the water. Using a conventional alum treatment seen in many water treatment facilities only 80% of the particles could be removed, even as aggregates indicate the potential for remaining in drinking water for a long time [111].

In a similar study the fate of SiO_2 nanoparticles in typical wastewater treatment streams [112]. It was observed that nonfunctionalized silica particles wouldn't settle in the typical primary water treatment process, indicating a need for a secondary process to ensure removal of the nanoparticle stream. Conversely, nanoparticles that were functionalized (in this case with the surfactant TWEEN) undergo flocculation and would likely be part of the sewage sludge which is often used as agricultural fertilizer [112]. This study indicates the difficulty in understanding the fate of nanoparticles in the wastewater treatment process, as they are highly dependent on the initial stream, the particles themselves, and the processes they undergo, all of which influence which downstream path the nanoparticles take in a wastewater treatment application. Treatment options for nanoparticles in wastewater streams can be achieved in a variety of methods. As previously mentioned conventional coagulation is less successful, but more advanced concepts such as electro-coagulation have been proven to be highly effective for certain nanoparticles [110]. Another interesting approach is the use of floatation, wherein dissolved air bubbles are added to the water stream and then collected with CTAB or other collector agents [110]. Newer processes like nanofiltration and ultrafiltration also could be used as secondary procedures by tuning pore sizes of membranes to screen for select sizes of nanoparticles [110]. In addition, to engineered nanoparticles, it is possible via a range of different mechanisms for

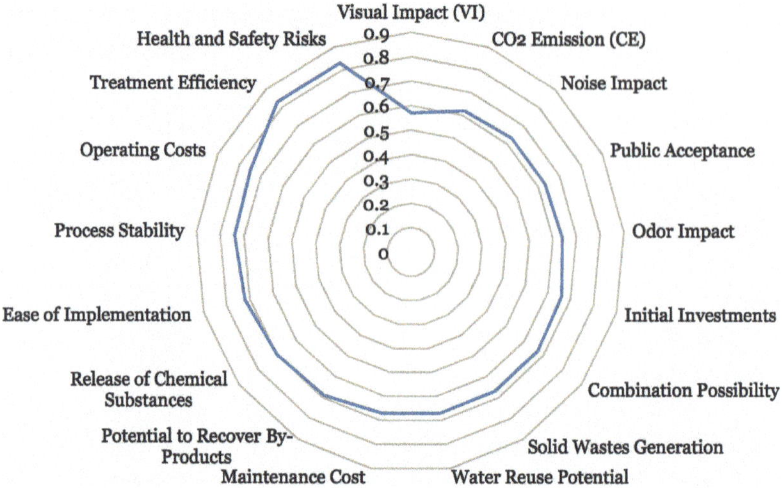

Figure 6.18 Weighted sustainability criteria for nanotechnology in industrial wastewater. *Reproduced with permission from M. Kamali, K.M. Persson, M.E. Costa, I. Capela, Sustainability criteria for assessing nanotechnology applicability in industrial wastewater treatment: current status and future outlook. Environment International 125 (2019) 261–276. https://doi.org/10.1016/j.envint.2019.01.055.*

nanoparticles to form naturally, but at this time there is very little known on the relative toxicity of natural nanomaterials to their engineered counterparts [113].

Less well-studied is understanding or weighing the benefits versus the associated risks. Recently, a set of sustainability criteria was developed specifically that could be applied to nanomaterials used in industrial wastewater treatment [114]. The criteria were based on surveying experts from around the world to develop a wide-ranging set of criteria to use for selection. The criteria were then weighted to determine which were of top importance, see Fig. 6.18. As can be seen, not surprisingly, the most important criteria were health and safety risks and treatment efficiency, with visual impact being the least impactful. This simple framework allows for at least simple weighting of impacts for different nanomaterials going forward, as well as further research. From the authors viewpoint the vast majority of the literature on integrating nanomaterials into different wastewater technologies have focused on the resulting treatment efficiency, for good reason. What is also needed from a research perspective is the release of these substances when used for water treatment, this is less well understood and will be equally important to understanding if they should be adopted.

6.4.5 Discussion/future work

Nanoparticles have increasingly become ubiquitous in all engineering applications because of their inherent alterations of physical, chemical, and optical properties. The use of nanoparticles as adsorbent or photocatalysts for water treatment has become a fairly well researched field. Coupling the nanoparticles with membranes has opened new avenues for enhanced and potentially economical water treatment due to the relative ease of recovery and reuse of nanoparticles. Additionally, the recent boom in solar-driven water treatment technologies has opened possibilities of further economic advantage through use of solar energy as heat source for water treatment processes. However, but the use of nanoparticles in solar-driven water treatment systems will require additional infrastructure to meet the regulations associated with water treatment. Looking forward, hybrid solar power generation and water treatment units where the two systems perhaps have a symbiotic link, could be an attractive area of research. Another area of future research could be in the simultaneous use of multiple nanomaterials to develop multifunctional capability within a single water treatment step. For instance, direct absorption membrane distillation where the nanoparticles provide solar absorption for thermal energy and photocatalysis for reducing biological fouling.

Nanomembranes, particularly molecular thick sheets of materials, represent a highly attractive avenue for many water-treatment options. The high selectivity and high permeability of nanomembranes makes them extremely promising candidates for a wide range of applications. While clearly novel for water treatment, the use of nanomembranes in solar applications is significantly less well understood. Potential application in solar PV can be envisioned where thin layers with controlled transport could be achieved. Within solar thermal thin layers of highly absorbing and highly conductive properties could be useful, but how this could be done to outperform existing solar selective coatings or creating additional benefits remains to be seen. Opportunities remain in the hybridization of nanomembranes with conventional solar processes to be either the driving force for the wastewater treatment, which could be particularly attractive if the energy requirements for treatment are much lower for nanomembranes.

Nanofibers represent a highly promising area for solar-driven water treatment as the fibers can easily be fabricated to have multifunctional properties such as adsorbent, photocatalytic, and high mechanical strength. Additionally, the relative ease in the fabrication either through chemical synthesis or through electrospinning allows for fairly large-scale fabrication. Further the nanofibers can result in highly porous networks that create strong pathways for water transport as well as light trapping. The recent boom in solar steam generation devices, has created a race for achieving the maximum photothermal conversion efficiency but largely these systems are tested under varying test conditions. As the vaporization rate is highly dependent on ambient air temperature, relative humidity, salt concentration, and wind speed, further work needs to focus on developing a consistent test standard for comparison, similar to what has been developed for testing photovoltaics. Additionally, the bulk of the research is focused on enhancing the evaporation rate. While this is clearly important, the energy cost associated with condensing the water back out of the vapor phase has been overlooked, although this is an important aspect as may become a limiting step. Notable challenges also include the formation of salt crystals on the solar absorber surface, particularly as the feedwater approaches saturation [105,106]. The traditional approach is to cycle the systems on and off sun, such that when "off sun" they become fully wetted and the salt dissolves back into the water. This inherently limits the on-sun time and productivity of the devices. Another key area to focus on is long-term stability. While many research works include testing over a number of cycles ($\sim 10-100$) [102,104,106], this number of cycles does not remotely approach the level of durability needed for a working device. Therefore some effort should be devoted to accelerated testing procedures to establish long-term durability.

In almost all of these areas, nanomaterials can have a positive impact on the solar harvesting ability, the water treatment ability, and the combined process. In many nanomaterial applications, the fate and toxicity of engineered nanoparticles is understood, especially for the most commonly investigated materials. The fate of nanomaterials in conventional wastewater streams, while highly complex, is also a well-studied area which indicates that typical primary wastewater treatment streams are not fully effective at nanoparticle removal. Ironically, some of the potential treatment options for removing nanomaterials will likely involve the increased use of nanomaterials. While there is a clear trend for the increased use of nanoparticles, nanofibers, and nanomembranes in various wastewater treatment strategies, further work is needed in balancing the improvements with the potential risks. In particular, the release of nanomaterials during the use phase or at the end-of-life can be in issue. For instance, the use of embedded nanoparticles in polymeric membranes is becoming common in research, but more study is needed on how long these nanoparticles remain bound in the polymer or how to dispose of these types of membranes effectively. Such areas of study will become more pressing as more nanomaterials are implemented in wastewater treatment processes.

6.4.6 Conclusion

Nanomaterials are providing new pathways for unlocking improved water treatment through a variety of different mechanisms. Additionally, nanomaterials can be designed to enhance thermal, optical, and electronic properties making them useful to a number of solar energy technologies. Not surprisingly, there exist a number of synergistic effects of combing both nanomaterials and solar energy in water treatment. For instance, a number of recent applications, where solar absorbing nanoparticles have been integrated into conventional polymeric membranes for thermal distillation which has been shown to minimize temperature polarization, limit the fouling of the membrane, and potentially allow for direct solar heating of the distillation process. In a similar approach to enhance photothermal absorption, solar nanofibers as floating foams are being designed as an effective way for highly efficient solar steam generation. Additionally, a number of photocatalytic materials exist that are easily synthesized as nanomaterials and could be utilized in a number of hybrid fashions with existing water treatment technology to provide additional benefits. Finally, one of the most promising areas for further development is the use of nanomembranes in different selective applications, like water treatment where a stacked layer of nanomembranes could selectively filter different solutes based on particle size or charge during the process. Nanomembranes have the ability to shift the paradigm in desalination from solution-diffusion process that requires high pressures to achieve high flux to a ballistic mode (via their dramatically reduced thickness) where the pressure requirements can be minimized, thus reducing energy use. However, further investigation is still critically needed on the long-term stability and environmental impacts of these nanomaterials within the proposed application, particularly with respect to releasing nanoparticles into the water stream in which they are being used for treatment.

References

[1] T.P. Otanicar, P.E. Phelan, R.S. Prasher, G. Rosengarten, R.A. Taylor, Nanofluid-based direct absorption solar collector, Journal of Renewable and Sustainable Energy 2 (3) (2010) 033102. Available from: https://doi.org/10.1063/1.3429737.

[2] O. Neumann, A. Urban, J. Day, S. Lal, P. Nordlander, N.J. Halas, Solar vapor generation enabled by nanoparticles, ACS Nano (2012). Available from: https://doi.org/10.1021/nn304948h.

[3] F.J. Beck, S. Mokkapati, K.R. Catchpole, Plasmonic light-trapping for Si solar cells using self-assembled, Ag nanoparticles, Progress in Photovoltaics Research and Applications 18 (7) (2010) 500−504. Available from: https://doi.org/10.1002/pip.1006.

[4] N. DeForest, A. Shehabi, J. O'Donnell, G. Garcia, J. Greenblatt, E.S. Lee, et al., United States energy and CO_2 savings potential from deployment of near-infrared electrochromic window glazings, Building and Environment 89 (2015) 107−117. Available from: https://doi.org/10.1016/j.buildenv.2015.02.021.

[5] V. Mlinar, Engineered nanomaterials for solar energy conversion, Nanotechnology 24 (4) (2013) 042001. Available from: https://doi.org/10.1088/0957-4484/24/4/042001.

[6] G. Chen, J. Seo, C. Yang, P.N. Prasad, Nanochemistry and nanomaterials for photovoltaics, Chemical. Society Reviews 42 (21) (2013) 8304. Available from: https://doi.org/10.1039/c3cs60054h.

[7] P.J.J. Alvarez, C.K. Chan, M. Elimelech, N.J. Halas, D. Villagrán, Emerging opportunities for nanotechnology to enhance water security, Nature Nanotechnology 13 (8) (2018) 634−641. Available from: https://doi.org/10.1038/s41565-018-0203-2.

[8] D. Jassby, T.Y. Cath, H. Buisson, The role of nanotechnology in industrial water treatment, Nature Nanotechnology 13 (8) (2018) 670−672. Available from: https://doi.org/10.1038/s41565-018-0234-8.

[9] ISO/TS 80004−2:2015(en), Nanotechnologies—Vocabulary—Part 2: Nano-objects. <https://www.iso.org/obp/ui/#iso:std:iso:ts:80004:-2:ed-1:v1:en> (accessed 17.2.21).

[10] C.S.C. Santos, B. Gabriel, M. Blanchy, O. Menes, D. García, M. Blanco, et al., Industrial applications of nanoparticles—a prospective overview, Materials Today: Proceedings 2 (2015) 456−465. Available from: https://doi.org/10.1016/j.matpr.2015.04.056.

[11] S.United States Choi, J.A. Eastman, Enhancing Thermal Conductivity of Fluids with Nanoparticles, Argonne National Lab., IL (United States), 1995.

[12] S. Gaikwad, S. Dhage, H. Potdar, V. Samuel, V. Ravi, Co-precipitation method for the preparation of nanocrystalline ferroelectric SrBi2Nb2O9 ceramics, Journal of Electroceramics 14 (1) (2005) 83−87. Available from: https://doi.org/10.1007/s10832-005-6588-y.

[13] F.E. Kruis, H. Fissan, A. Peled, Synthesis of nanoparticles in the gas phase for electronic, optical and magnetic applications—a review, Journal of Aerosol Science 29 (5-6) (1998) 511−535. Available from: https://doi.org/10.1016/S0021-8502(97)10032-5.

[14] A. Menbari, A.A. Alemrajabi, A. Rezaei, Heat transfer analysis and the effect of CuO/water nanofluid on direct absorption concentrating solar collector, Applied Thermal Engineering 104 (2016) 176−183. Available from: https://doi.org/10.1016/j.applthermaleng.2016.05.064.

[15] P.G. Jamkhande, N.W. Ghule, A.H. Bamer, M.G. Kalaskar, Metal nanoparticles synthesis: an overview on methods of preparation, advantages and disadvantages, and applications, Journal of Drug Delivery Science and Technology 53 (2019) 101174. Available from: https://doi.org/10.1016/j.jddst.2019.101174.

[16] A.V. Rane, K. Kanny, V.K. Abitha, S. Thomas, Methods for synthesis of nanoparticles and fabrication of nanocomposites, Synthesis of Inorganic Nanomaterials, Elsevier, 2018, pp. 121−139. Available from: https://doi.org/10.1016/b978-0-08-101975-7.00005-1.

[17] R. Taylor, S. Coulombe, T. Otanicar, P. Phelan, A. Gunawan, W. Lv, et al., Small particles, big impacts: a review of the diverse applications of nanofluids, Journal of Applied Physics 113 (2013) 11301. Available from: https://doi.org/10.1063/1.4754271.

[18] T.P. Otanicar, P.E. Phelan, J.S. Golden, Optical properties of liquids for direct absorption solar thermal energy systems, Solar Energy 83 (7) (2009) 969–977. Available from: https://doi.org/10.1016/j.solener.2008.12.009.

[19] M. Vivar, V. Everett, A review of optical and thermal transfer fluids used for optical adaptation or beam-splitting in concentrating solar systems, Progress in Photovoltaics Research and Applications 22 (2012) 612–633. Available from: https://doi.org/10.1002/pip.

[20] N. Goel, R.A. Taylor, T. Otanicar, A review of nanofluid-based direct absorption solar collectors: design considerations and experiments with hybrid PV/thermal and direct steam generation collectors, Renewable Energy 145 (2020) 903–913. Available from: https://doi.org/10.1016/j.renene.2019.06.097.

[21] Taylor, R.A.; Phelan, P.E.; Otanicar, T.P.; Tyagi, H.; Trimble, S. Applicability of nanofluids in concentrated solar energy harvesting (2011), 825–832. https://doi.org/10.1115/es2010-90055.

[22] Jain, P.K.; Lee, K.S.; El-Sayed, I.H.; El-Sayed, M.A. Calculated absorption and scattering properties of gold nanoparticles of different size, shape, and composition: applications in biological imaging and biomedicine, The Journal of Physical Chemistry B, 110 (2006) 7238–7248. https://doi.org/10.1021/jp057170o.

[23] Y. Xie, L. Carbone, C. Nobile, V. Grillo, S. D'Agostino, F. Della Sala, et al., Metallic-like stoichiometric copper sulfide nanocrystals: phase- and shape-selective synthesis, near-infrared surface plasmon resonance properties, and their modeling, ACS Nano 7 (8) (2013) 7352–7369. Available from: https://doi.org/10.1021/nn403035s.

[24] A. Bergerud, D. Nordlund, E.L. Runnerstrom, D.J. Milliron, S.D. Lounis, Influence of dopant distribution on the plasmonic properties of indium tin oxide nanocrystals, Journal of American Chemical Society 136 (19) (2014) 7110–7116. Available from: https://doi.org/10.1021/ja502541z.

[25] S. Kumar, C.L. Tien, Dependent absorption and extinction of radiation by small particles, Journal of Heat Transfer 112 (1) (1990) 178–185. Available from: https://doi.org/10.1115/1.2910342.

[26] S. Pillai, M.A. Green, Plasmonics for photovoltaic applications, Solar Energy Materials and Solar Cells, Vol. 94, North-Holland, 2010, pp. 1481–1486. Available from: https://doi.org/10.1016/j.solmat.2010.02.046.

[27] H.R. Stuart, D.G. Hall, Island size effects in nanoparticle-enhanced photodetectors, Applied Physics Letters 73 (26) (1998) 3815–3817. Available from: https://doi.org/10.1063/1.122903.

[28] D. Derkacs, S.H. Lim, P. Matheu, W. Mar, E.T. Yu, Improved performance of amorphous silicon solar cells via scattering from surface plasmon polaritons in nearby metallic nanoparticles, Applied Physics. Letters 89 (9) (2006) 093103. Available from: https://doi.org/10.1063/1.2336629.

[29] B.P. Rand, P. Peumans, S.R. Forrest, Long-range absorption enhancement in organic tandem thin-film solar cells containing silver nanoclusters, Journal of Applied Physics 96 (12) (2004) 7519–7526. Available from: https://doi.org/10.1063/1.1812589.

[30] H.A. Atwater, A. Polman, Plasmonics for improved photovoltaic devices, Nature Materials 9 (2010) 205–213. Available from: https://doi.org/10.1038/nmat2629.

[31] B. Eanest Jebasingh, A. Valan Arasu, A comprehensive review on latent heat and thermal conductivity of nanoparticle dispersed phase change material for

low-temperature applications, Energy Storage Materials, 24, 2020, pp. 52–74. Available from: https://doi.org/10.1016/j.ensm.2019.07.031.

[32] P. Manoj Kumar, R. Anandkumar, K. Mylsamy, K.B. Prakash, Experimental investigation on thermal conductivity of nanoparticle dispersed paraffin (NDP), Materials Today: Proceedings 45 (2020) 735–739. Available from: https://doi.org/10.1016/j.matpr.2020.02.798.

[33] P.K. Madathil, N. Balagi, P. Saha, J. Bharali, P.V.C. Rao, N.V. Choudary, et al., Preparation and characterization of molten salt based nanothermic fluids with enhanced thermal properties for solar thermal applications, Applied Thermal Engineering 109 (2016) 901–905. Available from: https://doi.org/10.1016/j.applthermaleng.2016.04.102.

[34] M.K. Saranprabhu, K.S. Rajan, Magnesium oxide nanoparticles dispersed solar salt with improved solid phase thermal conductivity and specific heat for latent heat thermal energy storage, Renewable Energy 141 (2019) 451–459. Available from: https://doi.org/10.1016/j.renene.2019.04.027.

[35] S.U.S. Choi, S. Li, J.A. Eastman, Measuring thermal conductivity of fluids containing oxide nanoparticles, Journal of Heat Transfer 121 (2) (1999) 280–289. Available from: https://doi.org/10.1115/1.2825978.

[36] S.M.S. Murshed, K.C. Leong, C. Yang, A combined model for the effective thermal conductivity of nanofluids, Applied Thermal Engineering 29 (11–12) (2009) 2477–2483. Available from: https://doi.org/10.1016/j.applthermaleng.2008.12.018.

[37] A.R. Moghadassi, S.M. Hosseini, D.E. Henneke, Effect of CuO nanoparticles in enhancing the thermal conductivities of monoethylene glycol and paraffin fluids, Industrial and Engineering Chemistry Research 49 (4) (2010) 1900–1904. Available from: https://doi.org/10.1021/ie901060e.

[38] Y. Devarajan, D.B. Munuswamy, Analysis on the influence of nanoparticles of alumina, copper oxide, and zirconium oxide on the performance of a flat-plate solar water heater, Energy and Fuels 30 (11) (2016) 9908–9913. Available from: https://doi.org/10.1021/acs.energyfuels.6b02264.

[39] J. Liu, Z. Ye, L. Zhang, X. Fang, Z. Zhang, A combined numerical and experimental study on graphene/ionic liquid nanofluid based direct absorption solar collector, Solar Energy Materials and Solar Cells 136 (2015) 177–186. Available from: https://doi.org/10.1016/j.solmat.2015.01.013.

[40] R.A. Taylor, T. Otanicar, G. Rosengarten, Nanofluid-based optical filter optimization for PV/T systems, Light: Science and Applications 1 (October) (2012) e34. Available from: https://doi.org/10.1038/lsa.2012.34.

[41] K. Simeonidis, S. Mourdikoudis, E. Kaprara, M. Mitrakas, L. Polavarapu, Inorganic engineered nanoparticles in drinking water treatment: a critical review, Environmental Science: Water Research and Technology 1 (2016) 43–70. Available from: https://doi.org/10.1039/c5ew00152h.

[42] G.P. Rao, C. Lu, F. Su, Sorption of divalent metal ions from aqueous solution by carbon nanotubes: a review, Separation and Purification Technology 58 (1) (2007) 224–231. Available from: https://doi.org/10.1016/j.seppur.2006.12.006.

[43] Y.H. Li, J. Ding, Z. Luan, Z. Di, Y. Zhu, C. Xu, et al., Competitive adsorption of Pb^{2+}, Cu^{2+} and Cd^{2+} ions from aqueous solutions by multiwalled carbon nanotubes, Carbon, Vol. 41, Pergamon, 2003, pp. 2787–2792. Available from: https://doi.org/10.1016/S0008-6223(03)00392-0.

[44] E.A. Deliyanni, D.N. Bakoyannakis, A.I. Zouboulis, K.A. Matis, Sorption of As(V) ions by akaganéite-type nanocrystals, Chemosphere 50 (1) (2003) 155–163. Available from: https://doi.org/10.1016/S0045-6535(02)00351-X.

[45] K.D. Hristovski, P.K. Westerhoff, T. Möller, P. Sylvester, Effect of synthesis conditions on nano-iron (Hydr)oxide impregnated granulated activated carbon, Chemical

Engineering Journal 146 (2) (2009) 237–243. Available from: https://doi.org/10.1016/j.cej.2008.05.040.
[46] K.D. Hristovski, H. Nguyen, P.K. Westerhoff, Removal of arsenate and 17α-ethinyl estradiol (EE2) by iron (Hydr)oxide modified activated carbon fibers, Journal of Environmental Science and Health, Part A 44 (4) (2009) 354–361. Available from: https://doi.org/10.1080/10934520802659695.
[47] Y.C. Chang, S.W. Chang, D.H. Chen, Magnetic chitosan nanoparticles: studies on chitosan binding and adsorption of Co(II) Ions, Reactive and Functional Polymers 66 (3) (2006) 335–341. Available from: https://doi.org/10.1016/j.reactfunctpolym.2005.08.006.
[48] J. Hu, I.M.C. Lo, G. Chen, Fast removal and recovery of Cr(VI) using surface-modified jacobsite ($MnFe_2O_4$) nanoparticles, Langmuir 21 (24) (2005) 11173–11179. Available from: https://doi.org/10.1021/la051076h.
[49] J. Hu, I.M.C. Lo, G. Chen, Performance and mechanism of chromate (VI) adsorption by δ-FeOOH-coated maghemite (γ-Fe_2O_3) nanoparticles, Separation and Purification Technology 58 (1) (2007) 76–82. Available from: https://doi.org/10.1016/j.seppur.2007.07.023.
[50] S.S. Banerjee, D.H. Chen, Fast removal of copper ions by gum arabic modified magnetic nano-adsorbent, Journal of Hazardous Materials 147 (3) (2007) 792–799. Available from: https://doi.org/10.1016/j.jhazmat.2007.01.079.
[51] J. Song, H. Kong, J. Jang, Adsorption of heavy metal ions from aqueous solution by polyrhodanine-encapsulated magnetic nanoparticles, Journal of Colloid and Interface Science 359 (2) (2011) 505–511. Available from: https://doi.org/10.1016/j.jcis.2011.04.034.
[52] Y.P. Chang, C.L. Ren, Q. Yang, Z.Y. Zhang, L.J. Dong, X.G. Chen, et al., Preparation and characterization of hexadecyl functionalized magnetic silica nanoparticles and its application in rhodamine 6G removal, Applied Surface Science 257 (20) (2011) 8610–8616. Available from: https://doi.org/10.1016/j.apsusc.2011.05.031.
[53] A. Alkhudhiri, N. Darwish, N. Hilal, Membrane distillation: a comprehensive review, Desalination 287 (January) (2012) 2–18. Available from: https://doi.org/10.1016/j.desal.2011.08.027.
[54] S. Al-Obaidani, E. Curcio, F. Macedonio, G. Di Profio, H. Al-Hinai, E. Drioli, Potential of membrane distillation in seawater desalination: thermal efficiency, sensitivity study and cost estimation, Journal of Membrane Science 323 (1) (2008) 85–98. Available from: https://doi.org/10.1016/j.memsci.2008.06.006.
[55] M.R. Qtaishat, F. Banat, Desalination by solar powered membrane distillation systems, Desalination 308 (2013) 186–197. Available from: https://doi.org/10.1016/j.desal.2012.01.021.
[56] M.N. Chernyshov, G.W. Meindersma, A.B. de Haan, Comparison of spacers for temperature polarization reduction in air gap membrane distillation, Desalination 183 (1–3) (2005) 363–374. Available from: https://doi.org/10.1016/j.desal.2005.04.029.
[57] T.Y. Cath, V.D. Adams, A.E. Childress, Experimental study of desalination using direct contact membrane distillation: a new approach to flux enhancement, Journal of Membrane Science 228 (1) (2004) 5–16. Available from: https://doi.org/10.1016/j.memsci.2003.09.006.
[58] N. Maximous, G. Nakhla, K. Wong, W. Wan, Optimization of Al_2O_3/PES membranes for wastewater filtration, Separation and Purification Technology 73 (2) (2010) 294–301. Available from: https://doi.org/10.1016/j.seppur.2010.04.016.
[59] T.H. Bae, T.M. Tak, Effect of TiO_2 nanoparticles on fouling mitigation of ultrafiltration membranes for activated sludge filtration, Journal of Membrane Science 249 (1–2) (2005) 1–8. Available from: https://doi.org/10.1016/j.memsci.2004.09.008.
[60] M.L. Lind, A.K. Ghosh, A. Jawor, X. Huang, W. Hou, Y. Yang, et al., Influence of zeolite crystal size on zeolite-polyamide thin film nanocomposite membranes,

Langmuir 25 (17) (2009) 10139–10145. Available from: https://doi.org/10.1021/la900938x.

[61] M.L. Lind, D.E. Suk, T.V. Nguyen, E.M.V. Hoek, Tailoring the structure of thin film nanocomposite membranes to achieve seawater RO membrane performance, Environmental Science and Technology 44 (21) (2010) 8230–8235. Available from: https://doi.org/10.1021/es101569p.

[62] M.L. Lind, B.H. Jeong, A. Subramani, X. Huang, E.M.V. Hoek, Effect of mobile cation on zeolite-polyamide thin film nanocomposite membranes, Journal of Materials Research 24 (5) (2009) 1624–1631. Available from: https://doi.org/10.1557/jmr.2009.0189.

[63] A.E. Kabeel, Z.M. Omara, F.A. Essa, Enhancement of modified solar still integrated with external condenser using nanofluids: an experimental approach, Energy Conversion and Management 78 (2014) 493–498. Available from: https://doi.org/10.1016/j.enconman.2013.11.013.

[64] M.J. Assael, C.F. Chen, I. Metaxa, W.A. Wakeham, Thermal conductivity of suspensions of carbon nanotubes in water, International Journal of Thermophysics 25 (4) (2004) 971–985. Available from: https://doi.org/10.1023/B:IJOT.0000038494.22494.04.

[65] T. Elango, A. Kannan, K. Kalidasa Murugavel, Performance study on single basin single slope solar still with different water nanofluids, Desalination 360 (2015) 45–51. Available from: https://doi.org/10.1016/j.desal.2015.01.004.

[66] X. Wang, Y. He, X. Liu, Synchronous steam generation and photodegradation for clean water generation based on localized solar energy harvesting, Energy Conversion and Management 173 (2018) 158–166. Available from: https://doi.org/10.1016/j.enconman.2018.07.065.

[67] E. Traver, R.A. Karaballi, Y.E. Monfared, H. Daurie, G.A. Gagnon, M. Dasog, TiN, ZrN, and HfN nanoparticles on nanoporous aluminum oxide membranes for solar-driven water evaporation and desalination, ACS Applied Nano Materials (2020). Available from: https://doi.org/10.1021/acsanm.0c00107.

[68] G. Zhu, J. Xu, W. Zhao, F. Huang, Constructing black titania with unique nanocage structure for solar desalination, ACS Applied Materials & Interfaces 8 (46) (2016) 31716–31721. Available from: https://doi.org/10.1021/acsami.6b11466.

[69] G.G. Jang, J.W. Klett, J. McFarlane, A. Ievlev, K. Xiao, J.K. Keum, et al., Efficient solar-thermal distillation desalination device by light absorptive carbon composite porous foam, Global Challenges 3 (8) (2019) 1900003. Available from: https://doi.org/10.1002/gch2.201900003.

[70] I.A. Said, T.R. Chomiak, Z. He, Q. Li, Low-cost high-efficiency solar membrane distillation for treatment of oil produced waters, Separation and Purification Technology 250 (2020) 117170. Available from: https://doi.org/10.1016/j.seppur.2020.117170.

[71] X. Wang, G. Ou, N. Wang, H. Wu, Graphene-based recyclable photo-absorbers for high-efficiency seawater desalination, ACS Applied Materials and Interfaces 8 (14) (2016) 9194–9199. Available from: https://doi.org/10.1021/acsami.6b02071.

[72] H.H. Mungondori, L. Tichagwa, D.M. Katwire, Removal of Pb^{2+} and Fe^{3+} from water using N-TiO_2 blended copolymer grafted asymmetric membranes, Water Science and Technology 73 (8) (2016) 1855–1864. Available from: https://doi.org/10.2166/wst.2016.024.

[73] M.M. Ibrahim, S. Asal, Physicochemical and photocatalytic studies of Ln^{3+}-ZnO for water disinfection and wastewater treatment applications, Journal of Molecular Structure 1149 (2017) 404–413. Available from: https://doi.org/10.1016/j.molstruc.2017.08.007.

[74] S. Sreeja, V. Shetty K., Photocatalytic water disinfection under solar irradiation by Ag@TiO_2 core-shell structured nanoparticles, Solar Energy 157 (2017) 236–243. Available from: https://doi.org/10.1016/j.solener.2017.07.057.

[75] H. Lee, J. Choi, S. Lee, S.T. Yun, C. Lee, J. Lee, Kinetic enhancement in photocatalytic oxidation of organic compounds by WO_3 in the presence of Fenton-like reagent, Applied Catalysis B: Environmental 138–139 (2013) 311–317. Available from: https://doi.org/10.1016/j.apcatb.2013.03.006.

[76] Z. Miao, S. Tao, Y. Wang, Y. Yu, C. Meng, Y. An, Hierarchically porous silica as an efficient catalyst carrier for high performance vis-light assisted Fenton degradation, Microporous and Mesoporous Materials 176 (2013) 178–185. Available from: https://doi.org/10.1016/j.micromeso.2013.04.009.

[77] S. Liu, X. Wang, W. Zhao, K. Wang, H. Sang, Z. He, Synthesis, characterization and enhanced photocatalytic performance of Ag_2S-coupled ZnO/ZnS core/shell nanorods, Journal of Alloys and Compounds 568 (2013) 84–91. Available from: https://doi.org/10.1016/j.jallcom.2013.03.149.

[78] J. Wang, Y. Yu, L. Zhang, Highly efficient photocatalytic removal of sodium pentachlorophenate with Bi_3O_4Br under visible light, Applied Catalysis B: Environmental 136–137 (2013) 112–121. Available from: https://doi.org/10.1016/j.apcatb.2013.02.009.

[79] V.K. Gupta, D. Pathania, S. Agarwal, P. Singh, Adsorptional photocatalytic degradation of methylene blue onto pectin-CuS nanocomposite under solar light, Journal of Hazardous Materials 243 (2012) 179–186. Available from: https://doi.org/10.1016/j.jhazmat.2012.10.018.

[80] S.F. Anis, R. Hashaikeh, N. Hilal, Functional materials in desalination: a reviewAvailable from: https://doi.org/10.1016/j.desal.2019.114077.

[81] A. Turchanin, A. Gölzhäuser, Carbon nanomembranes, Advanced Materials 28 (29) (2016) 6075–6103. Available from: https://doi.org/10.1002/adma.201506058.

[82] H.M. Nguyen, O. Seitz, W. Peng, Y.N. Gartstein, Y.J. Chabal, A.V. Malko, Efficient radiative and nonradiative energy transfer from proximal CdSe/ZnS nanocrystals into silicon nanomembranes, ACS Nano 6 (6) (2012) 5574–5582. Available from: https://doi.org/10.1021/nn301531b.

[83] W. Peng, S. Sampat, S.M. Rupich, B. Anand, H.M. Nguyen, D. Taylor, et al., Hybrid light sensor based on ultrathin Si nanomembranes sensitized with CdSe/ZnS colloidal nanocrystal quantum dots, Nanoscale 7 (18) (2015) 8524–8530. Available from: https://doi.org/10.1039/C5NR00334B.

[84] A. Kasry, M.M. Fadlallah, N.H. Voelcker, A.A. Maarouf, Experimental and theoretical demonstrations of ultraviolet absorption enhancement in porous nano-membrane graphene, Carbon N. Y 155 (2019) 65–70. Available from: https://doi.org/10.1016/j.carbon.2019.08.038.

[85] Z. Jakšić, O. Jakšić, Biomimetic nanomembranes: an overview, Biomimetics 5 (2) (2020) 24. Available from: https://doi.org/10.3390/biomimetics5020024.

[86] G. Jie, Z. Kongyin, Z. Xinxin, C. Zhijiang, C. Min, C. Tian, et al., Preparation and characterization of carboxyl multi-walled carbon nanotubes/calcium alginate composite hydrogel nano-filtration membrane, Materials Letters 157 (2015) 112–115. Available from: https://doi.org/10.1016/j.matlet.2015.05.080.

[87] D. Cohen-Tanugi, J.C. Grossman, Water desalination across nanoporous graphene, Nano Letters 12 (7) (2012) 3602–3608. Available from: https://doi.org/10.1021/nl3012853.

[88] S.C. O'Hern, D. Jang, S. Bose, J.-C. Idrobo, Y. Song, T. Laoui, et al., Nanofiltration across defect-sealed nanoporous monolayer graphene, Nano Letters 15 (5) (2015) 3254–3260. Available from: https://doi.org/10.1021/acs.nanolett.5b00456.

[89] S. Zheng, Q. Tu, J.J. Urban, S. Li, B. Mi, Swelling of graphene oxide membranes in aqueous solution: characterization of interlayer spacing and insight into water transport mechanisms, ACS Nano 11 (6) (2017) 6440–6450. Available from: https://doi.org/10.1021/acsnano.7b02999.

[90] Y. Yang, P. Dementyev, N. Biere, D. Emmrich, P. Stohmann, R. Korzetz, et al., Rapid water permeation through carbon nanomembranes with sub-nanometer channels, ACS Nano 12 (5) (2018) 4695–4701. Available from: https://doi.org/10.1021/acsnano.8b01266.

[91] T. Lee, B.-S. Kim, Two-dimensional designer nanochannels for controllable ion transport in graphene oxide nanomembranes with tunable sheet dimensions, ACS Applied Materials and Interfaces 12 (11) (2020) 13116–13126. Available from: https://doi.org/10.1021/acsami.9b20398.

[92] S. Karthick Kumar, S. Suresh, S. Murugesan, S.P. Raj, CuO thin films made of nanofibers for solar selective absorber applications, Solar Energy 94 (2013) 299–304. Available from: https://doi.org/10.1016/j.solener.2013.05.018.

[93] Q. Wali, Z.H. Bakr, N.A. Manshor, A. Fakharuddin, R. Jose, SnO_2-TiO_2 hybrid nanofibers for efficient dye-sensitized solar cells, Solar Energy 132 (2016) 395–404. Available from: https://doi.org/10.1016/j.solener.2016.03.037.

[94] M.N. Mustafa, S. Shafie, M.H. Wahid, Y. Sulaiman, Light scattering effect of polyvinyl-alcohol/titanium dioxide nanofibers in the dye-sensitized solar cell, Scientific Reports 9 (1) (2019) 14952. Available from: https://doi.org/10.1038/s41598-019-50292-z.

[95] Y. Liao, C.-H. Loh, M. Tian, R. Wang, A.G. Fane, Progress in electrospun polymeric nanofibrous membranes for water treatment: fabrication, modification and applications, Progress in Polymer Science 77 (2018) 69–94. Available from: https://doi.org/10.1016/j.progpolymsci.2017.10.003.

[96] J.-C. Kim, S.-I. Oh, W. Kang, H.-Y. Yoo, J. Lee, D.-W. Kim, Superior anodic oxidation in tailored Sb-doped SnO_2/RuO_2 composite nanofibers for electrochemical water treatment, Journal of Catalysis 374 (2019) 118–126. Available from: https://doi.org/10.1016/j.jcat.2019.04.025.

[97] Y. Shen, L. Li, K. Xiao, J. Xi, Constructing three-dimensional hierarchical architectures by integrating carbon nanofibers into graphite felts for water purification, ACS Sustainable Chemistry & Engineering 4 (4) (2016) 2351–2358. Available from: https://doi.org/10.1021/acssuschemeng.6b00030.

[98] D. Malwal, P. Gopinath, Efficient adsorption and antibacterial properties of electrospun CuO-ZnO composite nanofibers for water remediation, Journal of Hazardous Materials 321 (2017) 611–621. Available from: https://doi.org/10.1016/j.jhazmat.2016.09.050.

[99] L.N. Nthunya, M.L. Masheane, S.P. Malinga, E.N. Nxumalo, S.D. Mhlanga, V.K. Thakur, Environmentally benign chitosan-based nanofibres for potential use in water treatment, Cogent Chemistry 3 (1) (2017) 1357865. Available from: https://doi.org/10.1080/23312009.2017.1357865.

[100] P.S. Saud, B. Pant, A.-M. Alam, Z.K. Ghouri, M. Park, H.-Y. Kim, Carbon quantum dots anchored TiO_2 nanofibers: effective photocatalyst for waste water treatment, Ceramics International 41 (9) (2015) 11953–11959. Available from: https://doi.org/10.1016/j.ceramint.2015.06.007.

[101] T. Lavanya, M. Dutta, K. Satheesh, Graphene wrapped porous tubular rutile TiO_2 nanofibers with superior interfacial contact for highly efficient photocatalytic performance for water treatment, Separation and Purification Technology 168 (2016) 284–293. Available from: https://doi.org/10.1016/j.seppur.2016.05.059.

[102] T. Yang, H. Lin, K.-T. Lin, B. Jia, Carbon-based absorbers for solar evaporation: steam generation and beyond, Sustainable Materials and Technology 25 (2020) e00182. Available from: https://doi.org/10.1016/j.susmat.2020.e00182.

[103] F. Jiang, H. Liu, Y. Li, Y. Kuang, X. Xu, C. Chen, et al., Lightweight, mesoporous, and highly absorptive all-nanofiber aerogel for efficient solar steam generation, ACS Applied Materials and Interfaces 10 (1) (2018) 1104–1112. Available from: https://doi.org/10.1021/acsami.7b15125.

[104] Y. Jin, J. Chang, Y. Shi, L. Shi, S. Hong, P. Wang, A highly flexible and washable nonwoven photothermal cloth for efficient and practical solar steam generation, Journal of Materials Chemistry A 6 (17) (2018) 7942–7949. Available from: https://doi.org/10.1039/C8TA00187A.
[105] T. Gao, Y. Li, C. Chen, Z. Yang, Y. Kuang, C. Jia, et al., Architecting a floatable, durable, and scalable steam generator: hydrophobic/hydrophilic bifunctional structure for solar evaporation enhancement, Small Methods 3 (2) (2019) 1800176. Available from: https://doi.org/10.1002/smtd.201800176.
[106] C. Zhang, B. Yuan, Y. Liang, L. Yang, L. Bai, H. Yang, et al., Carbon nanofibers enhanced solar steam generation device based on loofah biomass for water purification, Materials Chemistry and Physics 258 (2021) 123998. Available from: https://doi.org/10.1016/j.matchemphys.2020.123998.
[107] E. Kabir, V. Kumar, K.-H. Kim, A.C.K. Yip, J.R. Sohn, Environmental impacts of nanomaterials, Journal of Environmental Management 225 (2018) 261–271. Available from: https://doi.org/10.1016/j.jenvman.2018.07.087.
[108] K.A. Dunphy Guzmán, M.R. Taylor, J.F. Banfield, Environmental risks of nanotechnology: national nanotechnology initiative funding, 2000 – 2004, Environmental Science and Technology 40 (5) (2006) 1401–1407. Available from: https://doi.org/10.1021/es0515708.
[109] D.G. Rickerby, M. Morrison, Nanotechnology and the environment: a European perspective, Science and Technology of Advanced Materials 8 (1–2) (2007) 19–24. Available from: https://doi.org/10.1016/j.stam.2006.10.002.
[110] Y. Liu, M. Tourbin, S. Lachaize, P. Guiraud, Nanoparticles in wastewaters: hazards, fate and remediation, Powder Technology 255 (2014) 149–156. Available from: https://doi.org/10.1016/j.powtec.2013.08.025.
[111] Y. Zhang, Y. Chen, P. Westerhoff, K. Hristovski, J.C. Crittenden, Stability of commercial metal oxide nanoparticles in water, Water Research 42 (8–9) (2008) 2204–2212. Available from: https://doi.org/10.1016/j.watres.2007.11.036.
[112] H.P. Jarvie, H. Al-Obaidi, S.M. King, M.J. Bowes, M.J. Lawrence, A.F. Drake, et al., Fate of silica nanoparticles in simulated primary wastewater treatment, Environmental Science and Technology 43 (22) (2009) 8622–8628. Available from: https://doi.org/10.1021/es901399q.
[113] V.K. Sharma, J. Filip, R. Zboril, R.S. Varma, Natural inorganic nanoparticles – formation, fate, and toxicity in the environment, Chemical Society Reviews 44 (23) (2015) 8410–8423. Available from: https://doi.org/10.1039/C5CS00236B.
[114] M. Kamali, K.M. Persson, M.E. Costa, I. Capela, Sustainability criteria for assessing nanotechnology applicability in industrial wastewater treatment: current status and future outlook, Environment International 125 (2019) 261–276. Available from: https://doi.org/10.1016/j.envint.2019.01.055.

CHAPTER 7

Issues and risks associated with generation I–IV solar-driven water purification technologies

Balaji Devarajan
Department of Mechanical Engineering, KPR Institute of Engineering and Technology, Coimbatore, India

7.1 Introduction

The high demand for water due to human activities necessitates the desalination of water. Water desalination has been established as a reliable alternative source of fresh water. Desalination, on the other hand, has had multiple impacts on the environment. The discharge of brine-loaded chemicals and greenhouse gas (GHG) emissions is under investigation. Some efforts have been made to reduce this impact [1]. Environmental impact (EI) mitigation and control strategies are the main topics addressed in current research. Many studies have collated the EI mitigation and control strategies of each desalination step, in preference to having them distributed among separate pieces of literature. The cited article tracks the flow from intake to outfall to explore how to reduce the impacts throughout the whole process. It begins with an intake system, passes through desalination technology, and ends with brine discharge. The EIs are thoroughly discussed with the introduction of mitigation and control strategies for every desalination process are explored. Studies have clearly demonstrated that specific design criteria and process improvements can have a substantial impact on substantially saving resources [2]. The feedwater source appears to have a significant influence on EIs. Additionally, the use of desalination has had a large impact on the properties of brine and energy consumption. Both the use of renewable and traditional desalination technologies have shown a reduction in energy consumption, while evidence also suggests that hybrids and emerging desalination methods have lower total EIs [3].

Desalination is essential to addressing the challenge of water scarcity. Desalination provides a primary water source for many countries where it

Solar-Driven Water Treatment
DOI: https://doi.org/10.1016/B978-0-323-90991-4.00004-9

can constitute more than 90% of the water supply for domestic purposes. Thermal and membrane desalination has become a technologically and financially viable method. New and current desalination methods are more productive and less energy intensive than previous desalination technologies [4]. As a result, desalination may cause many EIs (or problems). EI assessment (EIA) plays a vital part in decision-making on desalination mitigation and control measures (EI strategies) are essential. The elements of desalination were described using a plant-to- outfall approach, working through the feedwater until the discharge of brine at the outfall. First, it was addressed by few industrialists to eliminate and reduce the EIs, and desalination technology as a pretreatment, followed by energy consumption and brine disposal [5]. It was found that desalination techniques had a large impact on overall desalting capacity, and also the water source and quality, and energy source. the most critical task in dealing with EI, as well as intake and discharge sizes, brine disposal, energy consumption, and disposal for desalination is related to the proper selection of desalination technology. Conventional desalination systems, emerging desalination technologies, and the utilization of renewable energies, all reduced desalination methods, were found to substantially support the mitigation and control systems [6,7]. Fig. 7.1 illustrates a schematic diagram of solar-driven desalination and evaporation along with the mechanism and application of the technology.

7.1.1 Brine management

Desalination plays a significant role in meeting the rising demand for fresh water due to the increasing population, and desalting by multiple stages is basically a type of distillation that desalinates using the efficiency of reverse osmosis (RO). To date, the majority of inland desalination plants use RO. Also, conventional desalination technology has gained a strong advantage over its competitors, however there are several newer methods under consideration that could be utilized in the development of inland desalination. Regardless of the technology, there are problems associated with inland desalination, and many environmental concerns about desalination exist [9]. Once it is expelled from the desalination plant, seawater is able to access the soil, and its physical and chemical properties are changed. The brine is likely to seep into the groundwater as well as be washed away in it. As of yet, however, much work remains to be done in improving the efficacy of brine management. Desalination releases a large

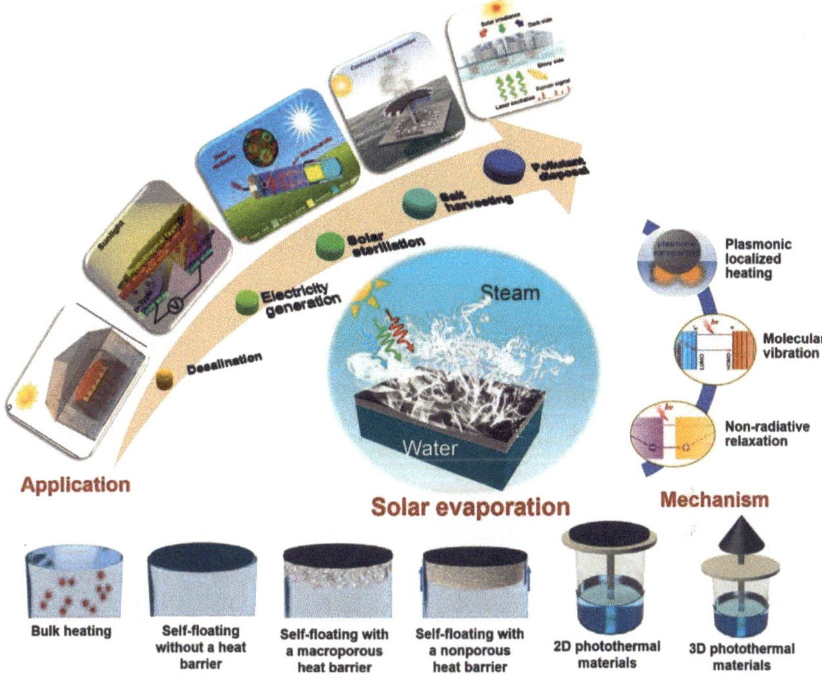

Figure 7.1 Solar-driven steam generation schematic [11].

amount of GHGs; however, it should be noted that these emissions may be offset by reforestation. The most common approaches aimed at reducing desalination energy consumption involve multiple approaches, all of which have been described in the literature. Sometimes processes or technologies are improved or alternatives found, while other times the focus is on making use of new energy savings and energy-saving technology; and the third way involves shifting to a process that results in a reduction in the consumption of energy while desalinating renewable resources [8,10].

The most simple, straightforward, and cost-effective method for disposing of brine from coastal desalination plants is to dump it into the ocean. Brine discharge to the ocean is harmful to the benthic (living on the ocean floor) ecosystem, while bathymetry, waves, and depth of the water column are crucial factors for the hydrodynamic impact, and are strongly dependent on environmental characteristics [11]. It is common for inland desalination plants to be located in areas distant from the discharge of brine into the sea, and it is not easy to use discharged seawater because of various concerns over environmental pollution, including cost,

EI, toxicity, and salinity. Thus, in turn, the major issues faced by inland desalination plants include finding ways to get rid of the brine (brine disposal issues). However, there are several options to resolve seawater rejection from inland desalination plants, and discharge into a well is currently the most common. Injection into deep aquifers, ocean water, and discharge of the liquid into a wastewater system are the most frequently used, and dispersion of the residual into land can affect the ground water also. Few other techniques for desalination includes the use of saline plants (halophytic crops) to irrigate or desalting plants and harvesting brine shrimp [12]. While these approaches work in moderate amounts of water, they have limited applicability for large quantities of brine solution.

7.1.2 Land use

Although the developments of desalination plants are likely to have harmful impacts on land use, they can be effectively mitigated. Most importantly, desalination plants (especially seawater desalinization plants) can keep pace with other useful products such as the value of drinking water and electricity. In this way, it could greatly impact recreational activities (like swimming, fishing, diving) as well as maritime operations (e.g., sailing) and tourist activities. Desalination plants are located almost entirely within a state or country's immediate coastline, and thus access to the supply is completely restricted for nearby regions. Land use and site availability of coastal land are typically considered to be linked. It is often connected to the amount of available undeveloped coastal real estate. Desalination plants are affected by multiple aspects of land use, and this is one of many negative issues that may arise. In desalination plant construction activities, these two factors interact in such a way that desalination has a deleterious effect on soil biota, as well as damaging the environment [13,14].

7.2 Policies for desalination

Desalination plants have a total capacity of 59.9 million m^3/day globally. Ever since 2008, there has been a 6.6 million m^3/day increase, making it the largest amount of desalination capacity added every year. The Shoaiba 3 project in Saudi Arabia, which was one of 700 new plants worldwide, is the largest plant, commissioned at 880,000 m^3/day. There are currently 14,451 desalination plants in service, with another 244 under contract or in the developmental stage. In addition, the installed capacity of seawater desalination plants has increased by a massive 35.9 million m^3/day, contributing to an almost 30% increase [15]. It does, however, provide an accurate picture of where the world stands in terms of desalination capacity. Below are descriptions of some of the major countries' desalination policies.

7.2.1 Kingdom of Saudi Arabia (KSA)

With 17% of global desalination water production, Saudi Arabia is the world's largest producer of desalinated water. With a growing population, Saudi Arabia will face a daily water requirement of an additional 6 million m^3 for the coming 20 years [16]. The cost of constructing this desalination capacity based on current process costs was estimated to be around $200 billion [17]. The two major processes are multistage flash distillation (MSF) and RO, with MSF accounting for approximately 64.2% of total installed desalination capacity in Saudi Arabia. The facility at Al-Jubail in Saudi Arabia produces around 2 million m^3 of desalinated water per day [18] and is the world's largest desalination plant. The country is attempting to control its water consumption in order to prevent aquifer depletion. In the future, the KSA may turn to other sources of energy, such as nuclear or solar, to meet its expanding demand for water and energy. A bold initiative to convert Saudi Arabia's desalination plants to solar-powered ones over the next 10 years has been announced [19].

7.2.2 The United States

The United States is the world's third-largest user of desalinators. Desalination plants have been constructed in every state of the United States. Desalinated water is used in both municipal and industrial applications. Brackish water and river water are the primary sources, both of which have low salinity levels, resulting in low energy consumption and cost. In addition, RO is the most common desalination approach in the United States. The next most popular method is nanofiltration (NF). In contrast to the Gulf countries, in the United States thermo-desalination has been used in a very limited capacity. Desalination and water recycling will be major contributors to the stability of an adequate water supply in the United States by the year 2020, according to the United States water body survey conducted in the year 2015. As a result, the US Bureau of Reclamation and Sandia have devised a strategy to address the country's water needs, with desalination as a key component. As a result, the

United States intends to invest heavily in the research and development of recycling of water and processes involving desalting [20–23].

7.2.3 Spain

The first water treatment plant in Europe was built 40 years ago in the Spanish Canary Islands. Desalinated water is used in Spain to support a thriving agricultural sector as well as an environmentally conscious tourism industry. Desalination production doubled between the years 2000 and 2004, with the government projecting another five-fold increase in the next 5 years [24,25]. During 2005, the country was hit by a drought. As a result, desalination was subsidized by the government, with the goal of making desalinated water as affordable as tap water. RO technology was responsible for 70% of desalinated water production, while multieffect distillation (MED) was responsible for another 23% of fresh water production [26].

7.2.4 China

In 2005, China had the world's 12th-largest desalination capacity, but by 2008, it had increased its capacity roughly 4.6-fold. It began to implement more desalination plants as a result of a lack of drinking water. It was then ranked fifth in the world for desalination, producing 4% of the world's desalinated water. MSF was first installed in 1989 on an industrial scale in two parallel installations, which supplied 6000 and 1200 m^3/day to Tianjin city, as well as an MED, which was installed in Qingdao city in 2004, experiencing a specific power consumption of 1.6 kWh/m^3 [27]. The first large-scale RO system was installed in 1997, with a yield of 500 m^3/day of fresh water from sea water, and seven more were installed by 2001, with the most recent having a yield of 18,000 m^3/day of fresh water from brackish water. China signed a $119 million contract in 2007 to build the world's largest seawater desalination plant in Beijing, which renders 100,000 m^3 of fresh water per day [28], and it commenced operation in August 2009 [29,30].

7.2.5 Kuwait

Kuwait is a significant desalination using country, with 4% of the world's capacity. It was the first country in the Gulf Cooperation Council to invest in desalination, beginning in 1953 with an MED plant which had a capacity of 9200 m^3/day. Between the 1950s and 2010, its desalination capacity grew at an almost exponential rate. MSF was introduced in large quantities in the 1970s, resulting in a desalination capacity of around 2.4 million m^3/day in 2008. This increase was estimated to be approximately 107% [31]. Due to the rapid progress of MSF technology, after 1960, all plants installed were MSF [32]. In Kuwait, all of the major desalination plants are dual-purpose cogeneration plants that use the MSF desalting system [33]. Several older units from the 1960s and 1970s have been retrofitted to extend their service life by another 10 years. MSF will be maintained in Kuwait for the next 20–30 years, according to a study of current trends. Developments in membrane science, technology maturation, and rising energy consumption may force the government to switch to RO. Furthermore, environmental concerns and advancements in renewable energy could lead to renewable energy convergence with the desalination industry [20]. Kuwait also yields more desalinated water each year than it can get from its renewable national water resources, and it recycles 20–40 m^3/c/year of wastewater. Furthermore, Kuwait uses between 50% and 70% of its water for agricultural purposes, with the majority of potential water going to domestic use. The Shuwaikh plant which has a capacity of 136,000 m^3/day, uses RO technology and is established in the year 2010. This plant is considered to be the most recent plant installed in Kuwait for the purpose of desalination [34–36].

7.3 Issues in desalination

7.3.1 Energy and environmental issues

Desalination has become a major source of drinking water for both domestic and industrial use as a result of the growing population and rising standards of living of people in urban areas, combined with unsustainable water use and pollution of available water sources. A renewable industrial process, as well as the design and implementation of water treatment, have the potential to have negative environmental consequences. The significance of these effects is determined by the technology employed [37]. The potential environmental damage caused by desalination processes has been summarized by various authors, which has highlighted the recognized issues and obtainable industrial strategies. Energy is given special attention because it is among the most serious environmental and significant costs of desalination. It is demonstrated that desalination processes consume very little energy when compared to the total national electricity consumption and also for transportation. The adverse effects of concentrate discharge into the sea are limited to a confined area around the discharge point. Some other potential environmental issues on the intake side, such as impingement and entrainment of marine life, are insignificant [38].

Desalination techniques are extremely effective in addressing municipal and industrial water supply demands. Despite the fact that desalination is regarded as an energy-intensive process, it has been demonstrated that the energy consumed by most desalination processes accounts for only a small portion of total energy consumed by the entire country. The energy required to produce 600 million m^3 of desalinated water per year, for example, is less than 1.4% of the entire energy consumption of Israel. Desalination technologies have raised many questions about their EIs. The critical areas of concern are air pollution and GHG emissions, entanglement of marine life, and the discharge of considerably high-temperature, high-salinity residues with higher chemical content. Nonetheless, there is a scarcity of research on these potential natural and environmental consequences, specifically when it comes to the long-term effects on the marine environment. Currently available data have suggested that high salinity and temperature affect only a specific area adjacent to the chemical disposal point, after which complete dilution with seawater diminishes any further effects. Priority should be given to developing renewable energy sources and water supplies that meet sustainability and environmental requirements in the near future [39].

7.3.2 Environmental issues in seawater reverse osmosis

The design and maintenance of intake systems, as well as the disposal of concentrate, have some environmental consequences for seawater RO (SWRO) desalination. The condensation and impingement of microorganisms in the water body is the primary effect of conventional open-ocean systems. These negative effects can be reduced by identifying the ingestion in an area with low oceanic productivity. Intakes with a velocity cap actually reduce the impacts by decreasing the number of fish trapped, and some forward screens may allow some oceanic organisms to survive. Prevention, such as habitat restoration or replenishment, can be an acceptable solution in cases of significant impacts on organisms. Although geologic intake systems prevent impaction and accumulation, they can have other less serious consequences such as restrictions to beach usage. If a poorly distilled output is flowing across the marine bottom, it can have a negative impact on benthic communities. Concentrate discharge impacts on benthic communities can be reduced by using carefully constructed diffuser systems that are designed and located based on current and flow modeling of the water bodies. EIs can be adequately minimized with suitable management of design based on a relatively detailed EIA analysis earlier to facility siting and design, according to the experiences of SWRO desalination to date [40].

Seawater desalination has become an important source of drinking water in many areas of the world from temperate to hyper-arid zones, especially in the Middle East and North Africa (MENA). Desalination has become more energy-efficient and economically viable as a result of the increase in use of the SWRO desalination process. The main focus is now on making SWRO as simple, efficient, and ecofriendly as possible. Desalination can be made "greener" now and in the future by employing a variety of techniques. Based on proper predesign, EIA research, and reduction of impacts when necessary, SWRO desalination can be done without having a significant impact on the environment. Nevertheless, a more internationally accepted approach to the environmental issues could be achieved by adopting a more conventional and environmental friendly practices. The scientific approach to evaluating the effects of impaction and accumulation of chemical wastes, which is currently done in a variety of ways that yield widely differing inferences, is significant. Additionally, when reviewing large storage SWRO plant discharges, the presumed environmental evaluation of concentrate discharge has to be unified and must include the most sophisticated near-field and far-field models. There is no agreed-upon threshold for what constitutes significant or unacceptable impacts [41].

While a zero-impact limit for any human influence may appear appealing on the surface, it is rarely feasible or financially viable. To design and deliver measures to safeguard the environmental quality of marine environments and minimize damage to the surrounding environment, sound science is usually undertaken, even while allowing for the economically efficient acquisition of the needed fresh water through SWRO. The marine ecosystem is most affected by standard shoreline and overseas open-ocean intakes. One of the most important aspects of these facilities' environmentally friendly design is to locate them where marine productivity is low, in order to have the least amount of impact. To establish a baseline for assessing potential impacts, predesign measurements of entrainable organism populations are required. To reduce the impacts of impingement and entrainment, passive wedge wire screens or seafloor intakes should always be used for exterior water intakes whenever possible, considering the economic feasibility of the process. When impingement or entrainment effects are unavoidable, contingency plans such as ecosystem creation or reestablishment, or refurbishing the sea with essential species of concern, can be used. Such initiatives were found to be economically viable, allowing desalination plants to operate without causing significant environmental damage [42].

By carrying out a detailed predesign inquest, which includes assessments of the living stock of aquatic creatures both in the water column and in the benthic environment, concentrate discharge in coastal SWRO systems can be done using eco-friendly methods. Natural seasonal fluctuations in populations, as well as potential effects caused by concentrate discharges, must be taken into account in proper predesign investigations. Data obtained during the process of an SWRO center can be perpetually assessed to determine whether any actual impacts occur by developing background data. Modifications to the residual effluent system may be necessary if impacts are discovered. Since numerous monitoring systems have been carefully designed and implemented, they cannot be effectively tied to true impact analysis. The key issue during SWRO operations is to perform monitoring to the extent that it is helpful in analyzing impacts. EIA is becoming more complicated as a result of global climate change, and therefore monitoring systems must use stations outside the impact zone to evaluate the changes in the ecosystem that are not related to the SWRO system's operation. Dilution of the chemicals close to the release point may reduce the varying density between the receiving and discharge water, as well as the employment of diffusor systems to stimulate rapid mixing for minimizing concentrate discharge impacts. Both methods necessitate a mixing area between the discharge point and a defined radius that permits mixing. Regulatory bodies must establish a reasonable "mixing zone" that is based on the real science rather than on a whim [43,44].

Diffusor systems can be formulated to suit virtually any combining zone boundary regulation requirement. Prevention strategies can be adopted to accommodate for the effects of concentrate discharge where they are considered relevant based on the actual feasible maintenance and implementation scenarios, either before or after the desalination site is operational. A similar generic strategy used to prevent impingement and entrainment can be applied here. In the case of benthic impacts, creating a habitat to the exterior of the discharge impact zone to expand the size of impacted organisms is a viable option. For affected swimming organisms, the introduction of benthic organism larvae or reorganizing may be successful in reducing EIs, but not for benthic organisms. The restocking strategy can be done on the basis of real-time data from the monitoring system. SWRO desalination in all capacity ranges along the world's largest shorelines can be constructed and maintained without causing major environmental effects, according to research results. However, proper EIA must be carried out prior to design, and continuous management must be carried out during operation to consistently evaluate environmental factors. This would allow for adjustments to be made to mitigate actual effects. Much research is therefore needed for SWRO systems proposed for use in interior locations to minimize the effects of concentrate discharge, particularly for zero discharge systems options [45].

7.4 Sociopolitical impacting factors

According to a growing day-zero condition, day-zero water shortage risks, greater salinity of conventional groundwater resources and the worldwide occurrence of these droughts have led to a heightened view of desalination as an important strategy to solve salt-water problems. Some sociopolitical factors can amplify the trend of desalination and distribution of clean water; however, the distribution and deployment of desalination methods are greatly influenced by a large number of factors not widely known yet. Though the greater part of earlier researchers did not given much consideration to the aforementioned factors, there have been several desalination technology developments which have met with varying degrees of success across the globe due to economic, environmental, and technical considerations, with a complete and critical examination of 15 societal and political parameters that had an impact on desalination and adoption/proliferation. An extensive literature review was conducted on these topics to analyze the relative significance of desalination-related relevance of these factors. Numerous examples were offered to enhance and strengthen the statement, most of which came from several different countries around the world. Research findings produced an additional finding, which led to the formulation of a goal for future research [23,27].

A detailed analysis of the political and economic factors was carried out, resulting in the conclusion that desalination is positively correlated with what was published in many studies. The four main variables of the research examined and the results of the SWOT analysis focused on various strengths, shortcomings, risks, and prospects, respectively. The article conducted an investigation into the ways in which desalination can positively impact aspects of society such as health, economic stability, industrial progress, and global security, by providing alternative water supplies for residents in numerous instances around the world. A few treatment systems can bring in the following strengths: desalination is scalable; it is possible to increase its output quickly; and deployment is relatively easy. Additionally, there was not much of a physical footprint during its operation. Desalination methods are helpful in resolving water issues, such as disputes, especially in areas that have political or ethnic tensions over access to resources. Desalination can have the ability to render ample amounts of water and even the ability to be expanded when needed to boost water availability around the globe. Desalination coupled with good distribution channels can also help the population and reduce the incidence of infectious water-spread diseases. There are many benefits that the desalination industry has to offer society. The various opportunities that are available to the locals while the desalination plant is being built and used are noted below. Furthermore, improved access to water in emerging nations would benefit women and give them time for greater access to work and education [12–17].

Many countries and communities have found that desalination has many important effects on their economy because it promotes other sectors, including tourism. Water desalination research studies also claim to identify benefits to the household water supply's consumers as well as desalinated water being used in other ways in this process. For instance, this effect was restricted by lower levels of destruction by water management and water-based equipment. Limitations associated with desalination plants (such as visual blight, excessive noise, and arguments about land use during desalination plant construction) were mostly highlighted by citizens. Much research has thrown light on many factors including the issue of remineralization and the impact of a depleted water supply on people's well-being. Desalination plants with higher capacities typically do not suffer from this problem, but very small and individual ones might, if they are located far from fresh water supply [26]. On the other hand, heavy dependence on desalination carries risks due to its unfavorable outcomes. With an exponentially expanding desalinated water supply as a prime example, it is expected that water consumption would increase domestically in areas such as the Gulf states. There is another kind of social conflict that could occur in desalination facilities: these facilities can be challenged by a small segment of the population due to concerns about the effect on the environment, and so they may face resistance. A small group with ardent, civilian protesters can have an impact on the economic, social, and development factors that determine how much it will be to build and how much the final project will cost. Natural disasters like tsunamis and floods can halt the expansion of desalination plants, while desalination plants are threatened also by other land uses and population growth, and can be affected by sea level rises, climatic factors like temperature and humidity changes, human development, population increases, and adverse weather conditions. Desalinated water has a tendency to expand, and with no additional purification steps it must be dealt with caution. Environmental factors including conflicts, natural catastrophes (e.g., hurricanes), oil spillages, and algal blooms can lead to the curtailment of river and waterway transportation activities. Desalination has had an impact on the lives of water-poor societies around the world and contributed to their growth and stability; however, no conclusive evidence has been established to support these claims. It is complex to take all the factors into account when trying to address the social and political facet of desertion. It is crucial to have a complete understanding of this subject area, and it is necessary to harness other sociopolitical elements to pave the way for desalination and to increase its prominence [33–35].

7.5 Economics and business issues with solar desalination

The economic issues are created with the aid of a patent analysis. The following graphs depict the existing status of solar desalination by country and year of publication. The source used for this patent analysis is the WIPO patent scope. It reveals the United States as the dominant country in patent filing, followed by PCT and Australia, respectively. This means that fresh water availability is lower in the United States than saline water. Therefore more conversion is required in countries like the United States and Australia compared to other countries. Figs. 7.2 and 7.3 illustrate the total number of patents filed by various countries over the past decade on an annual basis. The patent publication per year largely lies in the range of 12–29, giving an average of 21. The surprising data is in the year 2021, the count is 18. This reveals two major things, first that countries are moving toward drought (water scarcity) conditions and the second is that we are in need of saline water for conversion into fresh water. Form an economic point of view, there is potential for desalination to be

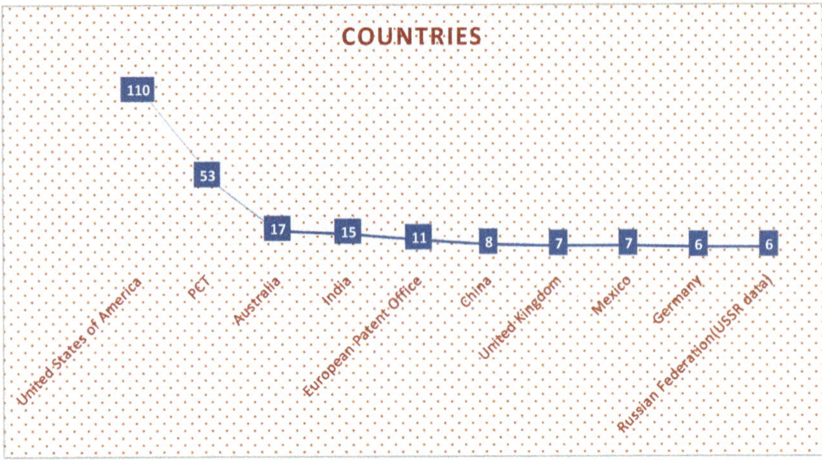

Figure 7.2 Patent analysis on solar desalination by country.

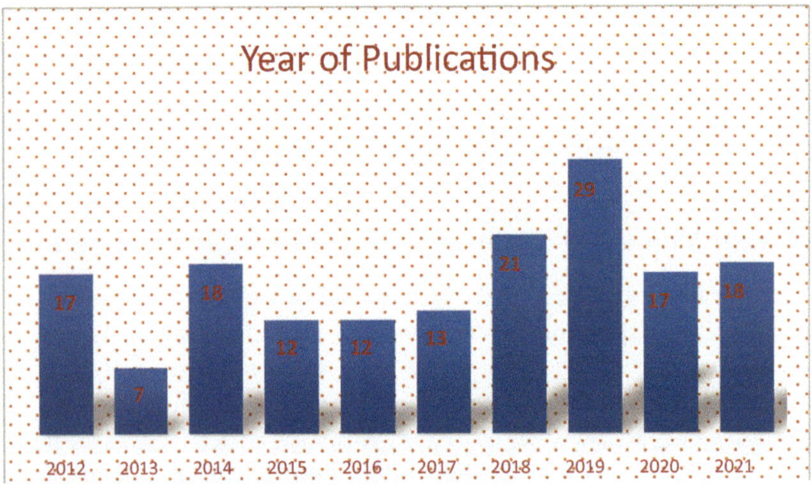

Figure 7.3 Patent analysis on solar desalination by year.

undertaken around the globe. Droughts are becoming more common around the world, not only in the United States and Australia. Therefore, where there is wide availability of saline water, desalination is seen as a potential business opportunity. The core analysis is done using renewable solar energy as the source, to avoid pollution. It is considered to be a potential global business with consistent growth over the years [46].

7.6 Australian status

The installed desalination plants in Australia provide only just over 1% of the production capacity, but regardless of this, there remains an outstanding need for a high volume of desalination throughout the country. Some researchers have discussed some of the most common seawater challenges arising in desalination projects in Australia: including, the advantages of RO and the characteristics of two desalination facilities in that area, and how to implement the technology in that context (Sydney and Perth). There are few but significant measures used to assess the possibility of desalination, such as ocean currents, seawater alkalinity, and temperature which have an effect on production expenses, capacity, and water quality, and may also be studied for efficiency and freshness of existing SWRO and implemented seawater reverse osmosis technologies (SWOT) that are in use for larger scale desalination plants in Australia. To date, the majority of the world's great capital cities have installed some form of desalination plants, mostly as alternatives to dams for water supply, because a lack of rainfall and this has necessitated the water conservation measures in those countries. Perth is a key example of a city making the switch to desalinated water in Australia, as its water supply was desalinated during the first decade of the 21st century, and Sydney followed soon after. There are currently plans to construct 13 other SWRO facilities for a variety of purposes, such as supplying drinking water to customers or wastewater to treatment plants [25,32,47]. The status of various Australian states is depicted in Fig. 7.4.

Although seawater desalination has become a more common practice despite the cost being greater for production. There is evidence that warmer weather due to global climate change has contributed to increasing water shortages in cities. Desalination accounts for the majority of desalting of small- and large-scale seawater desalting projects, and the sole technology preferred for emerging-scale seawater desalination plants. Developments and improvements of membrane technology are intended to help strengthen its use, to ensure that membranes provide more effective intake of seawater [48]. Chlorinated water treatment Adding or subtracting a small amount of sodium fluoride to adjust the pH d dual-media filtration Immediately after treatment with the water that was pre-exposed to lime. Membrane based desalination plant has been in operation for over 60 years. The first ever run of seawater desalination began over half a century ago. Once extensive shortages of natural water resources

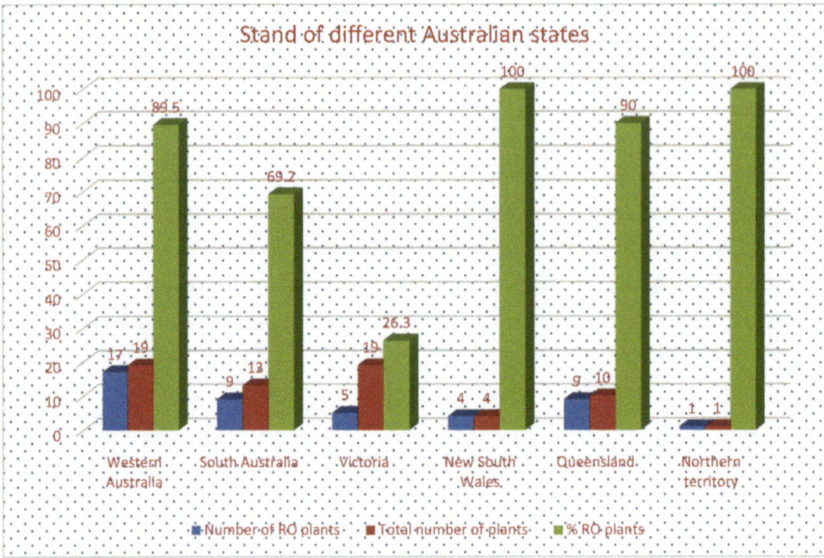

Figure 7.4 Status of different Australian states.

threaten the availability of water to a number of cities in Australia, seawater desalination will become highly attractive. The EI of desalination is specific to the project in which it is installed and depends on many factors, such as the quantity of the water that is used and the geographic location, and also: (1) habitats may be threatened; (2) sea level increase and erosion; (3) marine life and benthic communities may be adversely affected; and (4) lowering of the ocean floor [31,39,42].

SWRO is given low environmental risk ratings based on its impacts on the local ecosystem as well as the population habitats and populations, both of organisms already living there and those yet to be established. In general, desalination projects start with a concept, followed by contracts being drawn up and plant construction, after which contracts are completed and licenses are issued, which permit operation. This is slightly different from the usual procedure utilized throughout the desalination process of having four-stage contracts in each location (design, construct, maintain, and optimize vs. the standard locations) owing to the additional cost of desalination. Thus, the factors involved in selecting the location for a desalinization plant and the intake and outflow systems and also the pipeline maintenance are essential, as extra costs can be incurred due to the construction, equipment, and parts. Over the long term, desalination technology will gain importance due to decreasing precipitation rates in major population centers, but could see more or less, as an issue due to climate change, decreasing, if other things remain unchanged. Traditional water is more scarce than fresh water produced at high costs, which in turn will necessitate the use of desalination for various processes, especially in cities and countries of substantial size [26,28–31].

7.7 Fifty years analysis (1953–2013)

The years preceding World War II witnessed scientists investigating a variety of salinity-reducing methods, or salinization procedures, which had previously been overlooked. Electrodialysis (ED) was first established in the market by General Electric in the early 1950s. Based on its economic advantages, some publications claimed that ED might be used for desalting seawater and that it was much more affordable than maceration. More attention was given to using desalination to supply fresh water for municipalities that required a less expensive method, however, much more desalinating water was required. ED expanded into muddy and shallow water markets all over the world. Although this proved to be very effective, it was discovered that it could not be done with ocean water with maximum capability; that is, as the salt concentration increased, so did energy consumption. As is the case in stagnant or slow-moving water, however, ED excelled both chemically and mechanically in brackish water also. An obstruction on the membrane surface means that it takes much longer to set in place, and can be expelled and cleaned with a brush before being returned to service. ED paved way to the concept of membranes. ED, because of its implementation ED, decreased the risk of adjustment during processing significantly [48].

Almost all of these desalinating plants were located in the Middle East, and were constructed as part of a water and electricity generation station, which both were used by the facility close to an oil power plant. The pressure and salinity were greatly increased in order to raise the reprocess stream's temperature above the brine pressure. Then, the blend is heated to the maximum brine boiling point. Once the saturated steam is supplied to a bath stage maintained at slightly below the critical temperature of the resulting heated brine, it represents a portion of the original water's moisture content. The hot gas, which has passed through a mist eliminator, makes contact with the heat transfer on the facade of the piping. When hot water is added to the plate, the condensed liquid flows into it as a compound. This process begins with unprocessed brine, which converts the same to liquid and then converts liquid water to gas in a third stage, where sparge is added to produce additional liquid. Distillate flow is reheated on the flowing paths at phases and blows away through a coolant system and draining boilers till the mixed brine and water distillate are eventually taken out of the plant as processed water. The number of years for which membrane assurance was offered was not more than three but has gradually and has increased up to seven. Compared to conventional press formats, RO manufacturers started using the economical spiral wound, as they used fewer plates but greater packing density, resulting in reduced manufacturing costs, in the first decade of the 1990s. This was the next phase in the shift from cellulose acetate to polymer composites, which allowed the use of thinner film layers. Over the course of the entire lifetime of the project, one can look to modules which have had original membranes in place for over 12 years. These were all important steps toward developing the system by Dow and, in the late 1970s and 1980s, lower operating pressures, increased flux, and reduced consumption of salt, led to significant energy efficiency, and greater flux consumption led to enhanced overall energy efficiency. TFC membranes were not essential for the development of the modern RO system, according to Pearce, but the majority of these plants were not built until after the turn of the century. The introduction of energy recovery turbines had a dramatic effect on the operating costs of SWRO. Based on the original concept developed in Europe, typically RO can expand seawater at approximately 1.5 times the minimum pressure, incorporating commercially available components such as pumps and membranes to regain power [49,50].

The development of membrane structures was deemed useless due to its cost factor in earlier times and only after 2006 did their impact on the desalination technologies begin. They are now the most commonly accepted technology, and are used even more extensively for various applications other than desalination. Pretreatment with membrane reduces particle build-up on the microfiltration process prior to its use in higher capacity filters used for RO which have higher particle accumulation capabilities during operation. On an RO scale, there is some evidence that use this method demonstrates that membrane treatment is cheaper than conventional chemical cleaning, saves energy, lowers the cost, and boosts reliability. One paper in particular showed savings of over traditional scrubbing with lower emissions [51]. It was a major point of contention that bioaccumulation may not inherently follow the use of membrane treatment. These bioaccumulation issues need to be taken care of in order to maintain a clear fluid interface and research on new membranes has identified possible solutions to mitigate or prevent the occurrence of bio-fouling. Improving the dissolution organics with stabilizers is a salient phase of the treatment method, but has the potential to result in coagulant fouling if the treatment levels are excessive. Treatment plants that have seen great reductions in bio-fouling over the past few years with this type of process are exhibiting less than typical examples. Since newly installed or reconfigured NF softeners are capable of increasing all other system efficiencies and productivity, this shows that they are good for thermal systems [52].

7.8 During and after 2013

Based on the fact that around 74.8 million gallons of desalination energy generated around the world per day in 2012, it rose approximately to 125.5 million gallons in the year 2017. Approximately equal to 47.6 million m^3 in 2007, just 5 years earlier, but only a fraction of that of a cubic meter in 1963. Aggregate licensed capacity rose by 800,500,000 barrels to 80.5 million barrels. Over 16,000 modern (or total) desalination plants were established by August 2012, bringing the total locations in which desalination has been applied all around the globe to 632 since mid-2011. Whilst the expansion of desalinization was reduced during the initial part of 2012, while 2013 saw the greatest growth in new desalinization capacity. Just a year later, in 2014, the total capacity had fallen to 24.2 million barrels per day (bpd), or 49% of the previous figure. Around 16.5 million bpd in the year 1992 was the total capacity of the network during that time. In the last 20 years desalination has seen an increase of nearly 58 million bpd [50,53].

Most desalination processes take place in the Gulf region, which produces between 40% to 50% of the total supply, and in North Africa, the Middle East, Spain, and other European countries, there are more than 150 locations around the world where desalination plants are located. After the signing of a large-scale agreement for desalination projects in Peru and Chile, Latin America has gained an increased share of the desalination market. Also, Asia is now a major force in the desalination industry, and China has emerged as one of the most important new markets. A new 12-year on installing 2.2 million–2.6 million m^3 higher operating desalination capacity incorporated into the country's national plan, which is three to four times the planned capacity built in for the first 5 years [54].

The Task Force met for the final time in Abu Dhabi in August 2013 and validated its mission by reporting on water goals to be achieved by December 2013. In the latter half of 2014, the IDA followed up on its 2015 Environmental Task Force program with a conference on desalination that included the participation of the Environmental Task Force. The Task Force gave its recommended conservation plan to achieve the 20% energy reduction goal at the 2015 International Diversification Alliance's World Congress. A number of useful approaches were taken in terms of the kinds of reductions in energy consumption that the industry has implemented. Advances in desalination technology may have the side effect of decreasing desalination costs. There was an alternative approach that recommended retrofitting plants that were already in place. Studies showed that a possible reduction of about 200 MW could be achieved by performing basic value engineering tasks in the Middle East. Many advances have been made on the thermal side of plants that have been designed to distil 15 tons of steam per ton of MED using low pressure and that are now used in small or mini power plants. Reducing electrical power to run the MED process to energy requirements of 1.0−1.8 kWh/m of electrical energy for every square meter helps to extend the effective life of RO membrane until it needs to be replaced. The original technologies included low-energy MED and membrane technology for water desalination. The most recent membrane innovations are illustrated by the work that has already been successfully done in a small-scale commercial plant and is now being distributed to a larger commercial scale. The RO's yield was expanded to allow for an increase in the recovery rate by 20%. For those in the demonstration phase, bio-imaging, pressure-retardation membranes, carbon nanotubes, and biomimetics are all in development stages. In the meantime, processes using nanoporous membranes and graphene have increased the membrane efficiency by an extremely significant degree [55,56].

Other examples include desalination plants such as the one in Kwinana in Australia, which uses wind to offset all of its electricity demand. KSA also has planned for many avenues to follow for the installation of solar-powered desalination plants and to invest in them at a higher pace. RO desalination was accomplished using photovoltaic cells at a capacity of 136,000 m^3/day. This required the task of moving these technologies from the laboratory or prototype stage to the stage of commercial viability. The Earth's increasing temperature and sustainability issues are beginning to rise to prominence as issues over water, and its centrality to life, causing an increase in desalination and advanced water reclamation. Desalination will be seen as a staple water supply instead of a niche one, with great potential that needs a great deal of exploration in other places around the world. Increasing numbers of organizations and agencies will implement an approach similar to that in Singapore and focus on managing the overall portfolio of water resources with the notion of a natural disaster. At first, water reclamation technologies were met with public opposition, but this issue has been overcome and they are being used in many places across the country today. In countries where sewage treatment has traditionally been of high quality, the most important aspects for the transformation of seawater to drinking water have been economies of scale and a conversion method called sewer mining [26,57,58].

7.9 Conclusion

Solar-driven interfacial water purification technologies have been implemented over the past 10 years and have accomplished new heights in their use and production. A review of various solar absorber materials, brine management techniques, land use for the purification of water, use of salt water desalination methods, and policies at the country level has been made. In addition, environmental, economic, and socioeconomic impacts are discussed along with the status of Australia where the technology of desalination entered a new phase. The status of desalination and water purification technologies over the past half century has also been consolidated and the future prospects are also elaborated on. While using the desalination technologies, a freshwater production and electricity generation balance has to be taken into account. It is difficult to evaluate these systems, because they are applied in various situations and the efforts rendered to make such systems work could not be predicted. While these procedures and methods are acceptable, it is important to confirm the normal experimental set-ups, such as the room temperature, bulk water temperature, and humidity, in order to ensure the accuracy of studies. The deployment of large-scale desalination and purification technologies with a very low cost of production is also very important to considered when attracting potential customers, which is also a challenge in implementing these technologies. The stability of the system is also a big challenge, specifically when the systems are used in water sources such as lakes, river water, and sea water, and the long-term stability is of great research focus. Water purification can also be done with the application of steam generation, which requires attention on the system manufacturability. Upon addressing all the above issues and challenges next generation water purification technologies can be applied for various applications such as catalysis, photothermal ablation, therapy, and sensor other than water treatment.

References

[1] K. Elsaid, M. Kamil, E.T. Sayed, M.A. Abdelkareem, T. Wilberforce, A. Olabi, Environmental impact of desalination technologies: a review, Science of the Total Environment 748 (2020) 141528.

[2] Q. Khan, M.A. Maraqa, A.M.O. Mohamed, Inland desalination: techniques, brine management, and environmental concerns, in: Pollution Assessment for Sustainable Practices in Applied Sciences and Engineering, 2021, pp. 871–918.

[3] S.A. Abdul-Wahab, M.A. Al-Weshahi, Brine management: substituting chlorine with on-site produced sodium hypochlorite for environmentally improved desalination processes, Water Resource Managment 23 (12) (2009). 2437e2454.

[4] A.M.O. Mohamed, M. Maraqa, J. Al Handhaly, Impact of land disposal of reject brine from desalination plants on soil and groundwater, Desalination 182 (1) (2005). 411e433.

[5] M. Ahmed, W. Shayya, J. Al-Handaly, Investigation on the Use of Evaporation Ponds for Brine Disposal in Inland Desalination Plants, 14, The Newsletter of the Middle East Desalination Research Center, 2000, p. 5.

[6] D. Balaji, R. Saravanakumar, S. Sivalingam, V. Bhuvaneswari, Catalyst derived from wastes for biofuel production: a critical review and patent landscape analysis, Applied Nanoscience (2021). Available from: https://doi.org/10.1007/s13204-021-01948.

[7] C. Sommariva, Desalination and Advanced Water Treatment: Economics and Financing, Balaban Desalination Publications, Hopkinton, MA, 2010.

[8] 22ndGWIDesalData/IDA worldwide desalting plant inventory (November 8, 2009).

[9] Y. Ibrahim, H.A. Arafat, T. Mezher, F. AlMarzooqi, An integrated framework for sustainability assessment of seawater desalination, Desalination 447 (2018) 1–17.

[10] M. Ramesh, L. Rajeshkumar, R. Bhoopathi, Carbon substrates: a review on fabrication, properties and applications, Carbon Letters 31 (2021) 557–580. Available from: https://doi.org/10.1007/s42823-021-00264-z.

[11] V.-D. Dao, N.H. Vu, S. Yun, Recent advances and challenges for solar-driven water evaporation system toward applications, Nano Energy 68 (2020) 104324.

[12] D&WR (The International Desalination & Water Reuse Quarterly industry website), Total world desalination capacity close to 60 million m^3/d, 2009. Retrieved from, http://www.desalination.biz/news/news_story.asp?id = 5121.

[13] ESCWA (Economic and Social Commission for Western Asia), Role of Desalination in Addressing Water Scarcity 2009, 2009 November. Retrieved from, http://www.escwa.un.org/information/publications/edit/upload/sdpd-09-4.pdf.

[14] ETAP (Environmental Technologies Action Plan), Water desalination market acceleration, 2006. Retrieved from, http://ec.europa.eu/environment/etap/inaction/pdfs/watedesalination.pdf.

[15] U. Bardi, Peak water in Saudi Arabia, 2008 January 29. Message posted to, http://europe.theoildrum.com/node/3520.

[16] Sandia Laboratory, Desalination & Water Purification Technology Roadmap, Sandia Laboratory & Bureau of Reclamation, 2003.
[17] Canary Islands Coastline, Desalination in Spain, Tech. Review Inc. & Trade Commission, 2005. Retrieved from http://www.spainbusiness.com/icex/cma/contentTypes/common/records/viewDocument/0,,,00.bin?doc = 4119833.
[18] L. Zhang, L. Xie, H. Lin, C. Gao, Progress and prospects of seawater desalination in China, Desalination Environment 182 (1–3) (2005) 13–18.
[19] IsHiTech (Israel High-Tech & Investment Report), China signs desalination plant order, 2007 July. Retrieved from, http://www.ishitech.co.il/0707ar2.htm.
[20] GWI (Global Water Intelligence). (2009).
[21] D. Balaji, V. Bhuvaneswari, A.K. Priya, G. Nambirajan, et al., Renewable Energy Resources: Case Studies, IOP Conference Series: Materials Science and Engineering 1145 (2021) 012026. Available from: https://doi.org/10.1088/1757-899X/1145/1/012026.
[22] M. Ramesh, L. Rajeshkumar, D. Balaji, Influence of Process Parameters on the Properties of Additively Manufactured Fiber-Reinforced Polymer Composite Materials: A Review, Journal of Materials Engineering and Performance (2021). Available from: https://doi.org/10.1007/s11665-021-05832-y.
[23] S. Miller, H. Shemer, R. Semiat, Energy and environmental issues in desalination, Desalination 366 (2015) 2–8.
[24] T.M. Missimer, R.G. Maliva, Environmental issues in seawater reverse osmosis desalination: intakes and outfalls, Desalination 434 (2018) 198–215.
[25] L. Rajeshkumar, A. Saravanakumar, V. Bhuvaneswari, et al., Tribological behaviour of AA2219/MOS_2 metal matrix composites under lubrication, AIP Conference Proceedings 2207 (2020) 020005. Available from: https://doi.org/10.1063/5.0000042.
[26] K. Firoozirad, A. Alagumalai, E. Solomin, F. Fangfang, H.S. Ahn, O. Mahian, Study on thermophysical properties of alumina nanoparticles enhanced ionic liquids (NEILs): a modeling approach, Journal of Molecular Liquids (2021) 115827.
[27] I. El Saliby, Y. Okour, H.K. Shon, J. Kandasamy, I.S. Kim, Desalination plants in Australia, review and facts, Desalination 247 (1-3) (2009) 1–14.
[28] A. Bennett, 50th anniversary: desalination: 50 years of progress, Filtration + Separation 50 (3) (2013) 32–39.
[29] A. Alagumalai, O. Mahian, M. Aghbashlo, M. Tabatabaei, S. Wongwises, Z.L. Wang, Towards smart cities powered by nanogenerators: bibliometric and machine learning–based analysis, Nano Energy (2021) 105844.
[30] R. Mahendran, P. Rajkumar, L. Nirmal raj, S. Karthikeyan, L. Rajeshkumar, Effect of deep cryogenic treatment on tool life of multilayer coated carbide inserts by shoulder milling of EN8 steel, Journal of the Brazilian Society of Mechanical Sciences and Engineering 43 (2021) 378. Available from: https://doi.org/10.1007/s40430-021-03100-7.
[31] B. Liu, C. Wang, S. Bazri, I.A. Badruddin, Y. Orooji, S. Saeidi, et al., Optical properties and thermal stability evaluation of solar absorbers enhanced by nanostructured selective coating films, Powder Technology (2020).
[32] S.A. Mohamed, Application of geo-spatial analytical hierarchy process and multi-criteria analysis for site suitability of the desalination solar stations in Egypt, Journal of African Earth Sciences 164 (2020) 103767.
[33] F. Aydin, H. Sarptas, Spatial assessment of site suitability for solar desalination plants: a case study of the coastal regions of Turkey, Clean Technologies and Environmental Policy 22 (2) (2020) 309–323.
[34] M. Ramesh, K. Marimuthu, P. Karuppuswamy, L. Rajeshkumar, Microstructure and properties of YSZ-Al_2O_3 functional ceramic thermal barrier coatings for military

applications, Boletín de La Sociedad Española de Cerámica Y Vidrio (2021). Available from: https://doi.org/10.1016/j.bsecv.2021.06.004.
[35] M. Ramesh, C. Deepa, L. Rajeshkumar, M.T. Selvan, D. Balaji, Influence of fiber surface treatment on the tribological properties of Calotropis gigantea plant fiber reinforced polymer composites, Polymer Composites (2021) 1900400. Available from: https://doi.org/10.1002/pc.26149.
[36] Y. Lu, D. Fan, H. Xu, H. Min, C. Lu, Z. Lin, et al., Implementing hybrid energy harvesting in 3D spherical evaporator for solar steam generation and synergic water purification, Solar RRL 4 (9) (2020) 2000232.
[37] A.K. Singh, D.B. Singh, V.K. Dwivedi, G.N. Tiwari, A. Gupta, Water purification using solar still with/without nano-fluid: s review, Materials Today: Proceedings 21 (2020) 1700–1706.
[38] F. Zhao, Y. Guo, X. Zhou, W. Shi, G. Yu, Materials for solar-powered water evaporation, Nature Reviews Materials 5 (5) (2020) 388–401.
[39] V. Bhuvaneswari, M. Priyadarshini, C. Deepa, D. Balaji, et al., Deep learning for material synthesis and manufacturing systems: A review, Materials Today: Proceedings 46 (9) (2021) 3263–3269. Available from: https://doi.org/10.1016/j.matpr.2020.11.351.
[40] Q.-F. Guan, Z.-M. Han, Z.-C. Ling, H.-B. Yang, S.-H. Yu, Sustainable wood-based hierarchical solar steam generator: a biomimetic design with reduced vaporization enthalpy of water, Nano Letters 20 (8) (2020) 5699–5704.
[41] L.N. Rajeshkumar, K.S. Amirthagadeswaran, VARIATIONS IN THE PROPERTIES OF COPPER-ALUMINA NANOCOMPOSITES SYNTHESIZED BY MECHANICAL ALLOYING, Materiali in Tehnologije 53 (1) (2019) 57–63. Available from: https://doi.org/10.17222/mit.2018.122.
[42] O. Mahian, A. Kianifar, S.A. Kalogirou, I. Pop, S. Wongwises, A review of the applications of nanofluids in solar energy, International Journal of Heat and Mass Transfer 57 (2) (2013) 582–594.
[43] O. Mahian, M.R. Mirzaie, A. Kasaeian, S.H. Mousavi, Exergy analysis in combined heat and power systems: a review, Energy Conversion and Management 226 (2020) 113467.
[44] M.M. Ghafurian, H. Niazmand, E.K. Goharshadi, B.B. Zahmatkesh, A.E. Moallemi, R. Mehrkhah, et al., Enhanced solar desalination by delignified wood coated with bimetallic Fe/Pd nanoparticles, Desalination 493 (2020) 114657.
[45] I. Zahmatkesh, M. Sheremet, L. Yang, S.Z. Heris, M. Sharifpur, J.P. Meyer, et al., Effect of nanoparticle shape on the performance of thermal systems utilizing nanofluids: a critical review, Journal of Molecular Liquids (2020) 114430.
[46] G. Crabtree, E. Kocs, T. Alaan, Energy, society and science: the fifty-year scenario, Futures 58 (2014) 53–65.
[47] D. Balaji, M. Ramesh, T. Kannan, S. Deepan, et al., Experimental investigation on mechanical properties of banana/snake grass fiber reinforced hybrid composites, Materials Today: Proceedings 42 (2) (2021) 350–355. Available from: https://doi.org/10.1016/j.matpr.2020.09.548.
[48] A.B. Pouyfaucon, L. García-Rodríguez, Solar thermal-powered desalination: a viable solution for a potential market, Desalination 435 (2018) 60–69.
[49] M.A. Yamasoe, N.M. Évora do Rosário, S.N.S.M. Almeida, M. Wild, Fifty-six years of surface solar radiation and sunshine duration at the surface in São Paulo, Brazil: 1961–2016, Atmospheric Chemistry and Physics Discussions (2020) 1–32.
[50] M. Ramesh, C. Deepa, K. Niranjana, L. Rajeshkumar, R. Bhoopathi, D. Balaji, Influence of Haritaki (Terminalia chebula) nano-powder on thermo-mechanical, water absorption and morphological properties of Tindora (Coccinia grandis) tendrils

fiber reinforced epoxy composites, Journal of Natural Fibers (2021). Available from: https://doi.org/10.1080/15440478.2021.1921660.
[51] S. Singh, L.F. Inambao, Solar desalination: a critical review, International Journal of Mechanical Engineering and Technology 10 (8) (2019) 244–270.
[52] W. He, S. Amrose, N.C. Wright, T. Buonassisi, I.M. Peters, Field demonstration of a cost-optimized solar powered electrodialysis reversal desalination system in rural India, Desalination 476 (2020) 114217.
[53] M. Ramesh, C. Deepa, L. Rajeshkumar, M.R. Sanjay, S. Siengchin, Life-cycle and environmental impact assessments on processing of plant fibres and its bio-composites: A critical review, Journal of Industrial Textiles (2021) 152. Available from: https://doi.org/10.1177/1528083720924730.
[54] K. Alshail, Analysis of solar energy in desalination plants in Saudi Arabia, 2020.
[55] M. Ramesh, L. Rajeshkumar, C. Deepa, M. Tamilselvan, et al., Impact of Silane Treatment on Characterization of *Ipomoea Staphylina* Plant Fiber Reinforced Epoxy Composites, Journal of Natural Fibers (2021). Available from: https://doi.org/10.1080/15440478.2021.1902896.
[56] L. Rajesh kumar, K.S. Amirthagadeswaran, Corrosion and wear behaviour of nano Al_2O_3 reinforced copper metal matrix composites synthesized by high energy ball milling, Particulate Science and Technology 38 (2) (2020) 228–235. Available from: https://doi.org/10.1080/02726351.2018.1526834.
[57] M. Ramesh, C. Deepa, M. Tamil selavn, L. Rajeshkumar, et al., Mechanical and water absorption properties of Calotropis gigantea plant fibers reinforced polymer composites, Materials Today: Proceedings 499 (2021). Available from: https://doi.org/10.1016/j.matpr.2020.11.480.
[58] O. Prakash, Solar Desalination Technology, Springer Nature Singapore Pte Ltd., 2019.

CHAPTER 8

Solar-driven water treatment: the path forward for the energy–water nexus

Rajeshkumar Lakshminarasimhan
Department of Mechanical Engineering, KPR Institute of Engineering and Technology, Coimbatore, India

8.1 Introduction

When it comes to sustainable development, access to fresh water is a crucial parameter and a major issue that the international community is concerned about. The issue is that of water stress which is worsened by increasing global warming, because it goes hand in hand with both population growth and urbanization in cities [1,2]. Increasing the availability of functional water is another goal, and these are vital, but the most essential is dealing with the infrastructure as well as policy issues. In several cases, it will mean exploring further down the supply line to sea or wastewater. There are high energy and economic costs associated with implementing these tools, however, technologies exist to render these resources fit for use. Owing to the urgent demand, there is a need for new approaches to meet the challenge of increasing usage of renewable, low-cost energy, or developing new technologies for water treatment. Seawater essentially can be regarded as an inexhaustible resource. Thus, desalinization, also known as reverse osmosis, produces fresh desalinated water, and consumes significant amounts of energy. This intractably couples water production with its supply, which makes obtaining desalinated water unnecessarily expensive and more difficult [3,4]. The next problem commonly faced is the constraints posed by the thermodynamic energy over the removal of salt from the water and to overcome this problem, water purification techniques were chosen from carbon-less resources so that the purification load on the entire energy source could be partly reduced. Unsurprisingly, the sun is the most plentiful energy resource available to the world. Electrical energy can be used in an RO desalination plant, for example, to run various pumps such as those described above [5,6]. Another method for

Solar-Driven Water Treatment
DOI: https://doi.org/10.1016/B978-0-323-90991-4.00008-6

generating potable water is use of a solar still. Although a method including such distillation would then consume more energy, the implementation may be even more convenient with lower scale or less infrastructure, including in regions with less infrastructure which do not have sufficient resources to use the more complex method of volatile RO processes. Since the solution to water treatment relies on the use of solar power, the fouling factor might introduce a particularly significant problem in treating wastewater, in which fouling might be an even greater challenge than filtration [7]. There is an exception: water itself is inefficient at absorbing light, and materials must be used that are excellent at acquiring the entire spectrum as well as being subsequently able to transform it into heat.

In biomedical applications, such as photothermal therapy, previously, photothermal materials were only employed in solar evaporation. Prior to the use of the plasmonic nanoparticle dispersal method, scientists attempted to use the bulk solution as a means of solar heating as well as vaporization for evaporative cooling [8,9]. However, the majority of the generated heat is lost in the process of increasing the bulk temperature, and the evaporation efficacy is therefore limited. Instead of heating the complete water body, treatment surfaces are warmed, and energy problems could be minimized by focusing the heat only at the evaporation surface (air—water). This particular local region can be empowered with the aid of photothermal constituents that hover on the upper surface of the water body [10—15]. Whole evaporation efficacy can be assessed by observing the mass variation underneath the unchanged light irradiation.

The prompt efficacy is assessed as

$$\eta = \frac{(\dot{m} \times h)}{P}$$

where, $\dot{m}$ is the variation among the evaporation rates in light with dark vicinities; h is the specific enthalpy change of liquid water to vapor; and P is the power from light [16].

The evaporation efficacy is mostly affected by two things: light-to-heat (L to H) transformation, which is greatly agitated by reflection, and absorption, and also by other energy conversion techniques. This conversion is overlook with the aid of heat loss. These solar-based evaporator materials and architectural design have been well researched in a number of books and major peer-reviewed papers [12,13,15,17—19]. Herein, the primary concern is geared toward improving evaporator effectiveness and improving evaporator stability over the long term. It gives particular consideration to strategies for expansion during desalination; to that end, their efficacy as well as mechanistic concepts that are scarcely noted in the existing literature are elaborate on. Prior to that, a review on all the photosensitive materials that have been used so far, including carbon, nanoparticles, polymers, and other semiconductors has been carried out extensively. While light absorption is an important component of evaporation, there are other significant factors that also influence evaporation. By expanding on heat utilization besides evaporation, it is shown how and when to improve evaporation productivity through system design. Finally, we take a look at the biggest hurdle to the commercial adoption of solar evaporators, which is scalability, and provide several workable solutions (Figs. 8.1 and 8.2).

Solar-based evaporation is a developing method to attain freshwater from saline water or wastewater, in which photothermal resources play a pivotal role. Great effort is devoted to enabling energy conversion efficacy

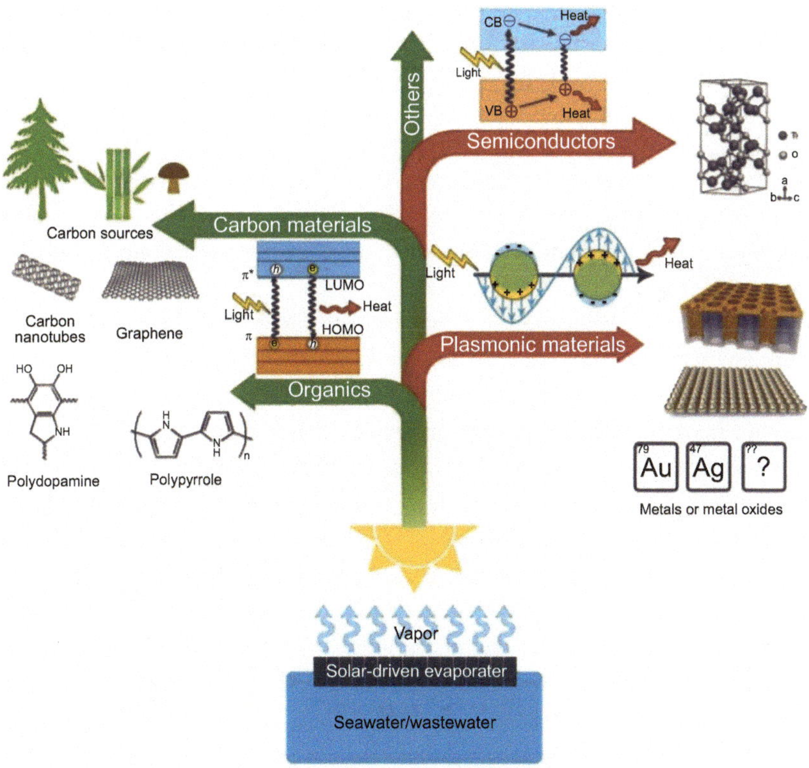

Figure 8.1 Typical photothermal constituents with L-to-H conversion [20].

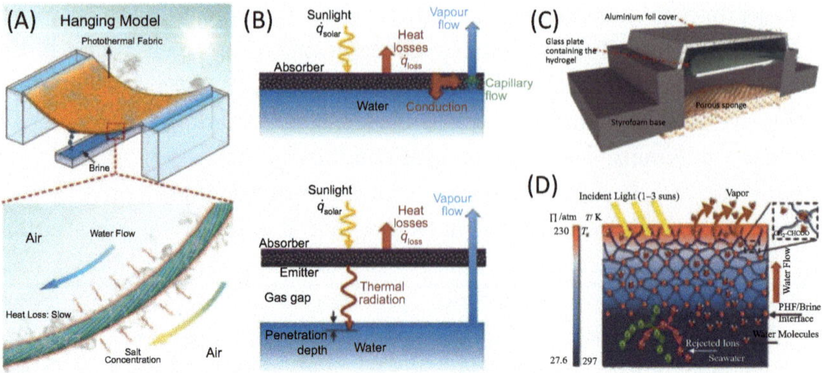

Figure 8.2 (A) Photothermal hanging technique; (B) direct method—contact as well as contactless absorption materials of thermal irradiation; (C) noncontact solar-based evaporator derived with super hygroscopic hydrogel; (D) salt-based rejection by extremely charged evaporators [20].

through the material and design of the device. The primary influences affecting long-term performance and evaporator efficiency, include selection of promotion of evaporation efficiency, photothermal materials, and solutions to the issue of scaling. Both the material components and structure will disturb factors such as absorption, and reflection of the incident light and the evaporation efficacy are enhanced by reducing heat loss, enlarging the surface area, and recycling latent heat. The scaling factor must be described by tailoring the properties of the surface along with the structure [20].

The recent development of evaporators fueled by the production of solar energy has moved researchers closer to becoming totally self-sufficient for water treatment technologies. The actions needed to best estimate how the efficiency and longevity of the solutions are affected will be described. The idea is to locate a strong light-absorbing thermal emitter that will serve all functions, since a broad-spectrum thermal material is required to make the ideal evaporator. The amount of light also depends on the morphology and arrangement of these materials, which broadens the absorption spectrum and increases the number of reflections. Heat management is essential to maintaining water's evaporative potential. An approach often used to further improve heat use is to extend the evaporation area, or introduce thermal barriers. As it is difficult to prevent scaling in desalination processes, evaporators have severe scaling issues and they will be more susceptible to this when used in long-term operations.

Surface or structural designs help mitigate this. Currently available materials allow for an increase in light harvesting of about 97%, and research shows little room for further gains, since materials currently allow for nearly 100% light absorption. Looking ahead, it is predicted that composite materials will be a major topic of interest because they can broaden the light range in next-generation solar radiation by promoting light utilization in the near-infrared spectrum. Another opportunity is their ability to be identified as photothermal materials suitable for industrial use, and their wide availability, low cost, and environmental safety. As with liquid water management, heat transfer via the heat exchanger has to be quick in order to ensure fast evaporation and low loss. As explained above, it is the relative humidity that determines the loss of latent heat through condensation. Diminishing the temperature of the interface aids in the evaporation, as the difference between evaporation as well as atmospheric temperature here lowers the humidity. The study of "cool evaporation" should gain momentum and could be done by expanding the evaporative cooling area [21–23]. While it may appear easy to understand that perhaps a relatively high surface temperature results in a higher photothermal efficiency by speeding up evaporation, recent research suggests that cool evaporation could even enhance heat utilization.

In addition, Asinari's group's suggested that the latent heat recycling design holds great potential in terms of a practical interface due to its efficient heat usage as well as easy-to-scale-up method. The majority of current investigations on the scaling issue generally aims at increasing the water supply in order to return the focused salt ions to something like the bulk. Growing water demand and supply, on the other hand, encourages not only mass transfer but also heat transfer that is detrimental to evaporation heat utilization. As a result, contactless evaporation appears to be a better option for dealing with this issue at a fundamental stage. Salt elimination devices, which prevent ions from reaching the evaporating interface, also appear to be promising. Scaling control, material design, and heat management must not be thought of in isolation. The wettability of photothermal material surfaces, for example, has an impact on scaling as well as heat transfer throughout evaporation. The concentration polarization caused by evaporation produces a saline gradient starting from surface to bulk that must be used to extract energy. Yang and his team merged a solar-based driving evaporation system with a contrary electrodialysis technique to develop both fresh water and clean energy, paving the way for future research that will take advantage of the salinity difference [24].

Furthermore, the interfacial evaporation-induced temperature gradient [25,26] and water flow [27] specific energy collectors could also be used to generate energy. Likewise, the increase in temperature and solute advancement could help photocatalysis perform better, so combining solar-driven evaporation with those other methodologies such as catalysis might be a new way to achieve better results.

Solar-powered evaporators seem to be on the verge of becoming practical desalination as well as wastewater treatment methodologies, rather than just a research intuition. Prospects of water treatment technologies through dispersed systems have become highly feasible which might be completely possible with large-scale RO treatment plants. Photothermal materials have found ever-widening potential applications, including such de-icing, because direct solar energy utilization offers sustainable development as well as potential financial advantages [28], diminishing petroleum viscosity for the transport system to save energy [29] as well as wastewater reuse [30–33]. While more researchers have focused on these intriguing systems, more applications to recognize a wide range of social needs are likely to emerge.

8.2 Research challenges to solar-driven water treatment

Concentrators must be used to increase solar irradiance in order to increase evaporation capacity. As solar thermal systems are widely used for the distillation of water, the amount of purified water solely depends on the amount of incident light energy. These systems work according to the saturated water vapour pressure and are characterized by non-homogenous temperature distribution within it [34]. In 2019, Dongare and his team, despite momentous advancements in material and structural developments, used salt deposition during continuous operation making it difficult to maintain and/or improve solar-driven interfacial saline water evaporation (SISE) performance. As a result, the SISE methodology is not yet ready for widespread use, particularly due to its lack of stability for longer periods of time.

8.2.1 Mechanism of crystalline salt deposition

Many factors influence SISE, including such a solar absorber's water supply ability, the composition of the salt water, interfacial efficiency, and environmental conditions [35,36]. The salinity upon its evaporator surface is essentially determined by the advection diffusion movement of salt ions combined with the evaporation effectiveness of salt water. To accommodate for the evaporated water, salt water containing salt ions would be transferred to the same evaporation surface via advection under solar illumination. As a consequence, the evaporation surface's ion concentration and the salinity slope between the evaporation surface and bulk water would both be boosted. The diffusive force, on the other hand, causes salt ions to move from the surface of the solar thermal absorber to the water bulk [37,38], lowering the salinity mostly on the evaporation substrate. As a result, with the balancing of the dynamics where high and low areas of salt ions meet, there is a continuous movement of salt into the absorber regions due to expansion of the solar surface and movement of the dissolved salts between the latter, determining the nucleation position of salt crystals. It is worth noting that as the salt concentration rises, crystallization of the active evaporation of absorber expansion decreases efficacy owing to enhanced reflection of the evaporation and it also increases during expansion, causing a decrease in efficiency. Worse yet, the crystallized salts obstruct the water supply [39–41]. The developed salts progressively either dissolve or diffuse back into brine water that is enclosed in the area of the solar absorber's microchannels at night. If the microchannel diameter of the solar absorber is large, water is transported rapidly over a long period of time; on the other hand, if the solar absorber is principally made from nanochannels, the channels can become blocked in the case of operation for a number of hours [42].

According to few experimenters, the size of the pore channels which is in the micrometer level is more suitable for water based on the capillary action compared with nanochannels [43]. This is due to the fact that the pore channels' nanoscale may increase the viscous confrontation of water, preventing water from being transported to the evaporative area. It is important to note that this point of view is widely held. Ni et al., (2018), for example, used this mechanism to demonstrate the efficient admittance along with the diffusion of numerous ions as well as molecules in channels at the microlevel. Furthermore, water fills the pores of a solar absorber with microchannels owing to capillary forces. A reduced pore size for a capillary channel should theoretically result in a capillary pressure which is higher. Thereby, the consequence is a smooth transition from various layer adsorption to capillary condensation, with reduced pores being further likely to be filled with water [44]. Capillary force can provide a water flux which is of a high level, and therein the flow resistance and capillary pressure always go hand in hand. The first is inversely proportional to pore diameter, whereas the latter is positively correlated. As a result, the capillary force and flow resistance are inversely proportional, owing to the thickness of the capillary force layer being dense enough to curtail conductive heat loss, and therein the trade-off is more protuberant in the SISE system. During such situations, the use of a dense insulation layer might protect the solar absorber from the saltwater [45], which results in a relatively lower evaporation rate when compared with theoretical calculations [46].

8.2.2 Destructiveness of salt deposit

The difference in pressure seen between saturated vapor close to the evaporation surface as well as the far-field vapor is connected to vapor transport. There are two main techniques of solute concentration influencing water evaporation through porous media: (1) expanding solution osmotic pressure to lower vapor pressure, and (2) mechanical obstruction of water passage by crystalline salts to lower the vapor diffusion coefficient [47,48]. As a result, it is reasonable to assume that ongoing salt accumulation will stifle vapor transport. Although solar absorbers could indeed consume all available solar energy, salt crystal deposition is immediately established as the evaporation efficacy in salt water is much inferior to that in clean water [42,49–51]. Kou et al. [51], for example, used a quick and easy dyeing approach to identify low-cost recyclable cotton carbon nanotube fabrics. These cotton fabrics were hydrophilic, flexible, and had high absorbency (approximately 95.7%) [52]. Cotton/carbon nanotube fabric was used to protect the evaporator, with the exception of the polystyrene foam surface. A rate of evaporation of seawater as high as 1.59 $kg/m^2/h$ has been observed in this configuration.

Solid salts have been observed on the surface of the cotton fabric when operating in saline water (NaCl, 17.5%). The evaporation rate dropped to 0.54 $kg/m^2/h$ after 8 hours of operation because more than 85% of the evaporation surface was enclosed. It was tested in the configuration shown in Fig. 8.3A then a bench-scale configuration was used, which showed 78% freshwater steam generation capacity, and delivered steam at the rate of 2 $kg/m^2/h$. A decrease in the salinity also has a considerable evaporation rate with lower concentration (1 $kg/m^2/h$) but it was kept sealed over many hours in concentrated salt water, thus limiting it to a greater extent (up to 7%) or with higher concentrations (15%), reducing the amount of water which evaporated from the mixture. The

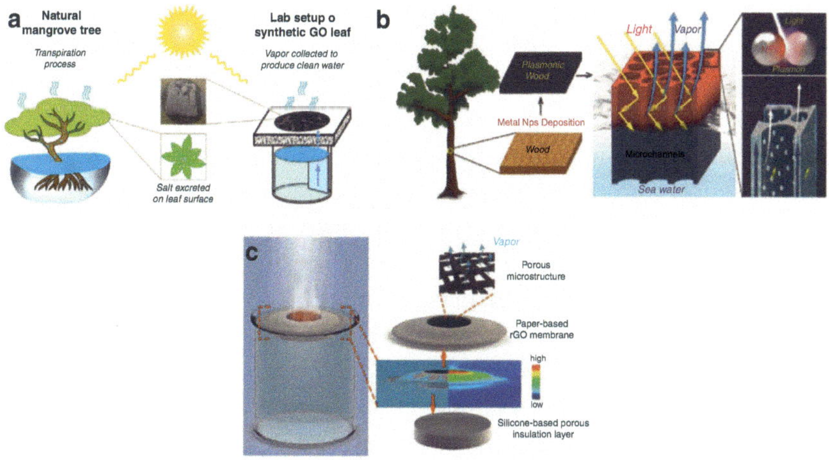

Figure 8.3 (A) Synthesized GO leaf by mimicking water transpiration-based solar evaporation [39]. (B) Plasmonic wood design [42]; (C) composite PrGO [50].

reduction in evaporation efficiency was as good as could be expected in seawater as long as vertically aligned carbon nanofibers had near black-body absorption, even though it was immersed in it [53]. The design shown in Fig. 8.3B achieved plasmonic pairing of plasmonic nanoparticles as well as microchannels in a wood matrix. This research successfully created a microchannel structure which preserves the wood's natural plasmon effects, with plasmonic interactions [42]. On the 6 days of operation under sun light, the composite material reached a power conversion efficacy of 85%, while the sunlight has been absorbed completely after 72 hours of its exposure and the production of useful energy started from that time. Despite the appearance of salt formation on the evaporation efficiency under the polychromatic light (sun) being lower than that predicted under one full sun, it decreased even more when exposed to a second light source which was not bright enough to permit evaporation, showing that the plastic's evaporation efficiency is only 69% in the vertical cell line schematic in Fig. 8.3C, where PrGO is hydrophilic and HLI water filter materials were used (PICV), allowing water to move to the periphery instead of the cell, moving up and not down (expanding horizontally in the schematic in Fig. 8.3C) This design totally eliminated the loss of heat energy caused by the evaporator as well as the upward conduction heat transfer to the water within the vaporizer. However, SISE's constant steady-state evaporation efficiency was approximately 89.7%, but when the evaporation tank was submerged in 3.5% sodium chloride solution, the percentage of NaCl increase resulted in a slight reduction in the percentage of evaporation.

There was recently published information on a new device by Hu et al. (to use an example from the authors) based on a silicon oxide/cellulosic nanofiber/carbon nanotube cloud hybrid solar absorber [54] with low tortuous porosity. This structure expanded in pure water, 1.09 kg for every 2 kg of a 3.5% solution of sodium chloride (NaCl) water. Despite this excellent performance, the desalination device was able to hold 100% of its capacity with sun illumination, and was desalted for 80% less than 100 hours. Finally, a conical biomimetic 3D structure was designed specifically for this project [55]. Among the 20 substances tested, the authors demonstrated the greatest evaporation rate at a concentration of 15% (or about 2.24 $kg/m^2/h$) but their experiment was still below the record when it came to pure water production at the rate of 2.63 $kg/m^2/h$, [56]. Although only NaCl was used in the simulation, this likely holds true for solid and inorganic water [55]. Most SISEs have contained potassium chloride because this compound acts as a salt inhibitor in the process of water uptake and escape. This, in addition to being a waste-removal factor, has also accumulated on the light-accepting surfaces. The white-coated surface would reduce light absorption capability, which would have an adverse effect on water pump performance and steam transmission, leading to a loss of water capacity and possibly water vapor evaporation, which has been studied by various experimenters. With the use of salinized water the membrane structure degraded, resulting in various losses over time periods longer than the maximum forecast of 3 years [57]. For cost-competitive solar evaporation, it is important to make sure the absorber is stable and durable.

8.3 Commercial challenges to solar-driven water treatment

The desalinization water technique is much more attractive in areas where there is a great shortage of fresh water due to distance from the ocean. Although water shortage and cost are key issues in desalination, the overall economy also plays a significant role. This report examines the costs and benefits of establishing water-only independence, and explores how this economic strategy could be used to determine if one was suitable to evaluate the viability of an additional water supply system. The calculations say that it costs 15 US dollars per m^3 to produce potable water using the compact unit, whereas it costs 18 dollars per m^3 to produce the same amount of water using the large one. The water-producing capacity of the membrane's life cycle and the life cycle of the plant are significant considerations to the total cost of water production. As the lifetime and/or membrane size continue to increase, the cost decreases. Compact as well as large solar-powered desalination systems for the supply of drinking water in arid regions have been looked into to see if they could be used profitably. Solar power still has a more negative impact on the economic viability of this system, and has proved to be more costly in comparison to other desalination processes. These compact as well as large systems' expected production costs have been compared to the norm, which was $15 and $18 per m^3 for the normal water supply cost. The results showed that there was a greater chance of increased reliability in membrane distillation, as well as an enhancement in plant lifespan, which would bring down the cost of produced water [58].

8.3.1 Challenges and opportunities for solar evaporation

Additionally, photothermal substances, thermal management, and interfacial designs have experienced marked efficiency improvements over the last few years, especially in sunlight. Current advancements in photothermal materials are focused on their photo-absorption. The different design principles, like thermal management, for enhancing evaporation, the transport of water, as well as antisalting structures, are all employed to some degree in the structural design to improve evaporation performance. Many great advances are being made in the potential fields of this appealing new area and are discussed here, including energy and environmental impacts. The current issues and researcher opportunities are now examined as well as long-term opportunities being presented, all with the aim of devising a course of action on ways to use solar evaporation going forward [59] (Fig. 8.4).

The development of photothermal and substrate materials, and also various structural engineering strategies involving thermal insulation, light absorption, interfacial properties, and water transportation, has made significant progress over the last decade. While there are some new applications which take full advantage of this fascinating technology, there are still a few challenges that are essential to be described before solar evaporation might be put into practice for the benefit of society more generally.

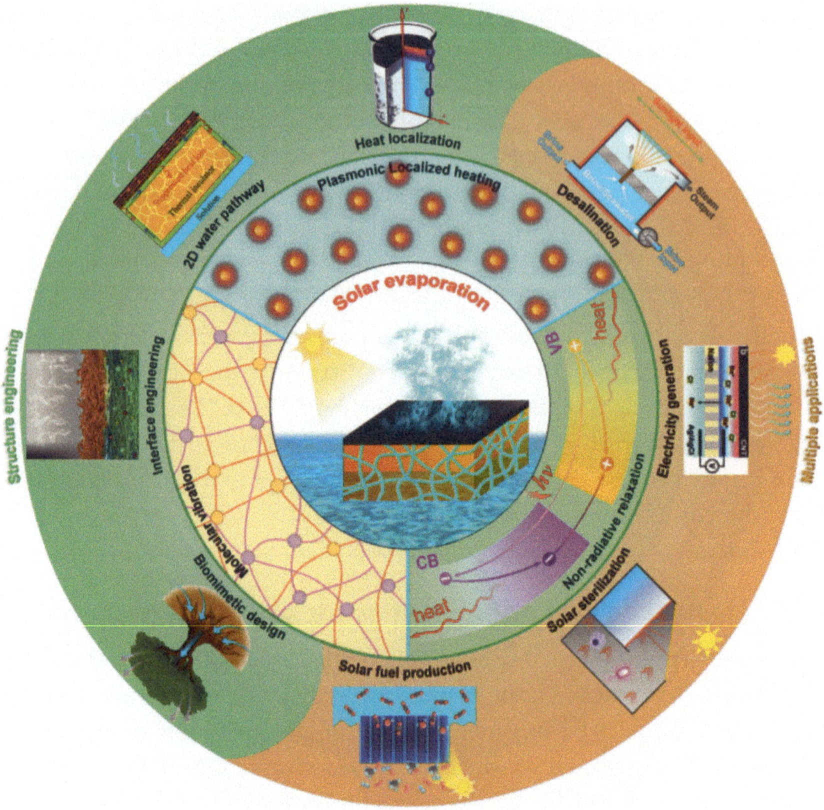

Figure 8.4 Challenges and opportunities for solar evaporation [59].

To create this clean, green, and sustainable technology as a reality, both fundamental research and practical explorations are required, including:

- Standardized measurements and evaluation;
- Multilevel stability against bacteria, salt, heat, weather, and water during extended term operation;
- Scalable and cost-effective production toward practical applications;
- Further understanding of the structure–property relationships of solar evaporation materials and structures;
- Discovering novel materials and structural design;
- Discovering novel functions and implementations.

Solar evaporation has been used for a wide range of applications ever since its inception, including solar–chemical fuel production, sterilization electricity generation, water purification, photocatalysis, and desalination.

These implementations, nevertheless, are already in their early stages of development, and many more functions as well as applications will need to be explored in the future. Rather than focusing solely on improving thermal efficiency, researchers should look into how to use the generated vapor for more practical purposes such as fuel generation, clean water, and power. In the coming decade, researchers will concentrate their efforts on integrating new functions into cutting-edge solar evaporation systems in order to expand the technology's applications. Increasing numbers of regulations, including about the direct discharge of brine systems, must be put in place.

Desalination that is driven by the sun, a sustainable and relatively inexpensive technique for removing minerals, and represents a promising implementation for treating water with food waste. This technology will likely see a great expansion as a result of recent discoveries in micro- and nanofluidics, recharging, electrocatalysis, and radiative cooling, as well as being linked to other emerging technologies, and complex systems like rechargeable batteries, and solar and catalytic conversion. Having abundant, renewable, and widely distributed solar energy resources is one of the most hopeful aspects in terms of clean water and fuel, and it is a proven approach for using concentrated solar power to generate heat for water and convert it into thermal energy. This green method does not leave behind any carbon residues and is easy to maintain, with no complex facilities required. It is anticipated that, over the next decade, the widespread use of environmentally friendly technology will combat both global water shortages and rising demand for clean energy. This system will be equally efficient, profitable, and adaptable over the long term if the strategic design effort focuses on bridging the technological evaporation deficit that exists between state-of-the-art solar systems and the future affordable and accessible alternatives [59].

8.4 Challenges to international water desalination

Access to water for drinking, agriculture, and industry will become an increasingly important challenging issue for several countries around the globe in the coming decades. Fresh water is presently in short supply and is being used in an unsustainable manner in the many parts of the world. Water covers is excess of 70% of the surface of Earth, but the majority of it is unfit for human consumption. As shown in Fig. 8.5, about 1.4 billion km^3, that is around 97.5% of the world's total water reserves, are in the oceans, while the remaining 2.5% is fresh water found in the atmosphere,

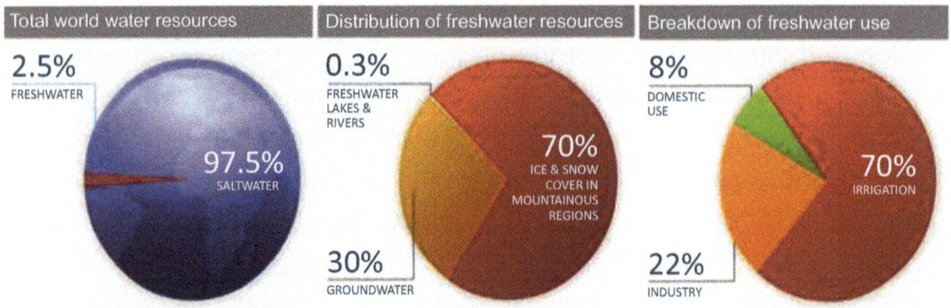

Figure 8.5 Global water resources [60,61].

freshwater lakes, glaciers, ground water, and rivers. Humans and other organisms have access to only about 0.014% of the total amount of water available [60].

Despite the scarcity of water, freshwater resources are allocated unequally around the globe. Every 20 years, the global water consumption rate doubles, outpacing population explosion by a factor of two. As a result, one in every three people across every continent is estimated to experience worsened water scarcity. Almost one-fifth of the global population lives in an area with water scarcity. The water requirement is expected to exceed consumption by 56% by 2025 as a result of persistent regional droughts and population shifts to urban coastal cities. Currently, approximately 3 billion people do not have access to a clean water source for drinking, and approximately 1.76 billion people live in regions where water scarcity is already severe. As competing demands for water grow owing to population growth, urbanization, the influence of greenhouse gases on the environment, and increases in household and industrial demands for water, the mismatch over the requirements for fresh water and their availability will worsen. Many other predictions evaluate that perhaps the world population will stabilize between 8 and 9.50 billion by 2050, with the majority of this growth taking place in the developing world. Population growth will have an impact on not only domestic water consumption, but also agricultural, industrial, other product, and energy consumption. The world's population had already tripled since the last millennium, resulting in a sixfold increase in global water usage. Many countries' rapid economic growth, especially China and India, has in recent times resulted in higher incomes, which has prompted increased consumption of goods and services. Massive amounts of water are required for the production of goods and services. One of the most difficult experiments in water management is decoupling income growth from water consumption.

Another demographic trend that contributes to increased water consumption is urbanization. The region's water resources can be put under tremendous strain as a result of such rapid growth. Deep wells are used to extract water from aquifers in many major cities today. In many developing nations, the discharge of untreated wastewater contributes to the pollution and deprivation of their already inadequate water sources. When water shortages occur, they impact both on the development of nations and the standard of living of their inhabitants. Although the use of water is obviously is vital to everyone, this resource needs to be a top priority on a global scale. Owing to the increased demand for fresh water, desalination technology could serve as a complementary method for supplying saline water, which is most often accomplished with fossil fuels that have volatile market prices that tend to rise and fall because of international financial status, thus requiring a lot of capital investment. Seawater is being used to meet the increased global water demand, along with fresh water and reclaimed river water. Also, due to the scarcity of fresh water and drought, desalination is rising as a necessity in the global market, and needs are now met by using seawater, river water, brackish water, and brine. The projected installed capacity was 60 million m^3/day in 2010, and it was expected to be 100 million m^3 by 2015. This facility will supply about 39 million m^3 per day to the Gulf Region, while the rest of the world will get 59 million m^3. In many areas with water resources similar to those described in this book, there has been long—and there will continue to be—significant progress with desalination technologies, with the Middle East, the Mediterranean and the Caribbean in particular, being important beneficiaries. A vast majority of new desalination plants with the ability to generate large quantities of energy have been built in the Middle East. The abundance of the world's largest oil reservoirs, and a severe scarcity of potable water resources in the Gulf accounts for 65% of the world's seawater desalination capacity. The countries with large desalination capacities include the United Arab Emirates, Saudi Arabia, Spain, and China, all of which have witnessed sizable increases in their water supply since 2002, and their abilities to desalinate seawater and fresh water have expanded greatly [62].

8.5 Opportunities across the globe for desalination

There is increasing recognition of the cost-effective nature of water purification across the globe, especially in countries with limited freshwater supplies. Purification of salt water is considered one of the most globally sound and inventive ways to meet both future and current water needs. Many communities and industrial sectors are dependent on water for their socioeconomic development. An estimated 14,000 purification plants are operational around the world, producing over one billion gallons of fresh water each day. This number includes the Middle East and other areas that use large- or traditional-scale thermal power stations for electricity, and account for 57% of the world's total power generation capacity. Additionally, due to the depletion of energy sources, current prices being higher, and harming of the environment with carbon dioxide, polluting the climate has become a more significant issue. Seawater can be found along the coast, but even in countries where there is plenty of coastline, many types of coastal plants cannot be afforded. Many countries experience energy crises and also electricity shortages and unreliability issues. Traditional approach costs rise, while agreements to improve and define renewable water desalination technologies are made, and sustainable energy integration strategies are laid out with the desalination techniques increasing in viability. In countries with a large total amount of sunshine, such as parts of the Middle East, North Africa, South Asia, and the Indian subcontinent, desalination could be a long-term alternative for the solution for obtaining fresh water from solar energy. The opportunities and challenges of solar water desalination around the world are discussed here. An in-depth examination is also given of the latest events in water desalination and solar desalination technologies, as well as incorporating the newest technologies for the most important efforts in the field of solar purification, such as economic and environmental considerations [62] (Fig. 8.6).

Water and energy are perhaps the most crucial components for long-term survival. Numerous countries around the world, especially developing nations, are hampered by a lack of water. Desalination of salt water is aimed to become a long-term source of water and to help alleviate the global water shortage. The large proportion of purification plants deployed across the globe are experiencing aquatic deficiencies are currently running on fossil fuels; though, because purification methods are energy intensive, they are increasingly expensive to operate and the emissions generated being recognized widely as very damaging to the environment. In some oil-rich countries, the thermal desalination method is often used entirely for seawater desalination, while the RO process has become the second most popular technique at a global level. In several developed

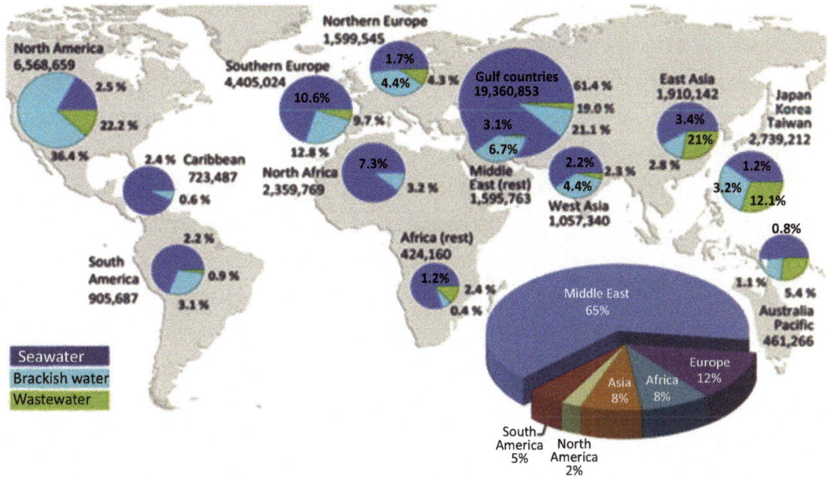

Figure 8.6 Global desalination scenario [63].

and developing nations, where traditional energy resources are limited, RO is currently the preferred option. Thermal distillation will, that being said, remain the dominant desalination industry throughout the Middle East for the predictable future, as the thermal combination of heat and power infrastructure prevails in the region with plenty of natural oil available. Desalination costs have been reduced significantly in recent years as a result of recent developments and improved methods in desalination technologies.

Desalination of saline and seawater will become more common as technology advances and the need for fresh water increases. Renewable energy and purification are becoming more dependable as a precisely established substitute to traditional desalination systems, particularly in areas with poor infrastructure, limited availability of conventional energy sources, and abundant renewable resources. Because of the increase in fossil fuel prices and the harmful effects of burning fossil fuels, combining desalination plants with renewable energy resources is becoming ever more desirable. Both solar and desalination technologies have already reached maturity as a result of recent advancements and enhancements. Consequently, using solar energy for water desalination in countries along the solar energy belt in the Middle East and Africa could be a viable method, helping to solve water scarcity issues while also decreasing carbon dioxide emissions through an ecologically responsible method. Solar distillation, on the other hand, may be beneficial for purification of salt water, as solar-MED and MSF are recommended for large-scale thermal purification, and other solar applications, such as photovoltaics and solar-thermal collectors must be explored, as they may be applicable for various desalination processes at an affordable rate and are very well appropriate for providing fresh water at remote locations.

The reverse osmosis film technique is the most cost-effective solar purification technique, with costs comparable to those of conventional desalination. As a result, combining solar PV cells with film aquatic purification methods could be the second most cost-effective option for desalinating salty water. Mitigation and adaptation research could perhaps concentrate on the development with reduced, reasonably priced, limitless, and clean solar energy technologies that will provide formally confirmed advantages and therefore can be assimilated with a variety of desalination processes. Although there have been visible advancements as well as enhancements in desalination processes, more research and development should be conducted to discover new purification films in terms of forward osmosis, as well as to decrease power consumption by using advanced pumping systems, heat, and retrieval of pressure systems. Meanwhile, similar research into the growth of low-cost solar energy technologies should be carried out in order to find low-cost materials (chemicals, membranes), low-cost solar collectors, and low-cost and long-lasting photovoltaic resources. The conclusion to use desalination as an alternative is based on a complex assessment of local circumstances, including the requirement of a water, available water source, availability of renewable energy source, finances, conservational, and societal influences.

In addition, ongoing investigations and projects are required to increase experience, knowledge, and faith in new environmentally friendly technologies, as well as state incentives in the form of new policies and economic support. These would almost certainly result in reduced power consumption and desalination water construction costs. Meanwhile, some benefits of wastewater purification over salt water purification should be emphasized and considered to be taken into account in any upcoming investigation for lower costs and lower energy consumption. Furthermore, because most collected wastewater is close to urban areas, transportation costs can be saved by using it for landscaping and industrial purposes. To combat water shortages and guard the aquatic sources worldwide, collaborative and complete investigation and development programs involving all investors—customers, local groups, administrations, businesses, academies, and research organizations—are essential to enhance the sustainable water purification

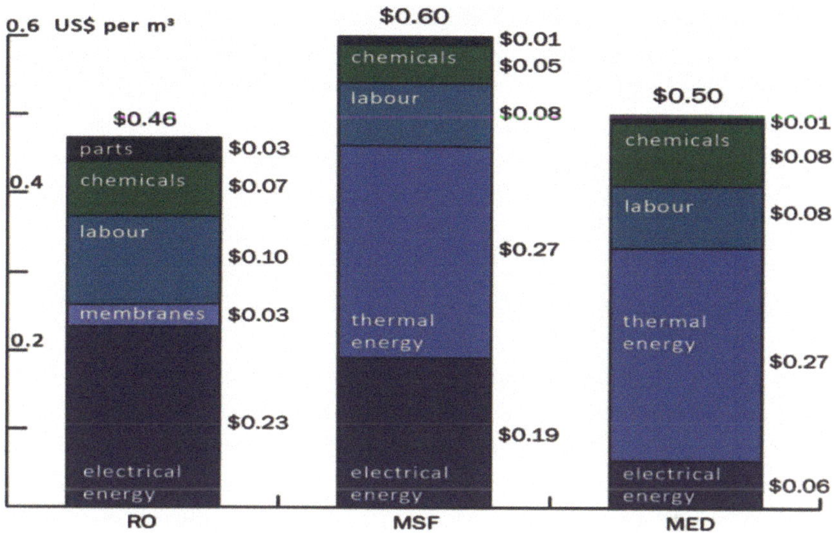

Figure 8.7 Cost analysis in US dollars [64].

policies and strategies. These works should comprise a complete, system-wide design process for a water recycling and purification technique, and also process streamlining to reduce excess recovery and energy consumption [61] (Fig. 8.7).

The cost of the engineering, procurement, and construction of an MS purification plant is approximately US $1235 per m^3 per day for the saline water to be desalinated. Since the MED and SWRO facilities have an average cost of US $916 and $641 dollars per day to operate, respectively, their capital costs will be around US $605 million and $342 million, which are both lower. The seawater production cost is estimated to be in the range of US $0.45–$0.60 per 1000 m^3 [63]. After paying for new components, the machine must be cleaned, that is, the unplanned operation of the water treatment facility, the feed water cleaning, and post water cleaning, must be applied, in addition to plant and other operations. Water supply costs may go up in the longer term due to currently rising energy prices, but desalination will remain a popular [62].

8.6 MENA region

Most seawater purification plants are found in the Gulf REA, with an entire purification capacity of around 12.1 million m^3 over 24 hours either or slightly less than half (44%) of global processing. The globalized sea water purification capability of countries is given as follows: Saudi Arabia 25%, the Gulf shore 11%, the Red Sea coast 12%, the United Arab Emirates 23%, and Kuwait 6%. Ninety-four percent of production is influenced by the Gulf region for thermal purification methods for water and power which combines both heat and power plants that use minimum value vapor and power from power plants as a heat source for purification. Eighty-one percent of MSF purification is carried out in Gulf countries. Compared to this method, both MED and reverse osmosis methods are not that very significant with values of 13% and 6%, respectively, for water generation [63] (Fig. 8.8).

In the Red Sea area countries like Egypt generates 3.6 MW of power each day which covers 13% of international volume. As in the Gulf, nearly all of the water is generated using cogeneration, located mostly on the Arabian coastline of Yanbu, Assir, and in the aforementioned two places, a remarkable water purification complex, capable of processing 1.6 million m^3 of ocean water per day, has been built. Also, more than twice as much water (2.6 million m^3/day) is produced from desalination plants in Saudi Arabia, with which it contributes to even more than three-quarters of the Red Sea's production, making up 78% of this total. Desalination is responsible for only 7% of the manufacturing (with

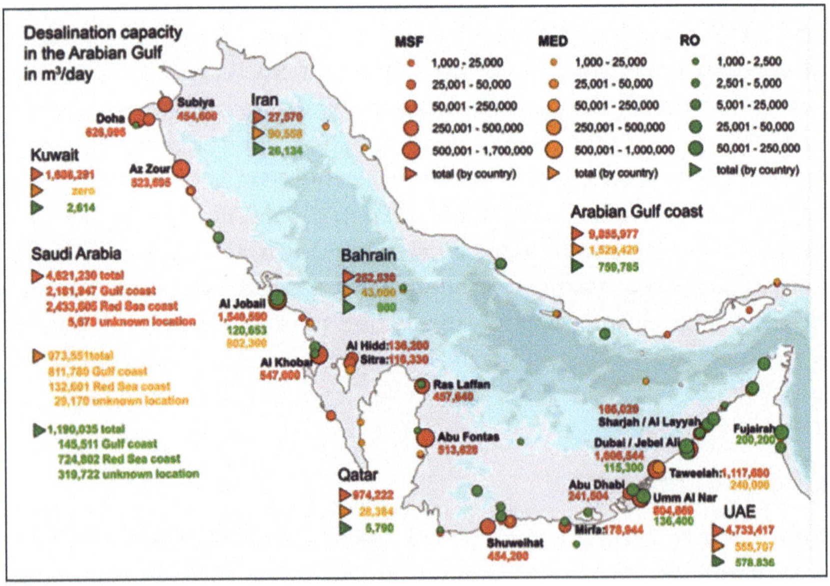

Figure 8.8 Middle East desalination [63].

0.2 Mm/day) of the Red Sea's water production, with 90% (about 6.8 Mm/day) heading from plants in the Sinai along the Red Sea [63].

In the area of the Mediterranean sea, slightly more seawater usage has been witnessed, and total usage is estimated to around 14% of the international capacity. Currently, the Spanish purification industry is about equal to that of the rest of the Western Hemisphere in terms of output, with 2.2 million m^3/day, giving it purification facilities equal to the entire area of the Americas. From 60% to 65% (1.4 million m^3/day) of capacity are found on the southern coasts of Spain and the Canary Islands, and from 30% to 35% (900,000 m^3/day) on the Balearic Islands and the Canary Islands. The program expanded desalination in order to meet water needs in the region by over 2.7 million m^3/day until 2010. Red sea serves as the major source of water for the Gulf regions while the Mediterranean coast gets the most out of Mediterranean sea. While other plants only appear in northern Europe and North Africa, there are plants found in larger numbers mostly along beaches of Libya and Algeria in this area. At the same time, at these locations, there are also new SWRO facilities, particularly in Latin America. Many great progresses are being made in Algeria, North Africa's leading desalination market, with the construction of a massive desalinization plant (or system) having been completed in February 2008, which is capable of producing 200,000 m^3/day [65]. It is the first in a sequence of desalination installations with capabilities of between 50,000 and 500,000 m^3/day, which was due to increase the country's total capacity to 4 Mm^3/day by the end of the year 2020 [66].

On the Israeli Mediterranean shoreline, the world's largest single-stage vacuum breaker is undergoing construction, and will be able to process 330,000 m^3. The Palmach SWRO project—the world's largest single-stage facility to date—is currently underway (83,000 m^3/day). They currently supply 8% of Israel's total water sources derived from purification facilities. As per the original estimates, this figure was supposed to increase to over 1 million $ft.^3$/day (28,000 m^3) by 2020 [67]. Although during late 2008 an original water demand target of 1 million m^3 per day which was then revised up to 2 million m^3/day in the latest crisis plan, other strategies may be able to increase the amount of purified water output per day to the revised target only by 2020 if the overall demand for water remains stable [68]. Due to numerous planned purification plants that will be developed with a total capacity of 274,000 m^3/day, in the Mediterranean along Israel's coastal borders, the future water provision will be in better shape than currently [69]. Additionally, it is estimated that more than two-thirds of the country's daily water resources will be utilized, expanding from the current 16,500 m^3 to approximately between 220,000 and 274,000 m^3 of water per day [68, 69]. Other methods of water use, legislated water rationing, increased fines, and investments in water conservation to enhance usage, and the capacity to lift the current level of 75% to 95% within 5 years bring in more effective ways of water treatement with more adaptable practices [68].

8.7 Conclusion

From the above discussions, it can be seen that the prevailing water shortage and the production of fresh water from seawater demands a huge amount of energy resources. In locations where large industries are located and in industries where high energy-consuming products are manufactured, the necessity for solar-driven desalination plants increases and, also, these desalination plants promise a high degree of water security across the globe in line with the global policies being developed. It is understood from the above discussions that the input to thermal power plants, which account for the largest energy-consuming plants for steam production, could use waste heat recovered or low-grade steam produced from the thermal plant and its coworking with solar-driven desalination plants to minimize the energy consumed by thermal power plants while also being an effective method to for the energy—water nexus. In doing so, the consumption of electricity in the operation phase and the energy—water nexus index in the controlling phase have to be given focus. It could also be concluded that the future of the energy—water nexus perspective will require due consideration be given to solar-driven desalination plants implementation in terms of regional supply chain management and energy consumption structure of the plant, which meets the need for energy and the drive for water conservation simultaneously.

References

[1] R.G. Taylor, B. Scanlon, P. Döll, M. Rodell, R. Van Beek, Y. Wada, et al., Ground water and climate change, Nature Climate Change 3 (4) (2013) 322—329.
[2] J. Schewe, J. Heinke, D. Gerten, I. Haddeland, N.W. Arnell, D.B. Clark, et al., Multimodel assessment of water scarcity under climate change, Proceedings of the National Academy of Sciences 111 (9) (2014) 3245—3250.
[3] B. Van der Bruggen, C. Vandecasteele, Distillation vs. membrane filtration: overview of process evolutions in seawater desalination, Desalination 143 (3) (2002) 207—218.

[4] L.F. Greenlee, D.F. Lawler, B.D. Freeman, B. Marrot, P. Moulin, Reverse osmosis desalination: water sources, technology, and today's challenges, Water Research 43 (9) (2009) 2317–2348.
[5] A.M. Delgado-Torres, L. García-Rodríguez, M.J. del Moral, Preliminary assessment of innovative seawater reverse osmosis (SWRO) desalination powered by a hybrid solar photovoltaic (PV)-Tidal range energy system, Desalination 477 (2020) 114247.
[6] C. Charcosset, A review of membrane processes and renewable energies for desalination, Desalination 245 (1-3) (2009) 214–231.
[7] A.K. Menon, I. Haechler, S. Kaur, S. Lubner, R.S. Prasher, Enhanced solar evaporation using a photo-thermal umbrella for wastewater management, Nature Sustainability 3 (2) (2020) 144–151.
[8] (a) O. Neumann, A.S. Urban, J. Day, S. Lal, P. Nordlander, N.J. Halas, ACS Nano 7 (2013) 42.
(b) O. Neumann, C. Feronti, A.D. Neumann, A. Dong, K. Schell, B. Lu, et al., Proc. Natl. Acad. Sci. U.S.A. 110, 11677.
[9] M.S. Zielinski, J.W. Choi, T. La Grange, M. Modestino, S.M.H. Hashemi, Y. Pu, et al., Hollow mesoporous plasmonic nanoshells for enhanced solar vapor generation, Nano Letters 16 (4) (2016) 2159–2167.
[10] D. Balaji, R. Saravanakumar, S. Sivalingam, et al., Catalyst derived from wastes for biofuel production: a critical review and patent landscape analysis, Applied Nanoscience (2021). Available from: https://doi.org/10.1007/s13204-021-01948-8.
[11] Z. Deng, J. Zhou, L. Miao, C. Liu, Y. Peng, L. Sun, et al., The emergence of solar thermal utilization: solar-driven steam generation, Journal of Materials Chemistry A 5 (17) (2017) 7691–7709.
M. Gao, L. Zhu, C.K. Peh, G.W. Ho, Energy & Environmental Science (12)(2019) 841–864.
[12] M. Ramesh, L. Rajeshkumar, R. Bhoopathi, Carbon substrates: a review on fabrication, properties and applications, Carbon Letters 31 (2021) 557–580. Available from: https://doi.org/10.1007/s42823-021-00264-z.
[13] Y. Zhang, T. Xiong, D.K. Nandakumar, S.C. Tan, Structure architecting for salt-rejecting solar interfacial desalination to achieve high-performance evaporation with in situ energy generation, Advanced Science 7 (9) (2020) 1903478.
[14] F. Zhao, Y. Guo, X. Zhou, W. Shi, G. Yu, Materials for solar-powered water evaporation, Nature Reviews Materials 5 (5) (2020) 388–401.
[15] D. Balaji, V. Bhuvaneswari, A.K. Priya, G. Nambirajan, J. Joenas, P. Nishanth, et al., Renewable Energy Resources: Case Studies, IOP Conference Series: Materials Science and Engineering 1145 (2021) 012026. Available from: https://doi.org/10.1088/1757-899X/1145/1/012026.
[16] C. Chen, Y. Kuang, L. Hu, Challenges and opportunities for solar evaporation, Joule 3 (3) (2019) 683–718.
[17] P. Wang, Emerging investigator series: the rise of nano-enabled photothermal materials for water evaporation and clean water production by sunlight, Environmental Science: Nano 5 (5) (2018) 1078–1089.
[18] V. Bhuvaneswari, M. Priyadarshini, C. Deepa, D. Balaji, et al., Deep learning for material synthesis and manufacturing systems: A review, Materials Today: Proceedings 46 (9) (2021) 3263–3269. Available from: https://doi.org/10.1016/j.matpr.2020.11.351.
[19] S.L. Wu, H. Chen, H.L. Wang, X. Chen, H.C. Yang, S.B. Darling, Solar-driven evaporators for water treatment: challenges and opportunities, Environmental Science: Water Research & Technology 7 (1) (2021) 24–39.

[20] Y. Shi, R. Li, Y. Jin, S. Zhuo, L. Shi, J. Chang, et al., A 3D photothermal structure toward improved energy efficiency in solar steam generation, Joule 2 (6) (2018) 1171–1186.
[21] X. Li, J. Li, J. Lu, N. Xu, C. Chen, X. Min, et al., Enhancement of interfacial solar vapor generation by environmental energy, Joule 2 (7) (2018) 1331–1338.
[22] M. Ramesh, K. Marimuthu, P. Karuppuswamy, L. Rajeshkumar, Microstructure and properties of YSZ-Al2O3 functional ceramic thermal barrier coatings for military applications, Boletín de La Sociedad Española de Cerámica Y Vidrio (2021). Available from: https://doi.org/10.1016/j.bsecv.2021.06.004.
[23] R. Mahendran, P. Rajkumar, L. Nirmal Raj, S. Karthikeyan, L. Rajeshkumar, Effect of deep cryogenic treatment on tool life of multilayer coated carbide inserts by shoulder milling of EN8 steel, Journal of the Brazilian Society of Mechanical Sciences and Engineering 43 (2021) 378. Available from: https://doi.org/10.1007/s40430-021-03100.
[24] Q. Shen, Z. Ning, B. Fu, S. Ma, Z. Wang, L. Shu, et al., An open thermo-electrochemical cell enabled by interfacial evaporation, Journal of Materials Chemistry A 7 (11) (2019) 6514–6521.
[25] L. Zhu, T. Ding, M. Gao, C.K.N. Peh, G.W. Ho, Shape conformal and thermal insulative organic solar absorber sponge for photothermal water evaporation and thermoelectric power generation, Advanced Energy Materials 9 (22) (2019) 1900250.
[26] M. Ramesh, C. Deepa, L. Rajeshkumar, M.T. Selvan, D. Balaji, Influence of fiber surface treatment on the tribological properties of Calotropis gigantea plant fiber reinforced polymer composites, Polymer Composites (2021). Available from: https://doi.org/10.1002/pc.26149.
[27] S.L. Wu, H. Chen, H.L. Wang, X. Chen, H.C. Yang, S.B. Darling, Solar-driven evaporators for water treatment: challenges and opportunities, Environmental Science: Water Research & Technology 7 (1) (2021) 24–39.
[28] S. Lyu, Y. He, Y. Yao, M. Zhang, Y. Wang, Photothermal clothing for thermally preserving pipeline transportation of crude oil, Advanced Functional Materials 29 (27) (2019) 1900703.
[29] H. Geng, Q. Xu, M. Wu, H. Ma, P. Zhang, T. Gao, et al., Plant leaves inspired sunlight-driven purifier for high-efficiency clean water production, Nature Communications 10 (1) (2019) 1–10.
[30] Q. Fang, G. Li, H. Lin, F. Liu, Solar-driven organic solvent purification enabled by the robust cubic Prussian blue, Journal of Materials Chemistry A 7 (15) (2019) 8960–8966.
[31] M. Ramesh, C. Deepa, K. Niranjana, L. Rajeshkumar, R. Bhoopathi, D. Balaji, Influence of Haritaki (Terminalia chebula) nano-powder on thermo-mechanical, water absorption and morphological properties of Tindora (Coccinia grandis) tendrils fiber reinforced epoxy composites, Journal of Natural Fibers (2021). Available from: https://doi.org/10.1080/15440478.2021.1921660.
[32] M. Ramesh, L. Rajeshkumar, D. Balaji, Influence of Process Parameters on the Properties of Additively Manufactured Fiber-Reinforced Polymer Composite Materials: A Review, Journal of Materials Engineering and Performance (2021). Available from: https://doi.org/10.1007/s11665-021-05832-y.
[33] P.D. Dongare, A. Alabastri, O. Neumann, P. Nordlander, N.J. Halas, Solar thermal desalination as a nonlinear optical process, Proceedings of the National Academy of Sciences 116 (27) (2019) 13182–13187.
[34] P. Tao, G. Ni, C. Song, W. Shang, J. Wu, J. Zhu, et al., Solar-driven interfacial evaporation, Nature energy 3 (12) (2018) 1031–1041.

[35] J. Zhong, C. Huang, D. Wu, Z. Lin, Influence factors of the evaporation rate of a solar steam generation system: A numerical study, International Journal of Heat and Mass Transfer 128 (2019) 860–864.
[36] M. Ramesh, L. Rajeshkumar, C. Deepa, M. Tamil selvan, et al., Impact of Silane Treatment on Characterization of *Ipomoea Staphylina* Plant Fiber Reinforced Epoxy Composites, Journal of Natural Fibers (2021). Available from: https://doi.org/10.1080/15440478.2021.1902896.
[37] D. Toksoz, I. Yilmaz, Influence of swelling clay content on ion migration and column performance in lime column treated soils, Geotechnical and Geological Engineering 38 (1) (2020) 813–832.
[38] C. Finnerty, L. Zhang, D.L. Sedlak, K.L. Nelson, B. Mi, Synthetic graphene oxide leaf for solar desalination with zero liquid discharge, Environmental Science & Technology 51 (20) (2017) 11701–11709.
[39] H. Ren, M. Tang, B. Guan, K. Wang, J. Yang, F. Wang, et al., Hierarchical graphene foam for efficient omnidirectional solar–thermal energy conversion, Advanced Materials 29 (38) (2017) 1702590.
[40] A. Saravanakumar, D. Ravikanth, L. Rajeshkumar, D. Balaji, M. Ramesh, Tribological Behaviour of MoS2 and Graphite Reinforced Aluminium Matrix Composites, IOP Conference Series: Materials Science and Engineering 1059 (2021) 012021. Available from: https://doi.org/10.1088/1757-899X/1059/1/012021.
[41] M. Zhu, Y. Li, F. Chen, X. Zhu, J. Dai, Y. Li, et al., Plasmonic wood for high-efficiency solar steam generation, Advanced Energy Materials 8 (4) (2018) 1701028.
[42] Y. Ito, Y. Tanabe, J. Han, T. Fujita, K. Tanigaki, M. Chen, Multifunctional porous graphene for high-efficiency steam generation by heat localization, Advanced Materials 27 (29) (2015) 4302–4307.
[43] L. Zhang, X. Hu, Q. Zhou, Sunlight-assisted tailoring of surface nanostructures on single-layer graphene nanosheets for highly efficient cation capture and high-flux desalination, Carbon 161 (2020) 674–684.
[44] G. Ni, S.H. Zandavi, S.M. Javid, S.V. Boriskina, T.A. Cooper, G. Chen, A salt-rejecting floating solar still for low-cost desalination, Energy & Environmental Science 11 (6) (2018) 1510–1519.
[45] J. Zeng, Q. Wang, Y. Shi, P. Liu, R. Chen, Osmotic pumping and salt rejection by polyelectrolyte hydrogel for continuous solar desalination, Advanced Energy Materials 9 (38) (2019) 1900552.
[46] D.L.M. Mendez, C. Castel, C. Lemaitre, E. Favre, Membrane distillation (MD) processes for water desalination applications. Can dense selfstanding membranes compete with microporous hydrophobic materials? Chemical Engineering Science 188 (2018) 84–96.
[47] V.A. Jambhekar, R. Helmig, N. Schröder, N. Shokri, Free-flow–porous-media coupling for evaporation-driven transport and precipitation of salt in soil, Transport in Porous Media 110 (2) (2015) 251–280.
[48] L. Rajesh kumar, K.S. Amirthagadeswaran, Corrosion and wear behaviour of nano Al_2O_3 reinforced copper metal matrix composites synthesized by high energy ball milling, Particulate Science and Technology 38 (2) (2020) 228–235. Available from: https://doi.org/10.1080/02726351.2018.1526834.
[49] Z. Wang, Q. Ye, X. Liang, J. Xu, C. Chang, C. Song, et al., Based membranes on silicone floaters for efficient and fast solar-driven interfacial evaporation under one sun, Journal of Materials Chemistry A 5 (31) (2017) 16359–16368.
[50] L. Rajeshkumar, S. Sivalingam, K. Prashanth, K. Rajkumar, Y. Huang, C. Yang, et al., Design, analysis and optimization of steering knuckle for all terrain vehicles, AIP Conference Proceedings 2207 (2020) 020003. Available from: https://doi.org/10.1063/5.0000044.

[51] D. Balaji, M. Ramesh, T. Kannan, S. Deepan, V. Bhuvaneswari, L. Rajeshkumar, Experimental investigation on mechanical properties of banana/snake grass fiber reinforced hybrid composites, Materials Today: Proceedings 42 (2021) 350–355. Available from: https://doi.org/10.1016/j.matpr.2020.09.548.
[52] Z. Yin, H. Wang, M. Jian, Y. Li, K. Xia, M. Zhang, et al., Extremely black vertically aligned carbon nanotube arrays for solar steam generation, ACS Applied Materials & Interfaces 9 (34) (2017) 28596–28603.
[53] R. Hu, J. Zhang, Y. Kuang, K. Wang, X. Cai, Z. Fang, et al., A Janus evaporator with low tortuosity for long-term solar desalination, Journal of Materials Chemistry A 7 (25) (2019) 15333–15340.
[54] L. Wu, Z. Dong, Z. Cai, T. Ganapathy, N.X. Fang, C. Li, et al., Highly efficient three-dimensional solar evaporator for high salinity desalination by localized crystallization, Nature Communications 11 (1) (2020) 1–12.
[55] P. Loganathan, G. Naidu, S. Vigneswaran, Mining valuable minerals from seawater: a critical review, Environmental Science: Water Research & Technology 3 (1) (2017) 37–53.
[56] A. Morelos-Gomez, R. Cruz-Silva, H. Muramatsu, J. Ortiz-Medina, T. Araki, T. Fukuyo, et al., Effective NaCl and dye rejection of hybrid graphene oxide/graphene layered membranes, Nature nanotechnology 12 (11) (2017) 1083.
[57] F. Banat, N. Jwaied, Economic evaluation of desalination by small-scale autonomous solar-powered membrane distillation units, Desalination 220 (1–3) (2008) 566–573.
[58] C. Chen, Y. Kuang, L. Hu, Challenges and opportunities for solar evaporation, Joule 3 (3) (2019) 683–718.
[59] UN water. http://www.unwater.org/statisticsuse.html (accessed 28.09.12).
[60] World resources. http://www.wrsc.org/attachimage/total-water-resources (accessed 26.09.12).
[61] M. Shatat, M. Worall, S. Riffat, Opportunities for solar water desalination worldwide, Sustainable cities and society 9 (2013) 67–80.
[62] S. Lattemann, M.D. Kennedy, J.C. Schippers, G. Amy, Global desalination situation, Sustainability Science and Engineering 2 (2010) 7–39.
[63] D.M. GWI, A Global Industry Forecast (CD ROM), Global Water Intelligence, Media Analytics Ltd, 2007.
[64] T. Pankratz, Hamma SWRO on-line, in: T. Pankratz (Ed.), Water Desalination Report, Media Analytics Ltd., Oxford, 2008. in cooperation with Global Water Intelligence, Houston, TX.
[65] NRC, Desalination: A National Perspective, Committee on Advancing Desalination Technology, Water Science and Technology Board, National Research Council (NRC) of the U.S. National Academies, 2008.
[66] Y. Dreizin, A. Tenne, D. Hoffman, Integrating large scale seawater desalination plants within Israel's water supply system, Desalination 220 (2008) 132–149.
[67] Media Analytics Ltd., Israel unveils drought-busting water strategy, Global Water Intelligence, Media Analytics Ltd., Oxford, 2008.
[68] GWI, DesalData.com by Global Water Intelligence, 2008.
[69] Media Analytics Ltd., Israel's brackish water challenge, Global Water Intelligence, Media Analytics Ltd., Oxford, 2008.

Index

Note: Page numbers followed by "*f*" and "*t*" refer figures and tables, respectively.

T

U

Kolibri 360 GmbH
Ferdinand-Jühlke-Straße 7
99095 Erfurt, Deutschland
produktsicherheit@kolibri360.de